Biosynthesis

Volume 7

A Specialist Periodical Report

Biosynthesis

Volume 7

A Review of the Literature Published
during 1979, 1980, and 1981

Senior Reporters
R. B. Herbert *Department of Organic Chemistry, University of Leeds*

T. J. Simpson *Department of Chemistry, University of Edinburgh*

Reporters
P. M. Dewick *University of Nottingham*

J. R. Hanson *University of Sussex*

E. Leete *University of Minnesota, Minneapolis, U.S.A.*

The Royal Society of Chemistry
Burlington House, London W1V 0BN

ISBN 0-85186-553-4
ISSN 0301-0708

Printed in Great Britain by
Adlard & Son Ltd, Dorking, Surrey

Foreword

This is the seventh volume in the series of Specialist Periodical Reports which covers the biosynthesis of natural products; it is also the last. Future surveys of this area will be included in the journal Natural Product Reports, which appears for the first time in 1984 and incorporates several Specialist Periodical Reports titles.

For various reasons, the publication of this volume has been delayed. It appears now with coverage of the literature for the period 1979 to 1981 inclusive, except in the case of Professor Leete's review, in which coverage is to the end of 1982.

The individual reviews which constitute this volume cover the same areas and appear under the same titles as previously. Unfortunately it was not possible to include reviews on the biosynthesis of triterpenoids, steroids and carotenoids, and on non-protein amino-acids, cyanogenic principles and glucosinolates; we hope to make good these omissions in early issues of Natural Product Reports as well as including new material in the burgeoning area which carries the title "Biosynthesis".

This volume has been put together against a very tight deadline, and it is a real pleasure to acknowledge the hard work and cooperation of the contributors to this volume. Without them, there would have been no volume 7 of "Biosynthesis".

Leeds and Edinburgh, June 1983

R B Herbert
T J Simpson

Contents

1

The Biosynthesis of Polyketides

BY T.J. SIMPSON

1 Introduction

This chapter covers the literature appearing between January 1979 and December 1981 and follows the format of the previous report.[1] It has been a particularly active period with a welcome, increasing trend towards studies aimed at elucidating the mechanisms of the intermediate steps in polyketide biosynthesis. The potential of ^{2}H-labelling mentioned previously[1] has been realised and extended in the review period with several studies using both direct methods, ie ^{2}H n.m.r., and indirect methods, ie ^{2}H α-isotope shifts and ^{2}H-^{13}C couplings in ^{13}C n.m.r. using doubly labelled [^{2}H,^{13}C] precursors. These methods have been reviewed.[2] A potentially more useful technique than the α-isotope shift method has been described.[3] In this, ^{2}H is placed β to the reporter ^{13}C nucleus in a doubly labelled precursor; an isotope shift is still observable but the unfavourable relaxation and nOe effects associated with ^{2}H directly attached to ^{13}C are avoided. This has been applied to only one study in the review period, but will clearly find much use. A related and also extremely useful technique makes use of ^{18}O-induced isotope shifts in ^{13}C n.m.r.[4] to detect the biosynthetic origins of oxygen by incorporating doubly labelled [^{18}O,^{13}C] precursors or by growing organisms in an atmosphere containing $^{18}O_2$ and subsequent ^{13}C n.m.r. analysis of the labelled metabolites. The number of studies using advanced intermediates continues to increase and ^{2}H-labelling has great potential in this area. A number of books which cover aspects of polyketide biosynthesis have appeared,[5,6] with Steyn's book on the biosynthesis of mycotoxins[7] being particularly valuable.

2 Fatty Acids

The stereochemical mechanism of enoyl reductase, the enzyme catalysing the final reduction in the cycle of condensation-reduction-dehydration-reduction that lengthens the fatty acid chain by one $-CH_2CH_2-$ unit at each turn of the cycle on fatty acid synthetase, appears to be species specific. As shown in Scheme 1, the enoyl reductase from yeast converts the enoyl thioester (1) to the acyl thioester (2) by an anti-addition of hydride from NADPH

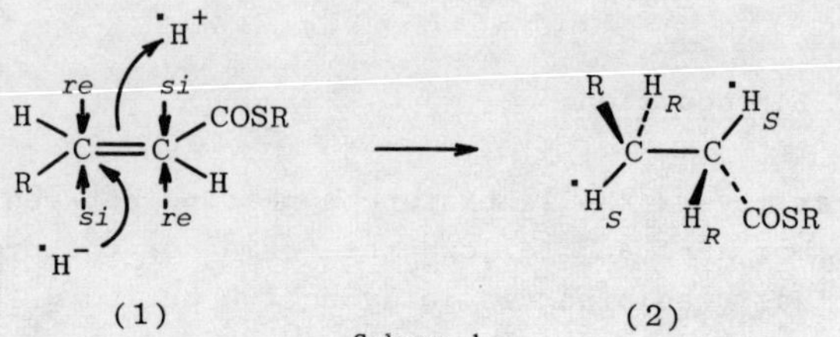

(1) (2)

Scheme 1

to the si face of the β-carbon with protonation of the α-carbon from the si face.[8] However with the reductase from both E. coli[9] and Brevibacterium ammoniagenes,[10] a syn addition of hydrogen via a 2-re, 3-si attack occurs, whereas the reductase from rat liver[10] also carries out a syn-addition, but this time via 2-si, 3-re attack. The stereochemistry of hydride donation from NADPH is related to the stereochemistry of addition, with the pro-4S hydrogen being used for 3-si addition and the pro-4R hydrogen being used for 3-re addition.[10,11]

Incorporation studies with $[2-^{13}C,^{2}H_3]$acetate and analysis of the ^2H α-isotope shifts in the simultaneously ^1H and ^2H noise decoupled ^{13}C n.m.r. spectrum has shown that palmitic acid (3) is biosynthesised in the alga Anacystis nidulans[12] with a gradation of ^2H retention along the acyl chain as shown in Scheme 2. The

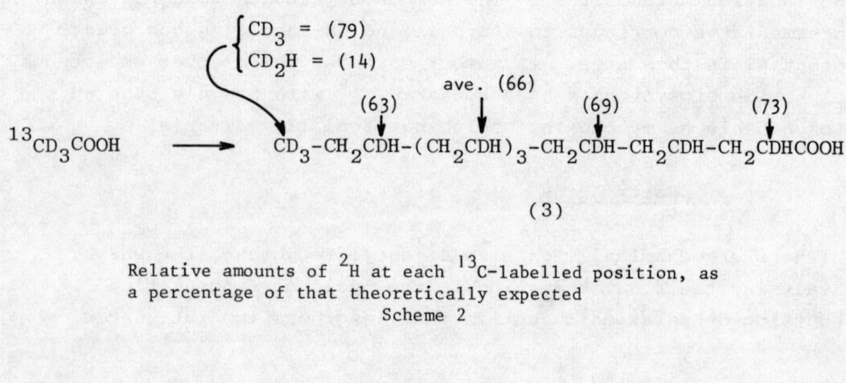

(3)

Relative amounts of ^2H at each ^{13}C-labelled position, as a percentage of that theoretically expected

Scheme 2

results are interpreted as being consistent with a 'post-malonate' exchange process, presumably associated with reversible transfer of the growing acyl chain from the acyl carrier protein to a cysteine residue of β-keto acyl ACP synthetase. Similar results have been obtained using [$^2H_3$]acetate in E. coli.[13]

The incorporation of ^2H from [$^2H_3$]acetate into lipoic acid (4) is consistent with its formation from octanoic acid with the loss of one ^2H label from C-8. The ^2H incorporated at C-6 of octanoic

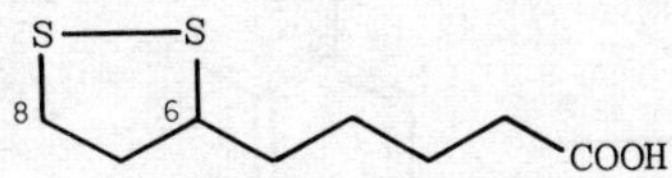

(4)

acid is retained, and since this ^2H is incorporated with the L-configuration during fatty acid biosynthesis but is known to have the D-configuration in lipoic acid, an inversion of configuration must occur at C-6 during sulphur insertion.[14] This suggested the involvement of hydroxylated octanoic acids as intermediates. However, feeding studies[15] with ^2H-labelled 6-hydroxy-, 8-hydroxy-, and 6,8-dihydroxyoctanoic acids gave negligible incorporations and so direct introduction of sulphur at the saturated carbons of octanoic acid seems likely.

An authoritative review of the structure of fatty acid synthetase has appeared.[16]

3 Tetraketides

Both ^2H n.m.r. spectroscopy and β-isotope shifts in ^{13}C n.m.r. have been used to measure the incorporation of ^2H from [$^2H_3$]- and [1-^{13}C,$^2H_3$]acetates into 6-methyl-salicylic acid (5) by Penicillium griseofulvum.[3] Both methods show that there is a

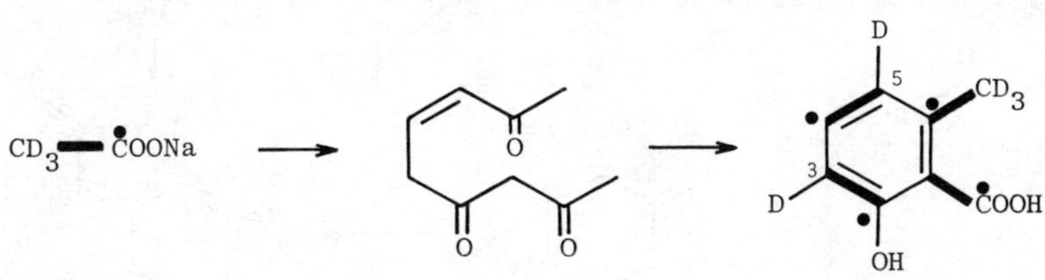

(5)

preferential incorporation into the methyl of the acetyl-CoA-derived starter unit and significantly more ^{2}H is retained at C-3 than at C-5. It is suggested that the non-uniform incorporation could arise from differing degrees of random exchange during the chain assembly process or, more interestingly, it could reflect the actual mechanisms of cyclisation and aromatisation of the precursor polyketide. More examples will be needed to test the validity of this observation.

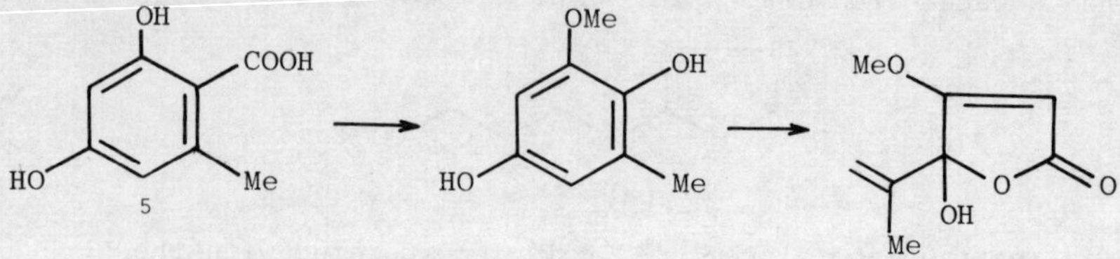

(6) (7) (8)

Addition of 5-chloroorsellinic acid to growing cultures of Penicillium cyclopium inhibits the biosynthesis of penicillic acid (8) and results in the accumulation of the previously indicated intermediates orsellinic acid (6) and 3-methoxytoluquinol (7) and its corresponding quinone.[17]

The post-gentisaldehyde part of the biosynthetic pathway to patulin (13) has been extensively investigated, using mutant strains of Penicillium urticae.[18] A patulin-minus mutant, J1, accumulates phyllostine (10) and isoepoxydon (11). Another patulin-minus mutant, J2, which is blocked immediately after gentisaldehyde (9),

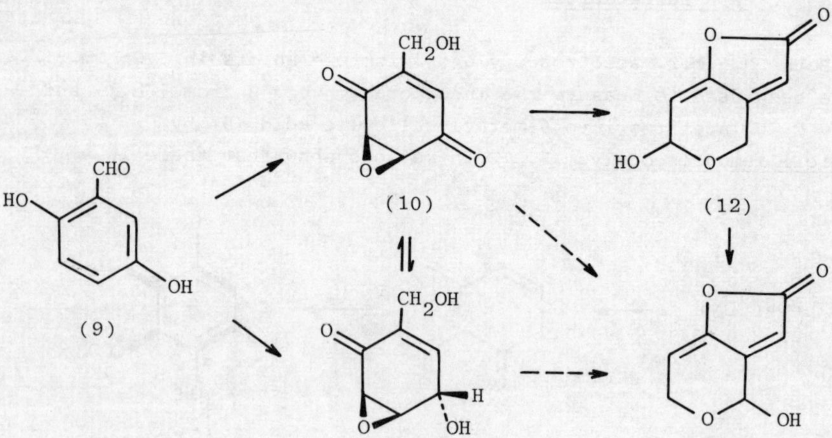

(9) (10) (11) (12) (13)

Scheme 3

converts both (10) and (11) to patulin in yields of 90% and 60% respectively. They are interconverted by a specific dehydrogenase and cell-free systems have been isolated from the mutants which carry out their interconversion and further transformation. A further mutant, S15, accumulates isopatulin (12) and immobilised cells of the wild-type strain convert phyllostine to isopatulin in good yield. Finally, a cell suspension of mutant P3, blocked between phyllostine and isopatulin, converts isopatulin to patulin.[19] These results are summarised in Scheme 3. The changes in oxidation levels occurring during these transformations are somewhat puzzling and labelling studies would be useful. This and earlier works are summarised in a review which also compares patulin biosynthesis with the pathways leading to a number of other fungal lactones.[20]

Elasnin (14), a novel inhibitor of human granulocyte elastase, has been isolated from <u>Streptomyces noboritoensis</u>. Incorporation of [$^{13}C_2$]acetate has shown it to be derived from twelve acetates[21] but it would appear to be best regarded as a tetraketide, as the

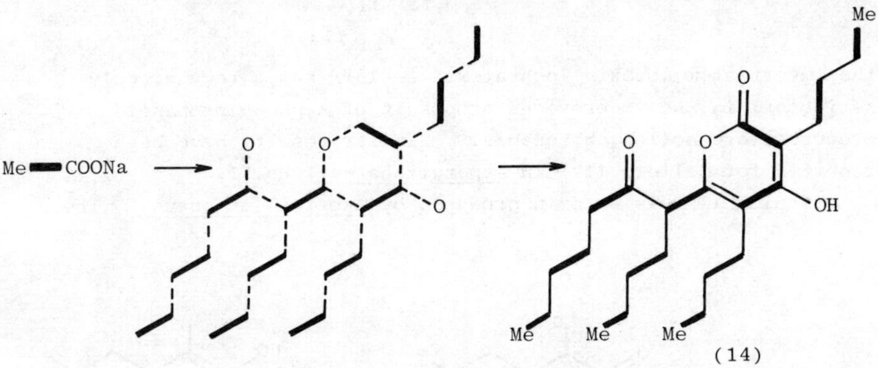

(14)

Scheme 4

most plausible route is extension of a hexanoate starter by three 2-butylmalonate units as indicated in Scheme 4.

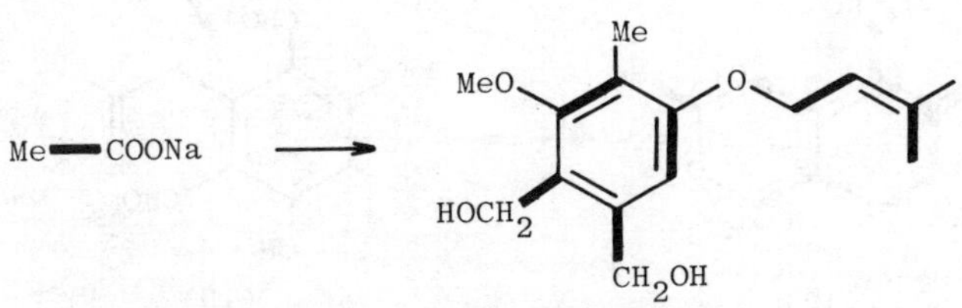

(15)

$[^{13}C_2]$Acetate is incorporated into zinniol (15) by cultures of Alternaria solani as shown.[22]

4 Pentaketides

Incorporation studies[23,24] with singly and doubly labelled $[^{13}C]$acetates have confirmed that the dihydroisocoumarin moiety of

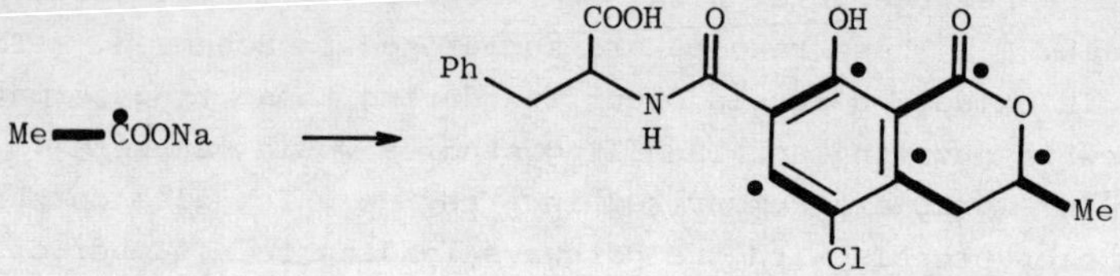

(16)

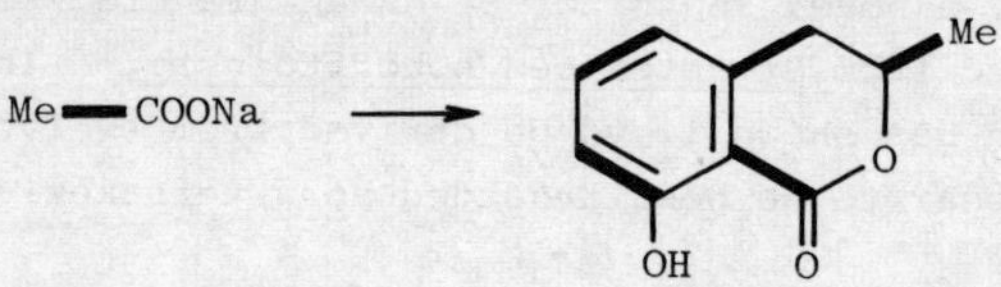

(17)

the important mycotoxin ochratoxin A (16) has a regular polyketide origin and so previous proposals of a phenylpropanoid precursor are no longer tenable. Similar results have been reported for mellein (17) in Aspergillus melleus.[25]

Austdiol (19) is a toxin produced by Aspergillus ustus. In-

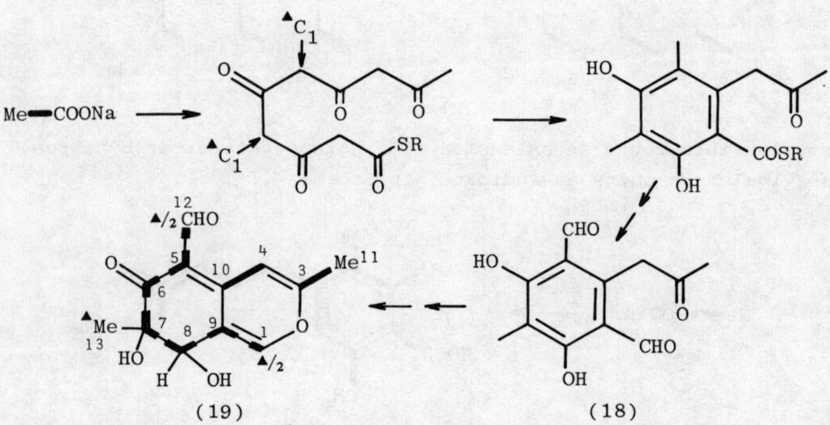

Scheme 5

corporation of $[^{13}C_2]$acetate resulted in two ^{13}C-^{13}C couplings being observed for carbons 5,6,7,8, and 9 while carbons 11,3,4,10,12, and 1 showed only one coupling. On feeding $[^{13}C]$methionine, C-13 was enriched approximately twice as much as C-1 and C-12. This labelling pattern is consistent with a biosynthetic pathway, shown in Scheme 5, where a methionine-derived methyl is oxidised and the polyketide carboxyl is reduced to give the symmetrical dialdehyde (18) as an intermediate.[26]

Full details of ^{13}C-labelling studies on aspyrone (20) have appeared.[25] Asperlactone (21) is a co-metabolite of aspyrone in A. melleus and has the same carbon skeleton. Computer-aided resolution enhancement of the ^{13}C n.m.r. spectrum of $[^{13}C_2]$acetate-enriched asperlactone shows a 2-bond ^{13}C-^{13}C coupling between C-2 and C-8. Pathways involving rearrangement of a linear polyketide (22), or rearrangement and cleavage of an aromatic precursor (23),

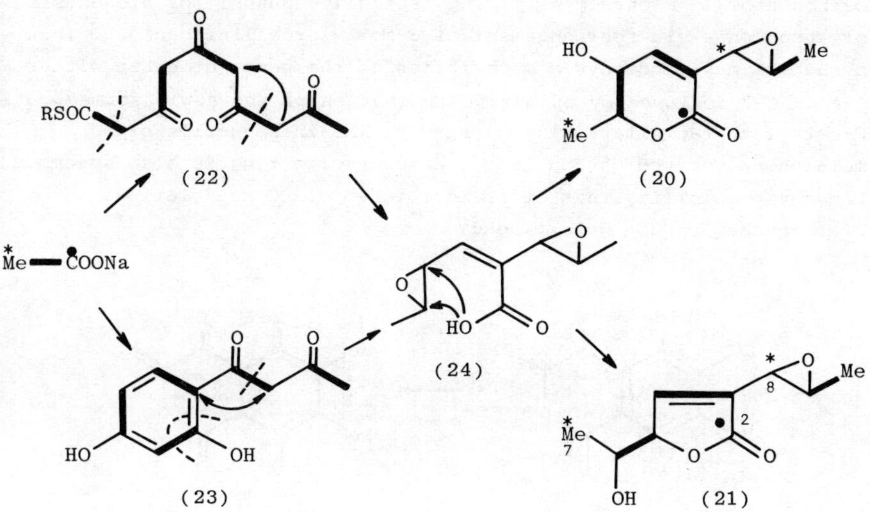

Scheme 6

were proposed.[27] However, on incorporation of $[2\text{-}^{13}C,^{2}H_3]$acetate and determination of the simultaneously ^1H and ^2H noise-decoupled ^{13}C n.m.r. spectrum, the resonance for the C-7 methyl showed two isotopically shifted signals, indicating that two acetate-derived hydrogens are retained on C-7, and so intermediates in which this carbon forms part of an aromatic ring are excluded.[28] On the basis of stereochemical differences between aspyrone and asperlactone, the epoxide (24) is proposed as a common intermediate; alternative modes of attack by the carboxylate on the epoxide would lead to (20) and (21) as shown in Scheme 6. Oxygen-18 labelling studies should yield further information on this point.

The biosynthesis of diplosporin (25), a toxic metabolite of Diplodia macrospora, has been studied, using ^{13}C-labelled acetates

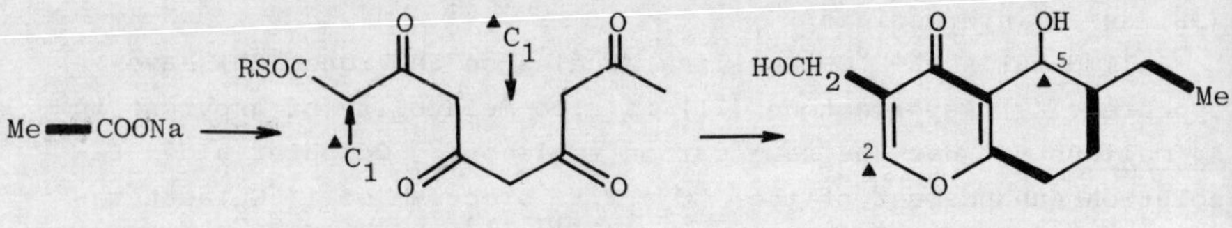

Scheme 7 (25)

and methionine.[29] The results indicate its derivation from a pentaketide chain, folded as shown in Scheme 7, with C-5 and C-2 derived from the C_1-pool. The presence of a methionine-derived carbon atom in a carbocyclic ring is highly unusual; cf biogenesis of tropolones via rearrangement of 3-methylorsellinic acid. The mechanism may proceed via methylation of the pentaketide at either C-4 or C-8 followed by oxidative activation of the newly formed C-methyl to facilitate ring closure. The introduction of a methionine-derived carbon into a heterocyclic ring is also uncommon and the possibility that it is introduced via O-methylation, cf rotenone, cannot be excluded.

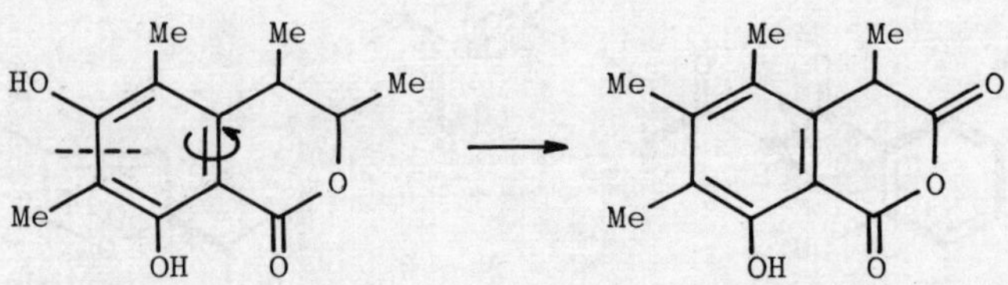

(26) (27)

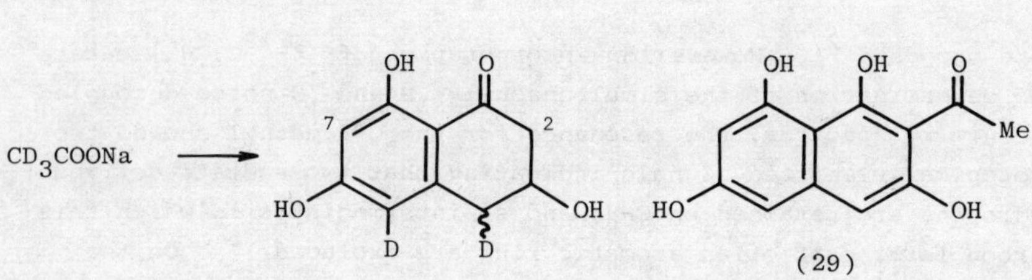

(28) (29)

Full details of ^{13}C, ^{2}H, and advanced precursor studies on sclerin biosynthesis in Sclerotinia sclerotiorum have appeared.[30] These are consistent with formation of sclerin (27) via ring cleavage and reorganisation of the carbon skeleton of sclerotinin A (26). A full paper has appeared on ^{13}C- and ^{2}H-labelling studies on scytalone biosynthesis in Phialaphora lagerbergii. To account for the lack of ^{2}H incorporation from acetate on C-2 and C-7 it is suggested that scytalone (28) may be formed via deacylation of a hexaketide-derived naphthol, eg (29). Such compounds are known - see O-methylasparvenone (46) below. Attempts to incorporate [^{13}C]-malonate to check for a 'starter' effect were unsuccessful in this study.[31] However, [2-^{13}C]malonate has been incorporated with high efficiency into scytalone and no 'starter' effect was observed.[32]

The antifungal metabolite citrinin (34), produced by Penicillium citrinum, has been the subject of intensive study by several research groups and notable use has been made of advanced precursors. The isocoumarin (35), labelled with ^{14}C at C-9, was specifically incorporated into citrinin whereas label from (36) was only incorporated after prior degradation to acetate, indicating that methylation of the polyketide precursor occurs before aromatisation.[33] However, in a study using a novel technique where P. citrinum was cultured in D_2O, incorporation of [1,2-$^{13}C_2$,$^{1}H_3$]-acetate and subsequent ^{13}C n.m.r. analysis allowed the origin of the hydrogens to be elucidated.[34,35] This indicated that the hydrogen on C-4 of citrinin was acetate-derived and so, although (35) is specifically incorporated, it cannot be an obligatory intermediate on the pathway. (This was confirmed by a ^{2}H-labelling study using [$^{2}H_3$]acetate and ^{2}H n.m.r.[35]). This study also revealed a marked difference in the protium content at C-1 and C-3, suggesting that the necessary reductions at these two sites are carried out at markedly different stages in the biosynthesis. Taken with the non-intermediacy of (35), this indicated that either the lactone (37) (reduction at C-3 but not C-1) or the aldehyde (30) (reduction at C-1 but not C-3) must be the first enzyme-free intermediate on the pathway. Both these compounds were synthesised with a single ^{2}H label on the C-11 methyl and fed to cultures of P. citrinum. ^{2}H n.m.r. analysis showed that only (30) was incorporated into citrinin.[36,37] The incorporation efficiency was 6.5%, with a dilution value of ca 62.5. In a further interesting experiment P. citrinum was grown in the presence of ethionine, which is known to inhibit methylation,

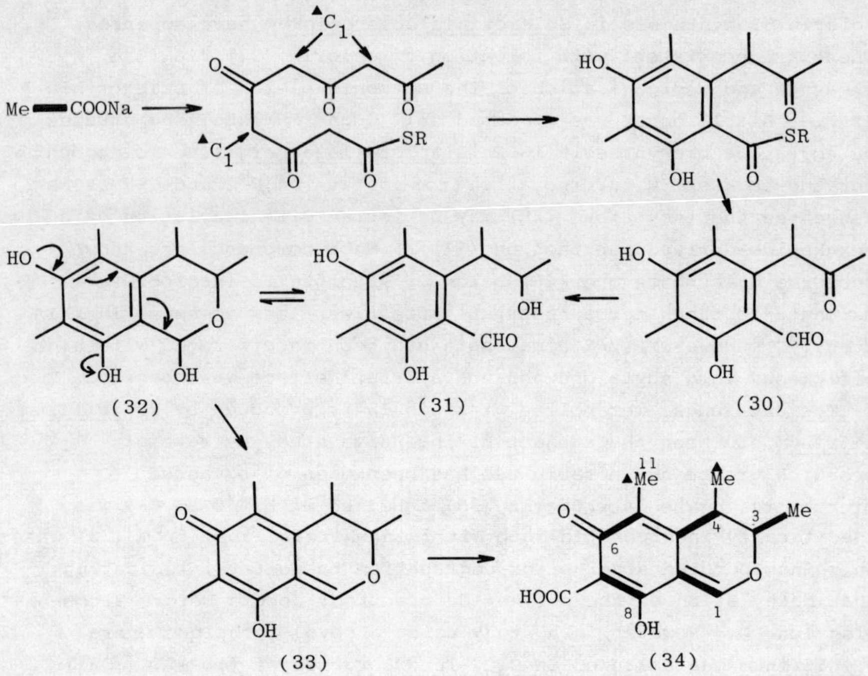

Scheme 8

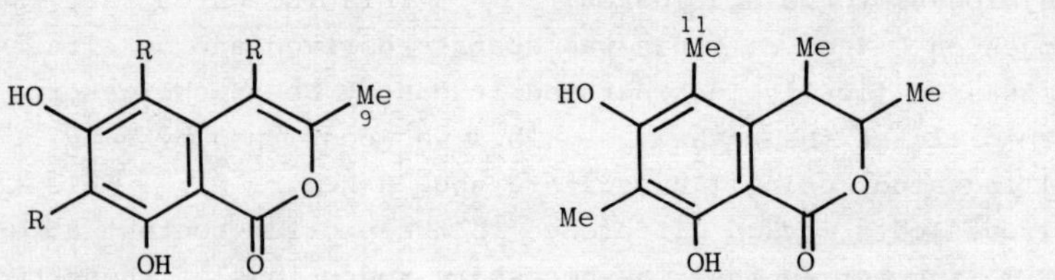

(35) R = Me
(36) R = H

(37)

resulting in suppression of citrinin production.[38] On incorporation of (30) labelled with ^{2}H at C-1 in the presence of ethionine in a replacement medium, a small amount of citrinin was isolated. Now, it was so highly enriched that the specific incorporation of (30) could be demonstrated by ^{1}H n.m.r. The incorporation rate was now 9.5% but the dilution was only 1.25. Parallel work by Scolastico and co-workers, using specifically ^{14}C-labelled precursors, has shown that both (30) and (32) are specifically incorporated,[39,40] and so the pathway shown in Scheme

8 is indicated for citrinin biosynthesis. Sankawa and co-workers have reported incorporations of [2-^{13}C,^{2}H$_3$]-, [1-^{13}C,^{18}O$_2$]-, and [1-^{13}C,^{17}O]acetates into citrinin.[41] The results are consistent with Scheme 8, and in the ^{13}C n.m.r. spectrum of [1-^{13}C,^{18}O$_2$]acetate-enriched citrinin, isotopically shifted signals are observed for the resonances due to C-3, C-6, and C-8, indicating origin of the attached oxygens from acetate, so that the quinone-methide structure must be formed by elimination of the hemi-acetal hydroxyl from (32).

5 Hexaketides

Incorporations of ^{13}C-labelled acetates and methionine[42] and of ^{14}C-labelled advanced precursors[43] into ascochitine (41), a

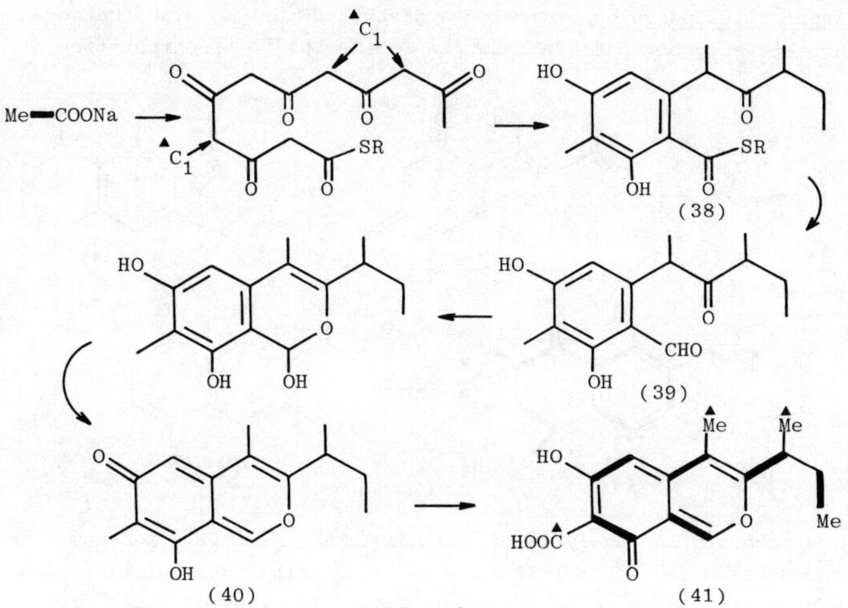

Scheme 9

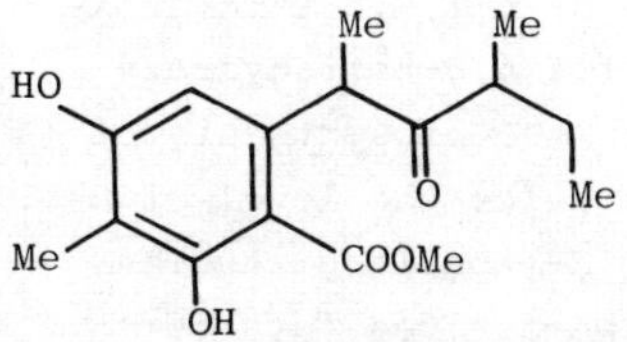

(42)

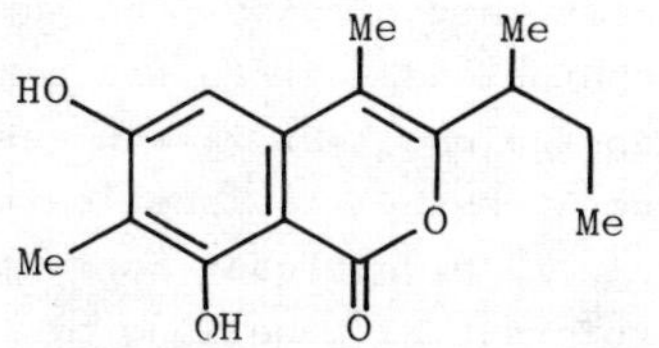

(43)

metabolite of the phytotoxic fungus Ascochyta fabae, have been reported. These show its derivation from a single hexaketide chain with introduction of three C_1 units from methionine to give a quinone-methide structure related to citrinin. The aldehyde (39) and quinone-methide (40) are specifically incorporated. The specific incorporation of the methyl ester (42) shows that the organism can convert it directly to the enzyme-bound thioester (38). The enol lactone (43) is also specifically incorporated, but 'enzyme trap' experiments show that it is not on the direct pathway, indicating that aldehyde (39) is formed by direct reduction of the thioester. Thus, the pathway shown in Scheme 9 can be proposed.

Incorporations of singly and doubly ^{13}C-labelled acetates into O-methylasparvenone (46), a dihydronaphthalene metabolite of Aspergillus parvulus, indicated a hexaketide origin with the novel acetate-assembly pattern shown in Scheme 10.[44] Incorporation of

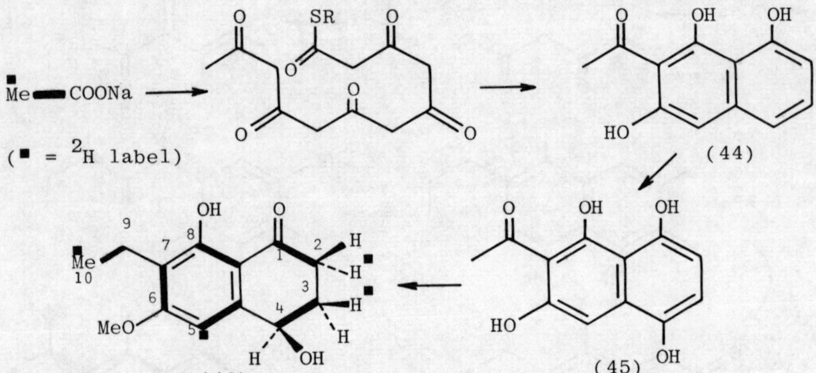

Scheme 10

$[^2H_3]$acetate and analysis of the resultant ^2H n.m.r. spectrum showed labelling of the 10-methyl, 5-, 2-axial, 3-axial hydrogens and significantly no labelling at C-4. The loss of label from C-4 and its appearance on C-3 can only be explained by an N.I.H. shift, which implies that hydroxylation of a 1,6,8-trihydroxy-naphthalene (44) to the corresponding 1,4,6,8-tetrahydroxy-naphthalene (45) is a necessary step in the biosynthesis of (46). The 10-methyl is labelled to less than twice the level of H-5. This is significant as, in a number of ^2H-labelling studies, preferential labelling of the acetyl-CoA-derived 'starter' position relative to positions derived from malonyl-CoA is observed. This therefore suggests that the necessary loss of ketide oxygen from C-9 occurs after aromatisation, allowing loss of label from C-10 relative to C-5 by exchange from an acetyl side-chain and via reduction to and dehydration of the resultant 1'-hydroxyethyl group during conversion to the ethyl side-chain. Support from this comes from a related study on the biosynthesis of the naphthoquinone (47), a metabolite of Hendersonula toruloidea.[45]

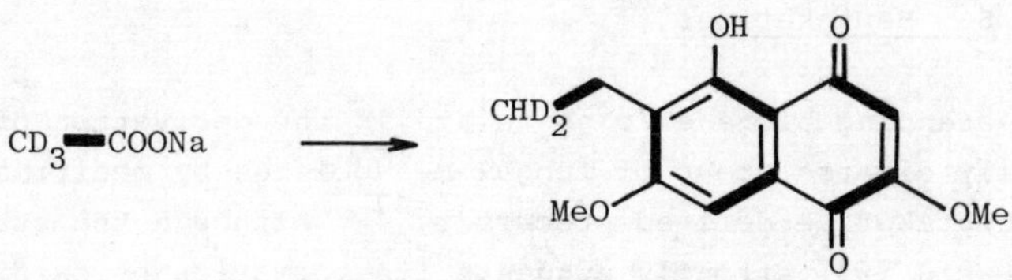

(47)

Incorporation of [$^{13}C_2$]acetate gave the assembly pattern shown and incorporation of [2-^{13}C,$^{2}H_3$]acetate indicated retention of only two acetate-derived hydrogens on C-10, consistent with its derivation from the acetate 'starter' and via the sequence $CH_3CO- \rightarrow CH_3CHOH \rightarrow CH_2=CH- \rightarrow CH_3CH_2-$.

Further details have appeared of biosynthetic studies on multicolanic, multicolic, and multicolosic acids (51; R = Me, CH_2OH, and CO_2H respectively) in Penicillium multicolor.[46] Specific incorporation of 6-pentyl[2-^{14}C]resorcylic acid (48), fed as the ethyl

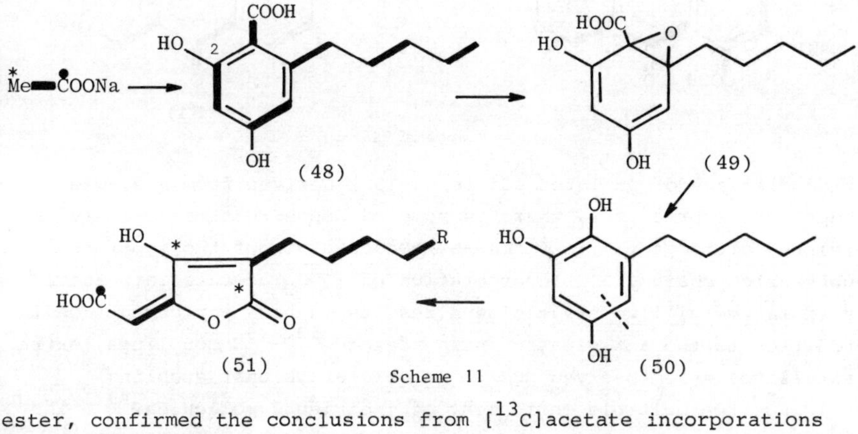

Scheme 11

ester, confirmed the conclusions from [^{13}C]acetate incorporations that these compounds are formed by oxidative cleavage of the aromatic ring of (48). The absence of randomisation of ^{13}C-^{13}C couplings in the [$^{13}C_2$]acetate-enriched metabolites excludes symmetrical intermediates, eg 5-pentylresorcinol, and, as shown in Scheme 11, the suggested pathway proceeds via the arene oxide (49) and the trihydroxyphenol (50). It seems likely that ω-oxidation of the pentyl side-chain occurs at a late stage in the biosynthesis.

6 Heptaketides

A long-standing biogenetic postulate is the derivation of a structurally diverse group of fungal metabolites by modification of a common heptaketide-derived precursor.[47] Although the structure of fulvic acid (53) strongly suggests its formation by oxidative

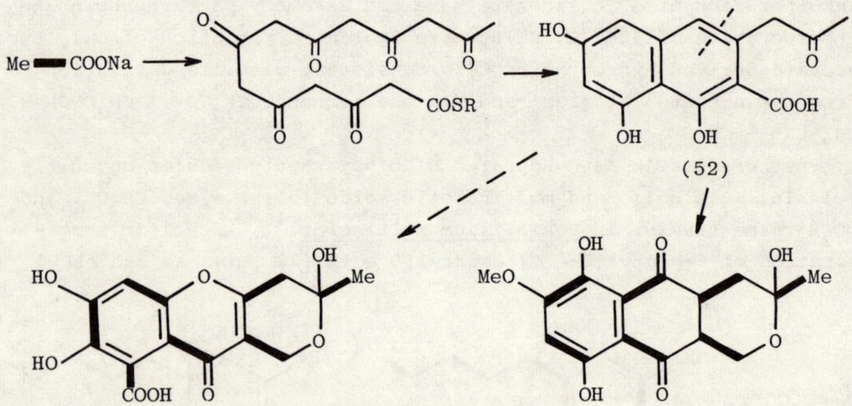

(53) (54)

Scheme 12

ring cleavage of an intermediate, eg (52), derived from a single heptaketide precursor, there is some evidence that the closely related citromycetin is formed by condensation of two separate polyketide chains.[48] Incorporation of [$^{13}C_2$]acetate into fulvic acid in *Penicillium brefeldianum* results in high enrichment and in addition to the anticipated *intra*-acetate ^{13}C-^{13}C couplings, extra satellites were observed due to *inter*-acetate unit coupling arising from multiple enrichment of individual molecules.[49] The ^{13}C-enrichment of the individual acetate-derived units was calculated to be *ca* 45% at the time of polyketide chain assembly in the presence of labelled acetate (subsequent dilution due to biosynthesis from endogenous acetate gives an overall enrichment of ca 2.6%). As the same level of inter-acetate couplings is observed between all the acetate units it is therefore suggested that a single chain origin is more likely than one in which separately formed chains come together. However, while the single-chain hypothesis is most attractive, in view of our poor understanding of the fundamentals of polyketide assembly processes at the enzyme level, these observations cannot be taken in any way as conclusive evidence. [^{13}C]Malonate or possibly ^{18}O-labelling studies would be more definitive, and ultimately advanced precursor studies are essential here. A potential precursor would be fusarubin, and the acetate assembly pattern of dihydrofusarubin (54) has been determined by feeding [$^{13}C_2$]acetate to *Fusarium solani* cultures.[50] This shows that (54) is biosynthesised via a heptaketide chain, folded as shown in Scheme 12. Deuterium from [2-^{13}C,$^{2}H_3$]acetate is incorporated only into the 11-methyl and the observed isotopically shifted signals show that up to three ^{2}H atoms are retained, thus proving its origin from the acetate 'starter' unit. Interestingly, parallel ^{2}H n.m.r. analysis shows the presence of some molecules containing ^{2}H but not ^{13}C. This suggests that during conversion of the enriched acetyl-CoA to malonyl-CoA, some ^{2}H is transferred to BCCP, where it does not exchange rapidly with the medium and is available for conversion of endogenous malonyl-CoA to ^{2}H-enriched acetyl-CoA.

Similar results are reported for incorporation of ^{2}H- and ^{13}C-labelled acetates into elsinochromes C and D (55) by *Pyrenochaeta terrestris*.[51] Both the C-14 and C-16 methyls incorporate up to

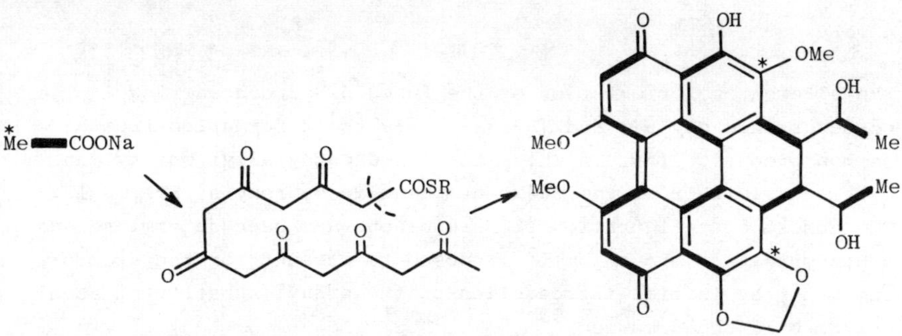

(55)

three ^{2}H atoms. Thus both are derived from acetate 'starter' units, excluding possible one-chain pathways, and so (55) is derived by dimerisation of two heptaketide-derived moieties, presumably by oxidative coupling after aromatisation. Note that in relation to the fulvic acid results above, extensive inter-acetate coupling was observed in [$^{13}C_2$]acetate-enriched (55) and there were no differences in the levels between and within the separate heptaketide-derived moieties.

Full details have appeared of biosynthetic studies with ^{13}C-labelled acetates and malonate on the phenalenone antibiotics deoxyherqueinone (58) and herqueichrysin (57) in Penicillium herquei.[52] Herqueichrysin is the only member of the group to have

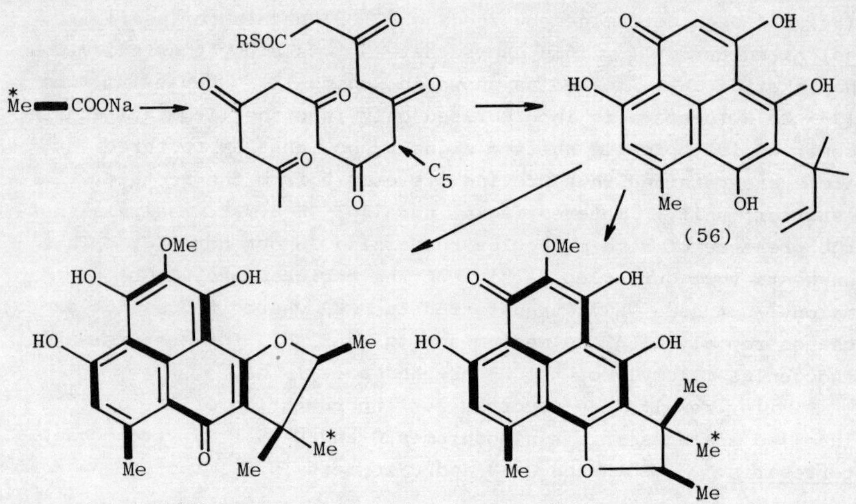

(57) (58)

Scheme 13

the alternative orientation of the fused dihydrofuran ring. The co-occurrence of (57) and (58) suggests their formation from a common precursor (56) in which the 1,1-dimethylallyl moiety can cyclise to either of the adjacent phenolic hydroxyls, Scheme 13. The absolute configuration of herqueinone has been determined and a biosynthetic mechanism is proposed to account for the stereochemistry by initial introduction of the prenyl substituent at a chiral bridgehead carbon.[53]

Further studies on incorporation of ^{2}H label from both acetate and the medium into griseofulvin by cultures of P. urticae are reported.[54]

7 Octaketides

The results of ^{13}C and ^{2}H n.m.r. analyses of [$^{2}H_3$]- and [$^{13}C_2$]-acetate-enriched tajixanthone (60), prepared by feeding experiments with Aspergillus variecolor, are reported.[55] These indicate derivation of the xanthone via ring cleavage of a precursor derived from an octaketide, folded as shown in Scheme 14.

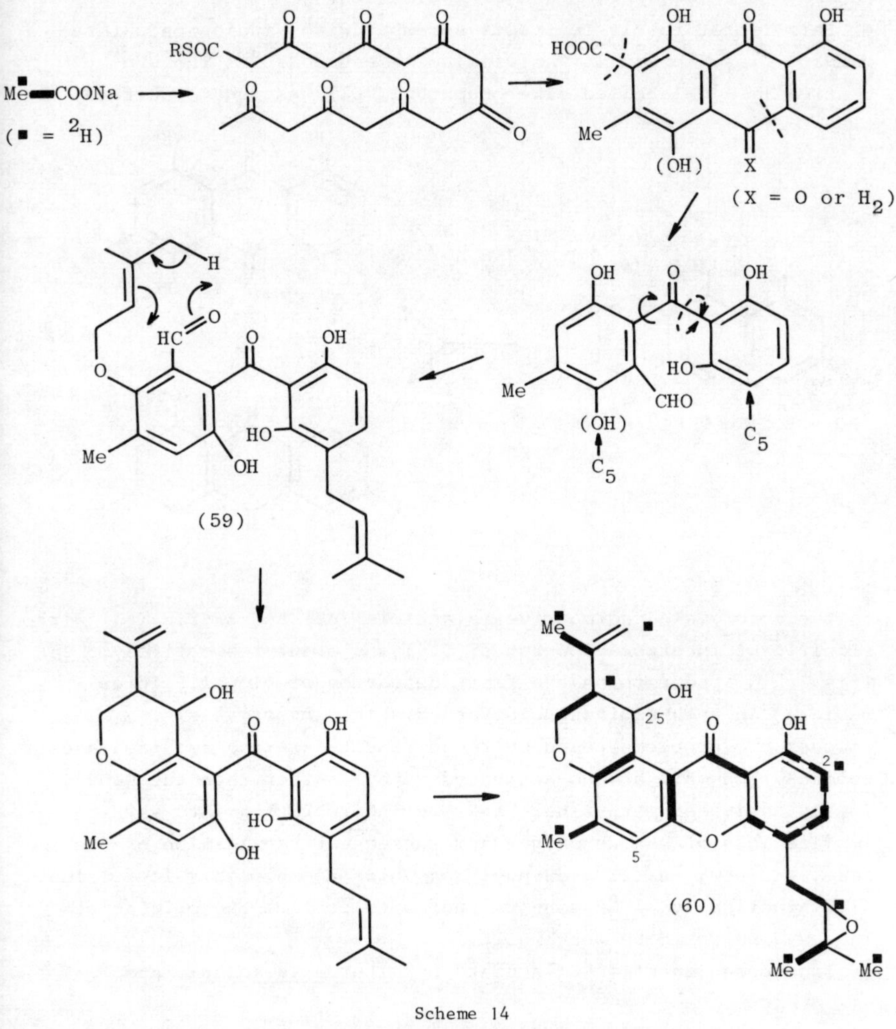

Scheme 14

Scrambling of label from [$^{13}C_2$]acetate in ring C indicates that ring cleavage must precede _C_-prenylation and the stereospecificity of ^{13}C-labelling in the dihydropyran ring moiety is consistent with cyclisation of the _O_-prenylaldehyde moiety in (59) by a concerted 'ene' reaction, suggesting that dihydropyran ring formation precedes formation of the xanthone ring system. The lack of ^{2}H label on C-25 implies cleavage of an anthraquinone rather than an anthrone as previously suggested; and the observation of ^{2}H on C-2 but not C-5 indicates that decarboxylation of the octaketide precursor occurs after cyclisation and aromatisation.

This latter result contrasts somewhat with studies on anthraquinone biosynthesis in _Penicillium islandicum_.[56] The synthetically ^{14}C-labelled diketonaphthol (61) was fed to surface

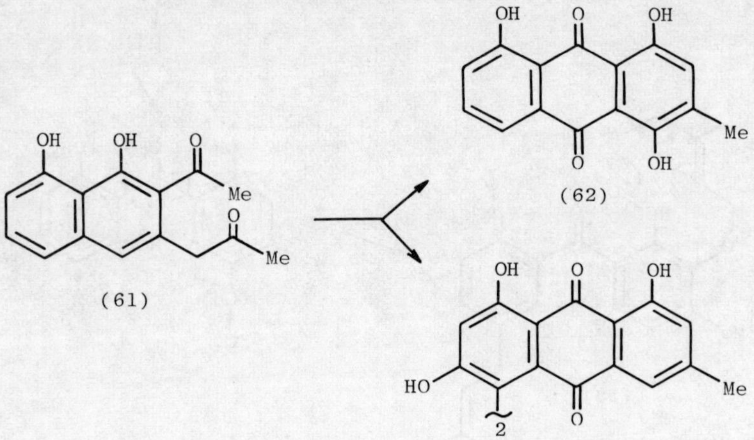

(63)

cultures to yield radioactive islandicin (62) and skyrin (63) with significant incorporation rates, 0.61 and 0.46% respectively. This would appear to be the first detection of bicyclic intermediates in anthraquinone biosynthesis in microorganisms. However, two problems posed by these results are the required aldol condensation onto a mono-activated methyl rather than the usual doubly activated methylene; and the incorporation into skyrin implies loss of ketide oxygen and subsequent reoxidation of the same, and now, unactivated position, which seems highly inefficient biosynthetically. Oxygen-18 studies to confirm the origins of the oxygens would be worthwhile.

Incorporation of [$^{13}C_2$]acetate by _Alternaria solani_ into

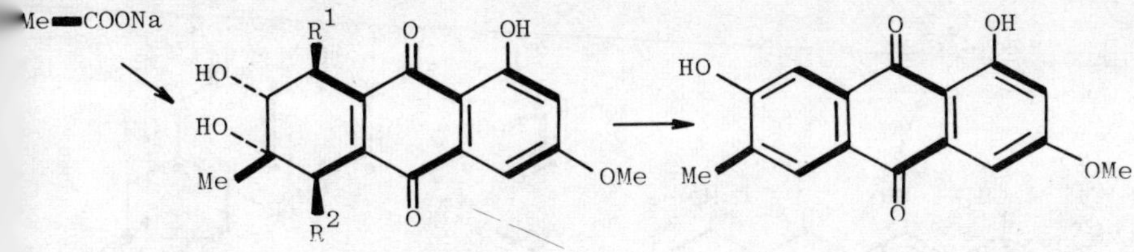

(64) $R^1 = R^2 = OH$
(65) $R^1 = R^2 = H$
(66) $R^1 = OH$, $R^2 = H$

(67)

altersolanol A (64) reveals the expected assembly pattern for an octaketide-derived anthraquinone.[22] On refeeding the ^{13}C-enriched altersolanol A to a mutant blocked for altersolanol A production, macrosporin A (67) was isolated and shown to be highly ^{13}C-enriched. Small amounts of altersolanol B (65) and (66) were also isolated. Thus, it is claimed that altersolanol A is metabolised by A. solani to altersolanol B and macrosporin A. However, the mechanisms are somewhat difficult to visualise.

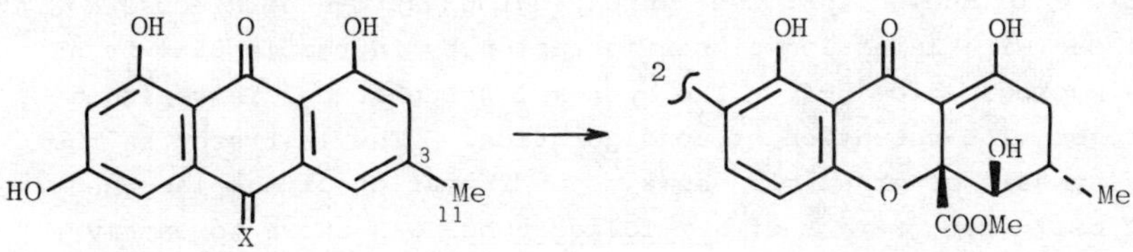

(68) X = O
(69) $X = H_2$

(70)

Concurrent incorporation of [3-^{14}C]emodin (68) and [11-^{14}C]-emodinanthrone (69) into secalonic acid D (70) by cultures of Penicillium oxalicum, and proving specificity of incorporation by Kuhn-Roth oxidation followed by Schmidt degradation of the resultant acetic acid, showed that the anthrone was incorporated 4.5 times better than the anthraquinone.[57] These results again necessitate loss of phenolic hydroxyl from the precursors.

The benzoisochromanequinone system is found in a large class of microbial metabolites. Several studies on the biosynthesis of this system have been reported. [1-^{13}C]- And [2-^{13}C]acetates are incorporated into granticin (72) by Streptomyces violaceoruber[58] as indicated in Scheme 15. Similar results are reported using Streptomyces olivaceus[59]. The remaining carbons are derived from

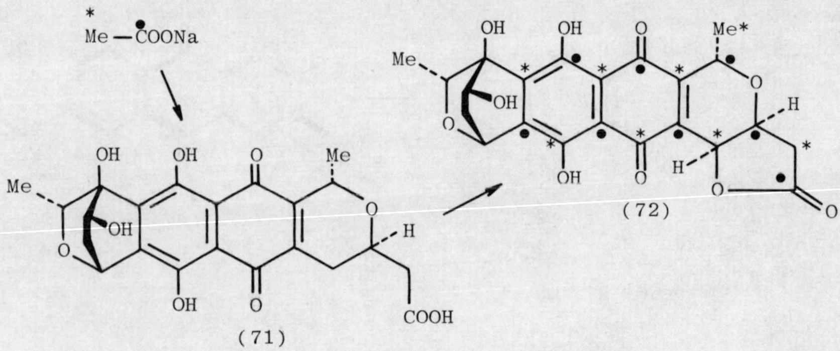

Scheme 15

glucose,[58] which is converted into a 2,6-dideoxyhexose and attached to the aromatic moiety by C-C linkages at C-1 and C-4. Conversion of glucose proceeds with retention of H-1, H-2, H-4, and the hydrogens at C-6 and loss of H-3 and H-5. Feeding (6R)- and (6S)-D-[4-^{2}H,6-^{3}H,6-^{14}C]glucose, followed by Kuhn-Roth oxidation, and determination of the chirality of the methyl of the resultant acetic acid showed that the hydroxyl group at C-6 of glucose is replaced with inversion of configuration by intramolecular transfer of a hydrogen from C-4. The hydroxyl group at C-2 is replaced by hydrogen with retention of configuration. The last step in the biosynthesis of granticin seems to be formation of the lactone ring, as a cell-free extract of S. violaceoruber was shown to catalyse

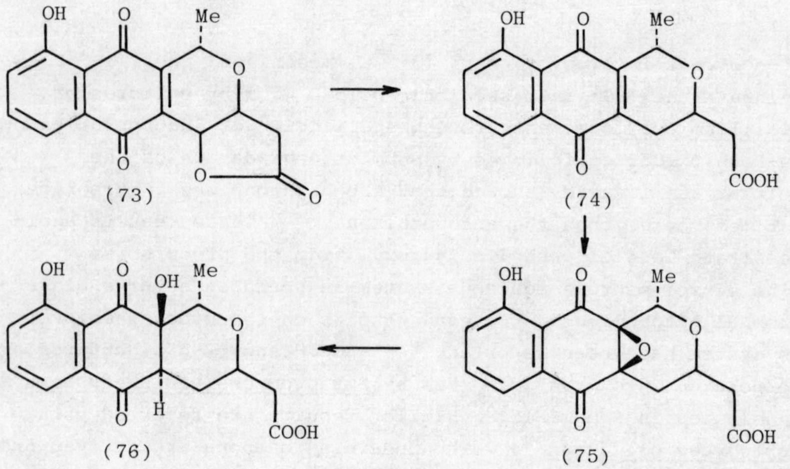

formation of (72) from dihydrogranticin (71) without incorporation of ^{18}O from $^{18}O_2$ to rule out a hydroxylation-lactonisation mechanism, and direct cyclisation of the carboxyl onto a quinone-methide intermediate is proposed. Contrasting results[60] have been found in biconversion studies of the nanomycins in cultures of Streptomyces rosa treated with cerulenin, which is known to inhibit the early stages of polyketide biosynthesis. These suggest that nanomycin D (73) is formed first and is converted to nanomycin A (74), then to E (75), and finally to B (76).

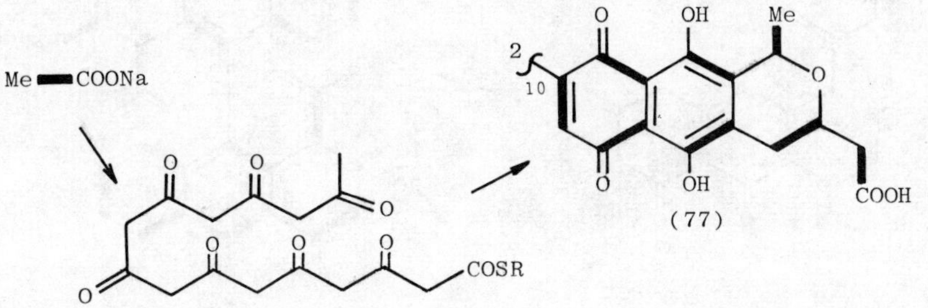

(77)

Scheme 16

A further benzoisochromane-quinone metabolite, actinorhodin (77), is elaborated by Streptomyces coelicolor. Its structure was known, apart from the point of dimerisation. Incorporation of [$^{13}C_2$]acetate established the acetate-assembly pattern shown in Scheme 16, and in addition the point of dimerisation was shown to be a carbon enriched from [2-^{13}C]acetate, ie C-10.[61]

8 Nonaketides

Incorporations of singly and doubly ^{13}C-labelled acetates and [2-^{14}C]- and [2-^{13}C]malonates by cultures of Pyrenochaeta

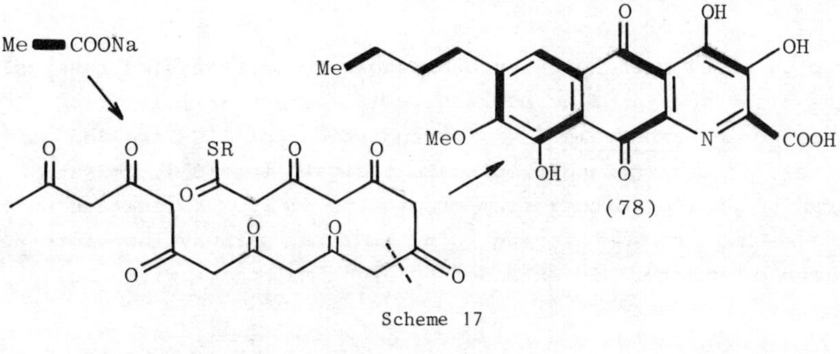

(78)

Scheme 17

<u>terrestris</u> indicate that the aza-anthraquinone phomazarin (78) is biosynthesised by condensation followed by oxidative ring fission of a nonaketide precursor, folded as indicated in Scheme 17.[62] Although a large number of metabolites appear to be formed by oxidative metabolism of anthraquinones, the cleavage of a benzenoid ring as in phomazarin biosynthesis, rather than the quinonoid ring, is highly unusual.

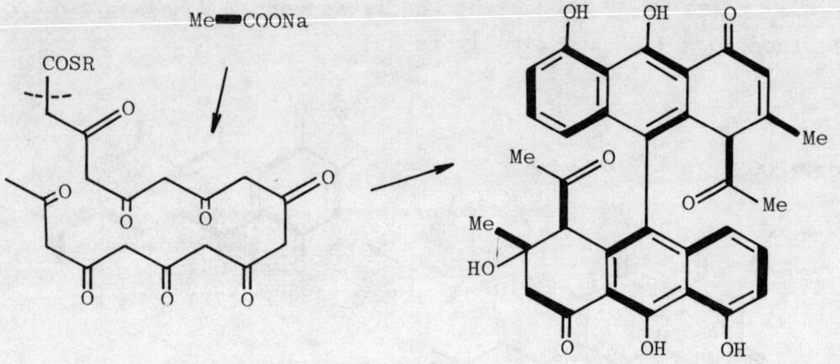

Scheme 18 (79)

Incorporations of [1-^{13}C]- and [$^{13}C_2$]acetates into the novel 9,9'-bianthryl antibiotic setomimycin (79) by cultures of <u>Streptomyces pseudovenezuelae</u> show that (79) is formed by oxidative coupling of two nonaketide-derived moieties, assembled as shown in Scheme 18 with loss of the terminal carboxyls.[63]

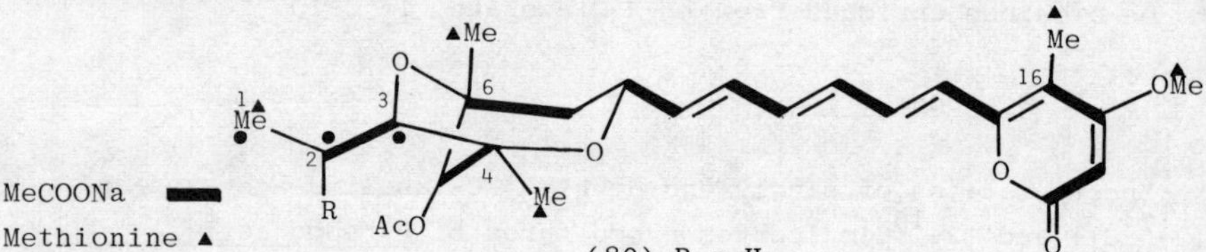

(80) R = H
(81) R = OH

Two parallel pathways have been shown to operate simultaneously in the biosynthesis of aurovertins B (80) and D (81) in <u>Calcarisporium arbuscula</u>.[64] Incorporations of ^{13}C-labelled acetates, methionine, and propionate indicate that C-1, C-2, and C-3 can be derived either from propionate or from an acetate unit plus a methyl from methionine. In the first pathway therefore a propionate 'starter' is extended by eight malonates, with the methyls attached to carbons 4, 6, and 16 derived from methionine. Incorporations of [2-^{13}C]malonate and [2-^{13}C,$^{2}H_3$]acetate reveal the absence of an acetate 'starter' unit and so the second pathway involves methylation of a C_{20} polyketide precursor at C-18, followed by loss of the chain-initiating acetate unit, C-19—C-20, with the other methyls again derived from methionine. This simultaneous operation of two independent pathways is unique amongst fungal metabolites.

9 Decaketides

The biosynthesis of the important mycotoxin aflatoxin B_1 (86) continues to attract considerable attention. The generally accepted pathway is summarised in Scheme 19, but there are still major gaps and many problems to be solved in this complex pathway. Further evidence that averufin (82) is an obligatory intermediate on the pathway is provided by experiments[65] in which averufin enriched biosynthetically from either [1-^{14}C]- or [$^{13}C_2$]-acetates is incorporated into aflatoxin B_1 with 18.5% efficiency (dilution value 5.1) by cultures of Aspergillus parasiticus ATCC 15517. Despite these excellent incorporation rates, the definitive study using a specifically labelled sample of averufin must be carried out.

The incorporation pattern resulting from incorporation of ^{13}C-labelled acetates into versiconal acetate (83) by dichlorvos-treated cultures of A. parasiticus is consistent with its origin from a single C_{20} polyketide and its proposed intermediacy in aflatoxin biosynthesis.[66] Two reports[67,68] have appeared on the isolation of cell-free enzyme systems capable of converting versiconal acetate to versicolorin A (84). Plausible mechanisms are proposed for this interesting conversion but their rigorous establishment awaits further study.

A kinetic pulse-labelling technique for the detection of transitory intermediates on the aflatoxin biosynthetic pathway has been described. It has been applied to both sterigmatocystin (85) formation in Aspergillus versicolor[69] and aflatoxin formation in A. parasiticus.[70] The results are largely in accord with the accepted sequence of intermediates. However, one ambiguity was the observation that radioactivity from [1-^{14}C]acetate appeared in the aflatoxins before it appeared in sterigmatocystin. This suggests that sterigmatocystin may not

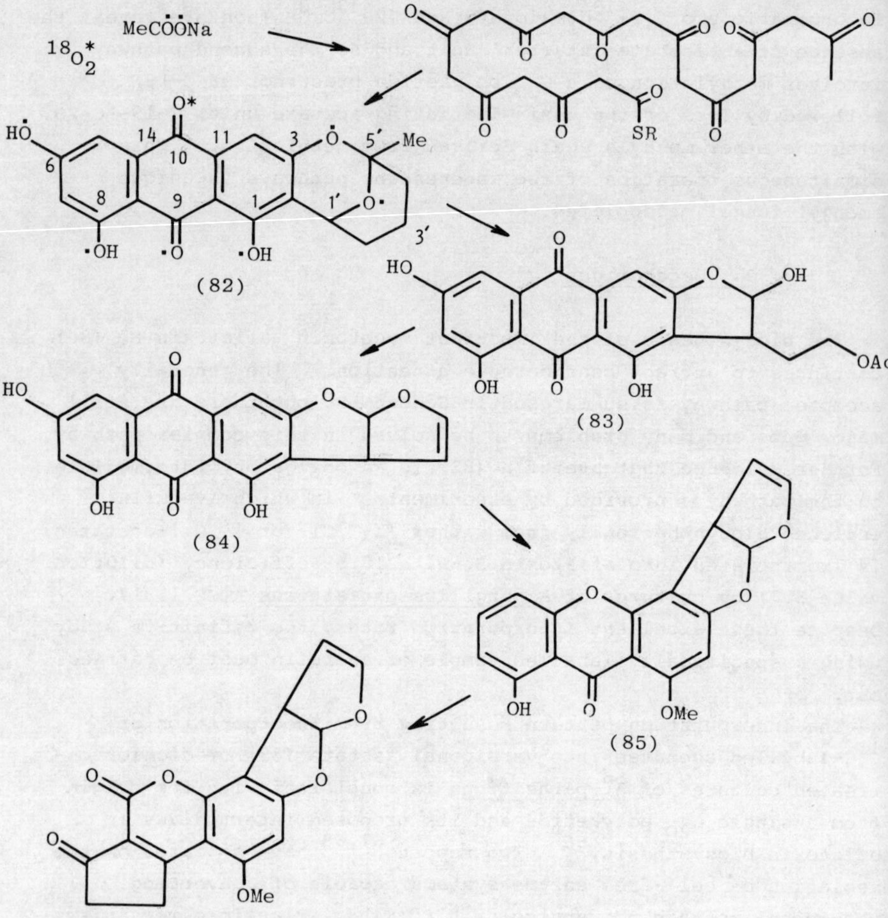

Scheme 19

actually be on the direct aflatoxin pathway and might explain, inter alia, why A. versicolor does not produce aflatoxins.

In one of the highlights of the review period, the first reported biosynthetic application of ^{18}O isotope shifts in ^{13}C n.m.r. has elucidated the origins of all the oxygen atoms in averufin.[71] [1-^{13}C, $^{18}O_2$]Acetate was added to cultures of A. parasiticus and the enriched averufin was analysed by ^{13}C n.m.r. As expected, carbons 1,9,8,6,11,14,3,1',3', and 5' were highly enriched. However, on expansion of the signals for carbons directly attached to oxygen, all except C-5' and C-10 showed isotopically shifted resonances (ΔδC 0.01-0.03 p.p.m.), indicating incorporation of ^{18}O, at C-1, C-3, C-6, C-8, C-9, and C-1'. Comparison of the integral ratios of the shifted and non-shifted signals suggested that about half the oxygen label was lost. In a second experiment, A. parasiticus was grown in an atmosphere highly enriched with $^{18}O_2$. This resulted in an isotopically shifted signal being observed for C-10 only.

10 Macrolides and Ionophores

The macrolide and ionophore group of antibiotics consist of long, usually polyoxygenated, carbon chains derived by combination of acetate, propionate, and butyrate units. Several groups of workers have reported studies using mainly ^{13}C-, ^{2}H-, and ^{18}O-labelled precursors, aimed at establishing the mechanistic details involved in the biosynthesis of these compounds, in particular with a view to establishing the extent to which the obvious similarities to classical fatty acid biosynthesis are in fact applicable to these functionally and stereochemically far more complex compounds.

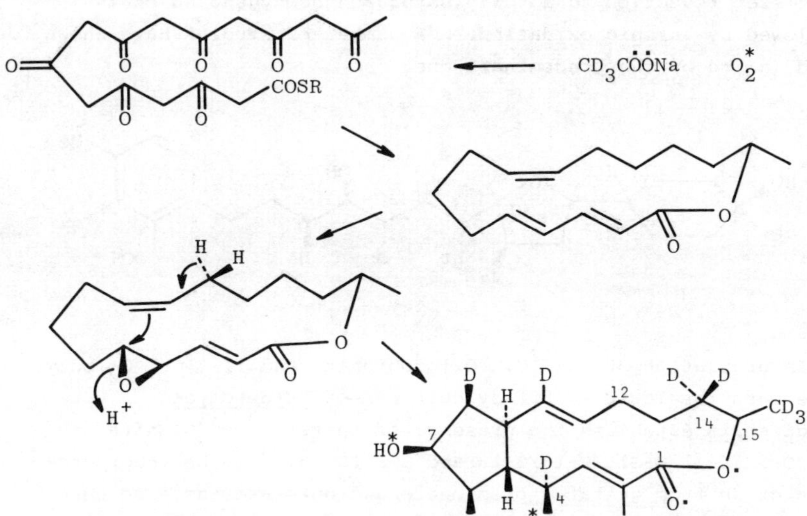

(87)

Scheme 20

One of the simpler systems to be studied has been brefeldin A

(87), a metabolite of Penicillium brefeldianum. Oxygen-labelling studies have established that the C-1 and C-15 oxygens are labelled equally by [^{18}O]acetate and that the C-4 and C-7 oxygens are labelled by two different molecules of oxygen.[72] Further studies using [^{14}C,^{3}H]-, [2-^{13}C,$^{2}H_3$]-, and [$^{2}H_3$]acetates have established that acetate-derived hydrogen is incorporated into brefeldin A with the regiospecificity[73] and stereospecificity[74] shown in Scheme 20. The results are fully discussed in terms of their similarities and differences to fatty acid biosynthesis. Noteworthy features are the unexplained lack of acetate-derived hydrogen on C-12 and the retention of both acetate-derived hydrogens on C-14, suggesting that the initially formed acetoacetyl-enzyme is rapidly reduced to β-hydroxybutyryl-enzyme before significant loss by exchange can occur. A proposed[72] mechanism for brefeldin A formation via an epoxide-initiated cyclisation is outlined in Scheme 20; cf monensin below.

Three possible mechanisms can be postulated for formation of the oxygen-bearing centres in the macrolides and ionophores: (a) retention or direct reduction of the keto group in the growing β-ketoacyl chain; (b) reduction followed by dehydration and stereospecific rehydration of the resulting enone; or (c) multistep reduction to a fully saturated deoxygenated chain followed by aerobic oxidation. A number of studies have shown path (a) to be the predominant one.

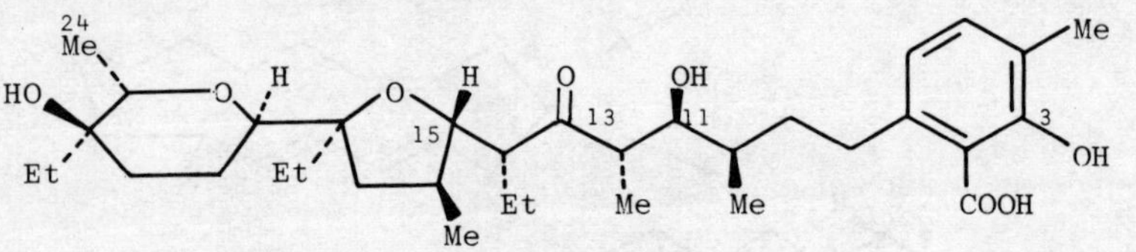

(88)

Incorporation of [1-^{13}C,$^{18}O_2$]propionate and [1-^{13}C,$^{18}O_2$]butyrate into lasalocid A (88) by cultures of Streptomyces lasaliensis establish the presence of intact ^{13}C-^{18}O units at C-3/O-3, C-11/O-11, C-13/O-13, and C-15/O-15.[75] The incorporation of ^{2}H from acetate, propionate, and butyrate has also been studied and the somewhat complex results are again analysed in relation to fatty acid biosynthetic processes.[76] All three ^{2}H labels from [1-^{13}C,$^{2}H_3$]acetate are incorporated into the propionate- and butyrate-derived methyls but oddly the apparent acetate 'starter'-derived methyl, C-24, of (88) retains no ^{2}H labels.

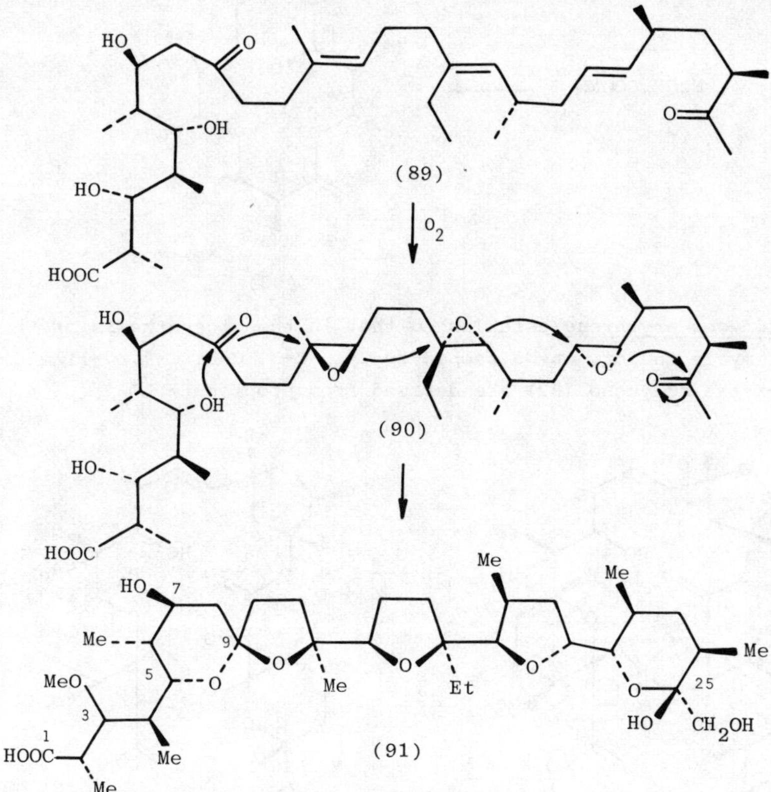

(91)

Scheme 21

Incorporation of ^{13}C-labelled precursors established the biosynthetic origins of all the carbons of monensin (91), an important antibiotic produced by <u>Streptomyces cinnamoniensis</u>.[77] When [1-^{13}C,$^{18}O_2$]propionate was fed, the oxygens attached to C-1, C-3, and C-5 were enriched, and [1-^{13}C,$^{18}O_2$]acetate enriched the oxygens attached to C-7, C-9, and C-25. It is likely that the remaining oxygens are derived from molecular oxygen. These results are therefore consistent with the derivation of monensin by formation of the triene (89), which can be converted to monensin via cyclisation of the triepoxide (90) as shown in Scheme 21.

Similarly, incorporation of [1-^{13}C,$^{18}O_2$]propionate by

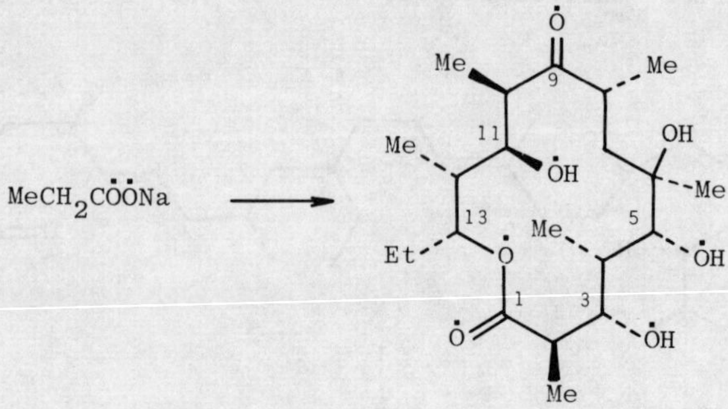

(92)

Streptomyces erythreus established that in the biosynthesis of erythromycin the oxygens attached to C-1, C-3, C-5, C-9, C-11, and C-13 of the aglycone (92) are derived from propionate.[78]

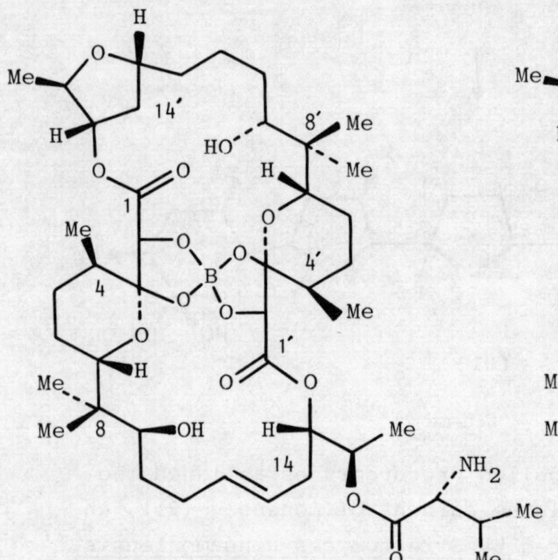

(93)

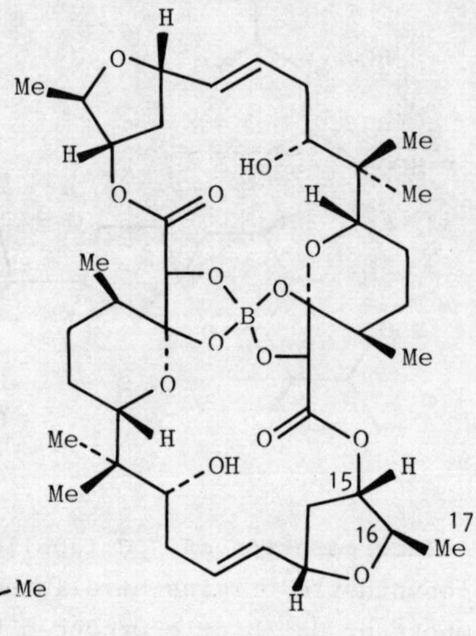

(94)

The biosynthesis of boromycin (93) has been studied[79] by feeding experiments with ^{13}C-labelled malonate and methionine and [2-$^{2}H_2$]valine in Streptomyces sp. MA 4423 followed by ^{13}C n.m.r. and mass spectral analysis. The results show that the carbon skeleton is derived from 14 acetate/malonate units, providing carbons 1-14 and 1'-14'; methionine gives rise to the methyls at C-4 and C-4' and the gem-dimethyl groups at C-8 and C-8'. D-Valine, rather than the L-isomer, is the immediate precursor of the D-valyl moiety of (93). Parallel studies[80] using [1,3-$^{13}C_2$]glycerol with the closely related aplasmomycin (94), produced by Streptomyces griseus, have shown that glycerol is the intact precursor of the 3-carbon 'starter' unit of the polyketide chain, C-15, C-16, and C-17. It is proposed that glycerol is converted via methylglyoxal to give lactate as the chain-initiating moiety.

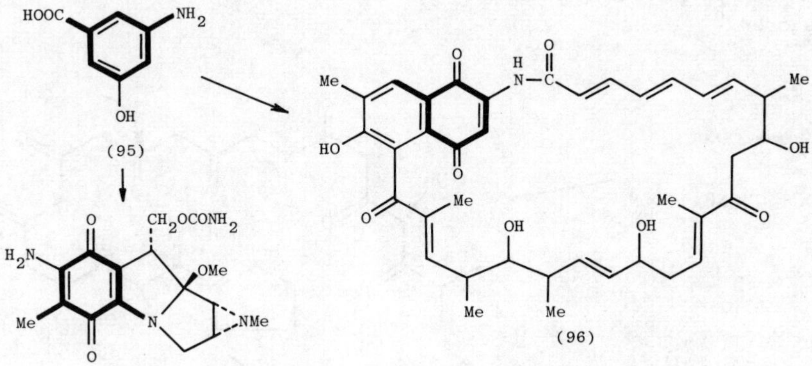
(95) (96) (97)

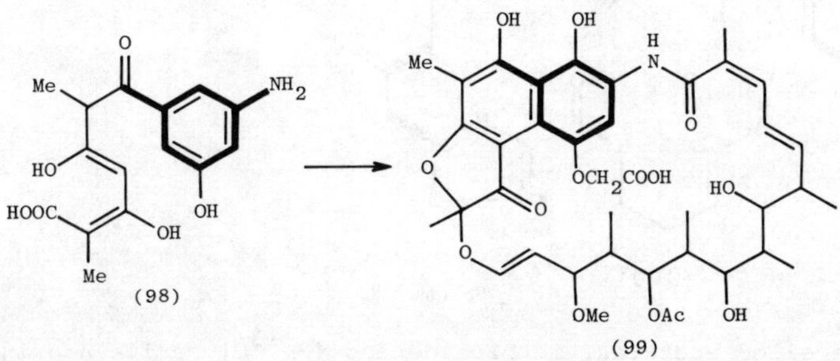
(98) (99)

The ansamycin group of antibiotics are formed by elongation of a 'C_7N' starter unit by acetate and propionate. The exact nature of the 'C_7N' unit was unknown but two independent studies have identified it as 3-amino-5-hydroxybenzoic acid (95). [Carboxy-^{14}C]-(95) was specifically incorporated by Streptomyces sp. E/784 into the novel ansamycin antibiotic actamycin (96).[81] Subsequent work showed that it was also incorporated into mitomycin (97) and h.p.l.c. analysis showed the presence of the free acid in culture filtrates of Streptomyces verticillatus; and its presence was also demonstrated by an isotopic dilution experiment using ^{13}C-labelled (95).[82] In an independent study using mutants of the rifamycin (99) producer Nocardia mediterranei it was found that a mutant P14 produced compound (98), which contains the 'C_7N' starter and the first three acetate/propionate chain-extending units of the ansa chain.[83] On cofermentation of P14 with a second mutant A8, which produced no rifamycin and instead accumulated shikimate, or on addition of (95) to A8, the normal rifamycin-producing capability of the parent strain was restored.[84]

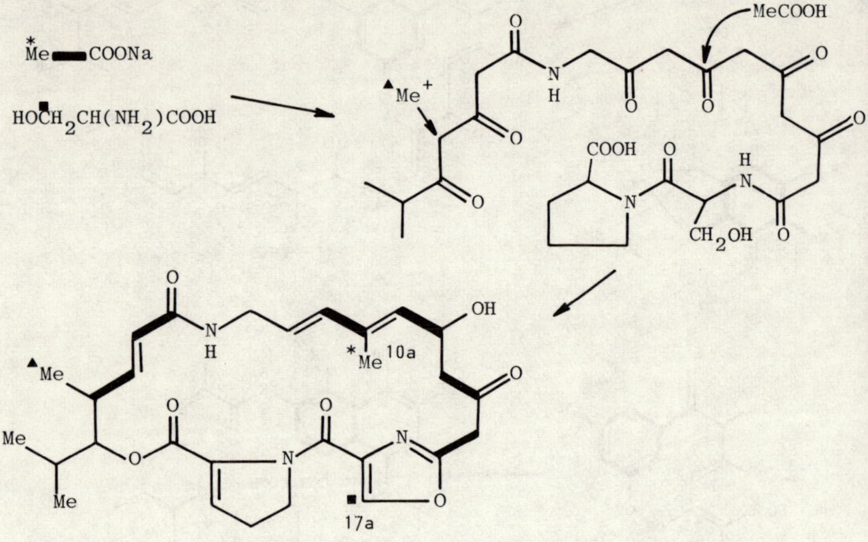

(100)

Scheme 22

^{13}C-Labelled acetates, methionine, and [3-^{13}C]-serine are incorporated into virginiamycin M (100) by Streptomyces virginiae as indicated in Scheme 22.[85] Noteworthy is the origin of the 10a-methyl group from an external acetate. Serine enriches C-17a consistent with the formation of the oxazole ring from an acylserine precursor. Isobutyryl CoA derived from valine is the likely 'starter' unit.

^{14}C- And ^{13}C-labelling studies have shown that the polyketide chain methyl groups in streptolydigin (101), a metabolite of Streptomyces lydicus, are derived from propionate and not from methionine.[86] The origins of the remaining carbons are yet to be

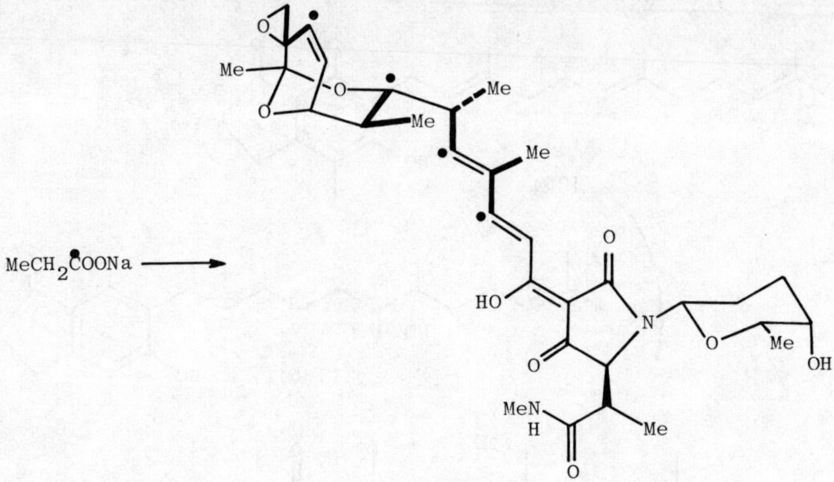

(101)

established but β-methylaspartic acid probably provides the remainder of the acyltetramic acid moiety.

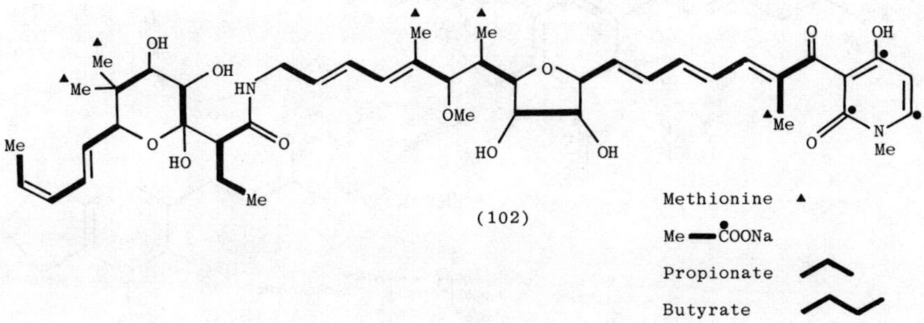

(102)

^{13}C-Labelling studies have shown that methionine, acetate, propionate, and butyrate provide most of the skeleton of aurodox (102), a metabolite of <u>Streptomyces goldiniensis</u>,[87] but the origins of the pyridone moiety are obscure. Whereas C-2, C-4, and C-6 are enriched by C-1 of acetate, C-2 of acetate failed to enrich C-3 and C-5 of (102).

11 Meroterpenoids

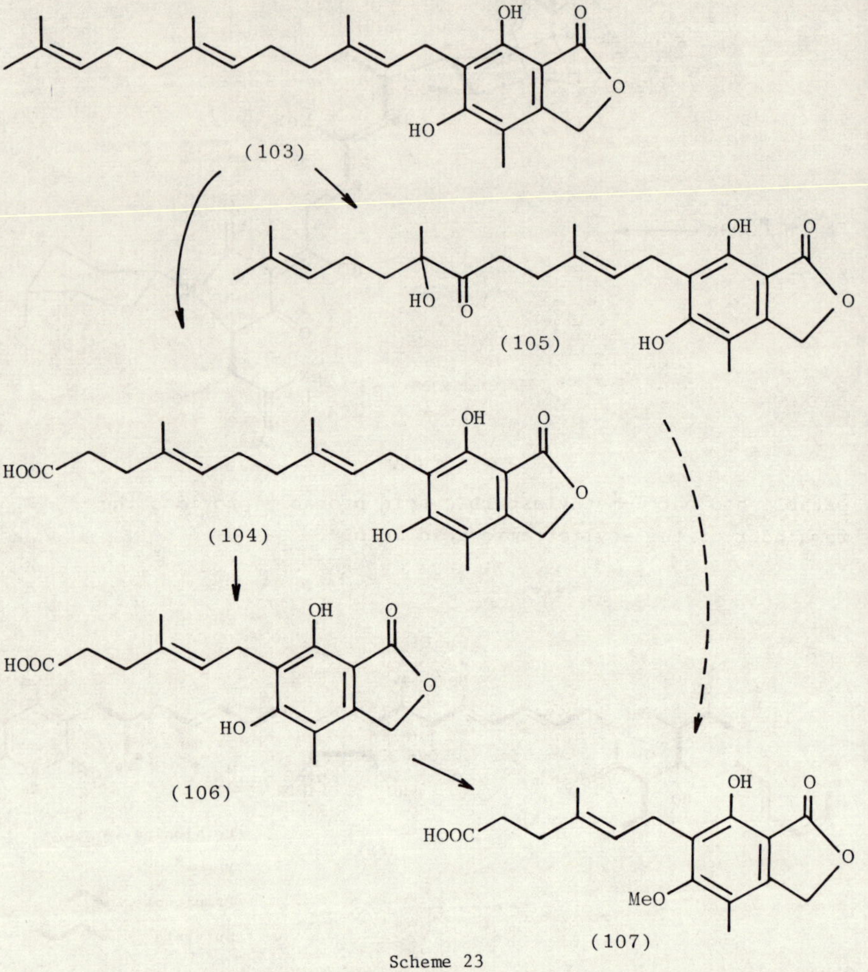

(103) (104) (105) (106) (107)

Scheme 23

Two separate studies of mycophenolic acid biosynthesis have shown that the prenylogue (104) is converted to mycophenolic acid (107) by whole cells of Penicillium brevicompactum.[88,89] Cell-free extracts have been shown to convert the phthalide (103) into the acyloin (105), which in turn is converted to mycophenolic acid by whole cells; Scheme 23. However, combined radiogas chromatography-mass spectrometric analysis of P. brevicompactum cultures indicate that both (103) and (104) but not the acyloin (105) are active in mycophenolic acid biosynthesis.[89]

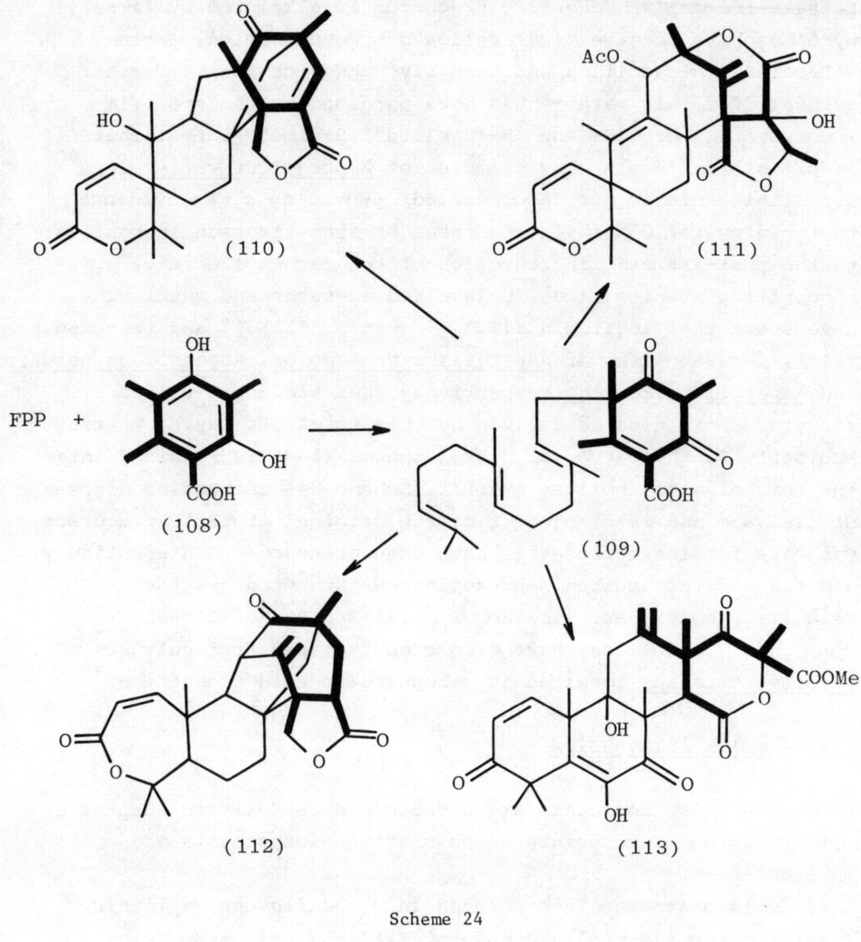

Scheme 24

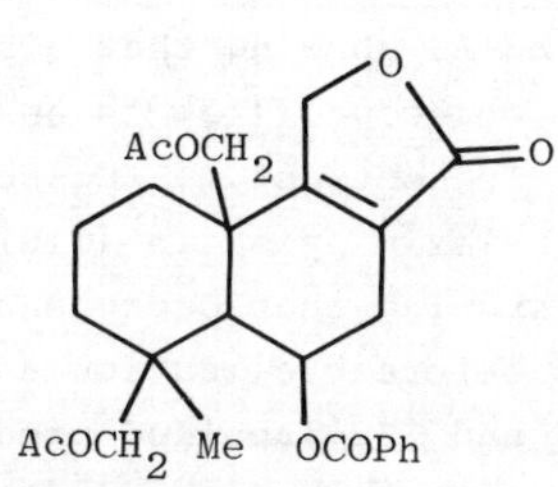

(114)

Previous ^{13}C-labelling studies had indicated that andibenin B (110) was formed by a novel pathway in which a bis-C-methylated tetraketide-derived phenolic precursor is alkylated by farnesyl pyrophosphate to give (109) followed by cyclisation, intramolecular cycloaddition, and oxidative modification. Further evidence for this pathway has been obtained by the specific incorporation of ^{14}C- and ^{2}H-labelled 3,5-dimethylorsellinate (108) into andibenin B by cultures of *Aspergillus variecolor*.[90] Orsellinic acid is not incorporated, providing clear evidence that biological C-methylation precedes aromatisation, in contrast to the post-aromatic introduction of the farnesyl moiety. Incorporation studies with ^{13}C-labelled acetates and methionine have shown that anditomin (112),[91] austin (111),[92] and terretonin (113),[93] metabolites of *Aspergillus variecolor*, *Aspergillus ustus*, and *Aspergillus terreus* respectively, despite their varied structures, can also be derived by this novel pathway. Whereas in andibenin B, the carbocyclic skeleton of (108) is retained intact, the remaining metabolites exhibit (Scheme 24) increasing degrees of cleavage and rearrangement of the original aromatic precursor. Pathways for these processes have been proposed. Interestingly, a mutant of the andibenin-producing culture produces the sesquiterpenoids, astellolides A (114) and B.[94] Closely related compounds, pebrolides, have also been isolated from cultures of *P. brevicompactum* impaired in mycophenolic acid production.[95]

12 Flavonoids

Though these compounds are discussed more fully in chapter 2, some features of relevance to polyketide biosynthesis are included here.

[$^{13}C_2$]Acetate was incorporated into the flavone apigenin (116) and the flavonol kaempferol (117) by cell suspension cultures of parsley, *Petroselinum hortense*, with randomisation of ^{13}C-^{13}C couplings in ring A, showing that a symmetrical intermediate, presumably the chalcone (115), is an intermediate in their biosynthesis.[96] In contrast, [$^{13}C_2$]acetate is incorporated into ring A of the phytoalexin pisatin (118) in *Pisum sativum* without randomisation, showing that deoxygenation of the polyketide precursor occurs before cyclisation and aromatisation.[97]

Incorporation of ^{14}C- and ^{13}C-labelled precursors into chloroflavonin (119) by cultures of *Aspergillus candidus* revealed

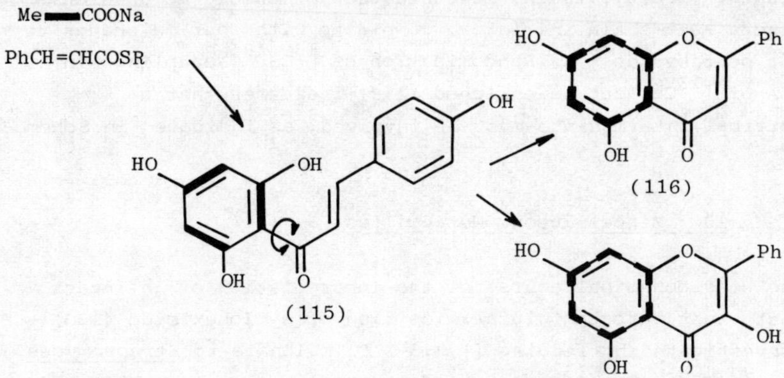

(115)

(116)

(117)

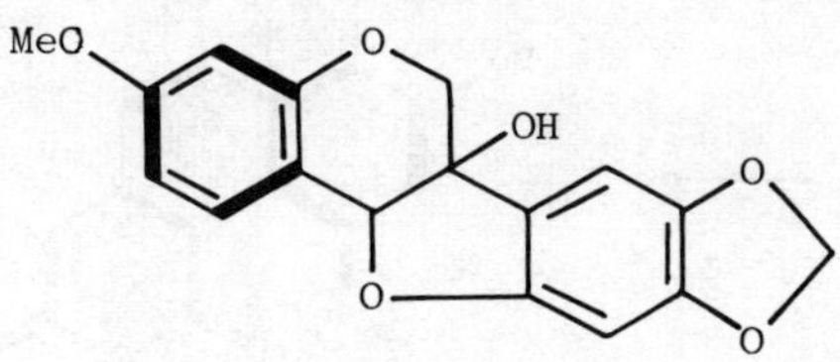

(118)

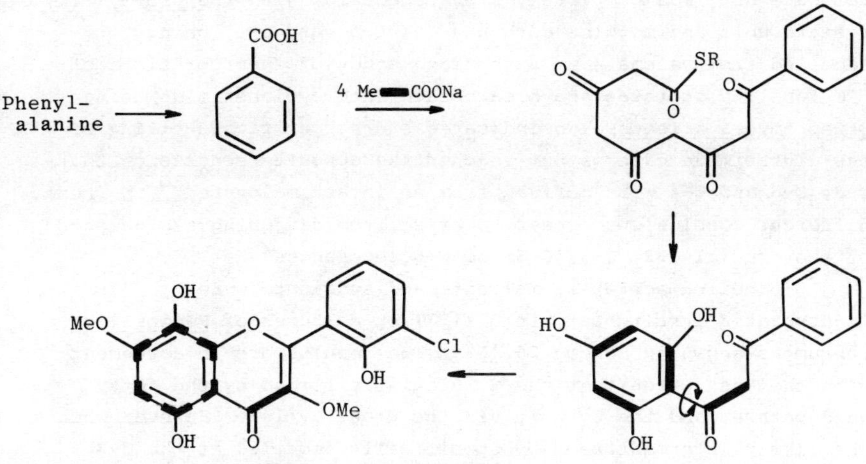

(119)

Scheme 25

a distinctly different route to the flavonoid skeleton compared to plants. Phenylalanine is converted to benzoic acid, which then acts as a chain initiator, combining with four malonates to give a pentaketide.[98] Randomisation of ^{13}C-^{13}C couplings in ring A of [$^{13}C_2$]acetate-enriched (119) indicates that a symmetrical intermediate must be involved, as indicated in Scheme 25.[99]

13 Miscellaneous Metabolites

The stereochemical course of the incorporation of an intact malonate unit into the glutarimide ring of cycloheximide (120) was investigated by feeding [1,2,3-$^{13}C_3$]malonate to Streptomyces naraensis.[100] The ^{13}C n.m.r. of the enriched cycloheximide and its derivatives indicated that the incorporation of an intact

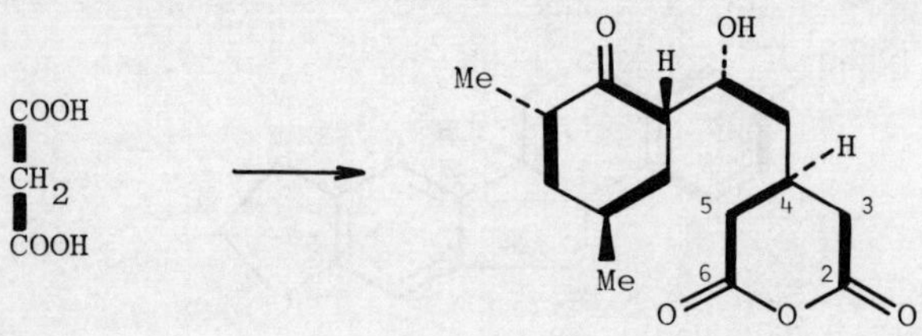

(120)

malonate unit was completely stereospecific, with the pro-S acetate unit and methine carbon (C-4, C-5, and C-6) being labelled from malonate. A separate study of incorporations of ^{13}C-labelled acetates and bicarbonate into cycloheximide, using Streptomyces griseus, also indicated stereospecific labelling of the glutarimide carbons but reached the opposite conclusion that C-2, C-3, and C-4 were derived from an intact malonate.[101] These different conclusions appear to arise from differing assignments of the crucial C-2, C-3, C-5, and C-6 resonances.

^{13}C-Labelled acetates, malonate, and succinate were incorporated into rubratoxin B (122) by cultures of Penicillium rubrum as shown in Scheme 26.[102] The results are in agreement with the assumption that the C_{10}-chain is formed by the fatty acid pathway and the C_3-unit via the Krebs cycle. However, when the likely intermediate (121), specifically labelled at the C-3 methyl, was fed, randomisation of label was observed, so its

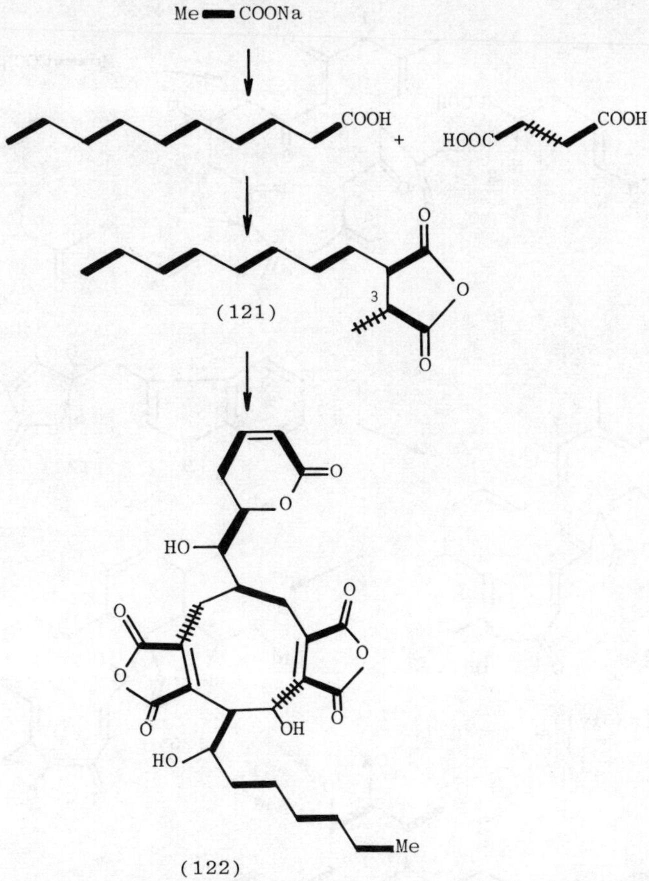

Scheme 26

status as a biosynthetic intermediate is not established.

Incorporation studies using ^{2}H- and ^{13}C-labelled acetates and methionine and ^{2}H-, ^{14}C-, and ^{15}N-labelled tryptophan indicate the pathway shown in Scheme 27 for the biosynthesis of chaetoglobosin A (126) and 19-O-acetylchaetoglobosin A (127) in Chaetomium globosum.[103] It is suggested that a C_{18} polyketide combines with L-tryptophan, probably forming an amide linkage first, and subsequently closing the lactam ring to form the tetramic acid (123). This is then transformed by reduction and dehydration to (124), which can undergo an internal Diels-Alder cyclisation to form

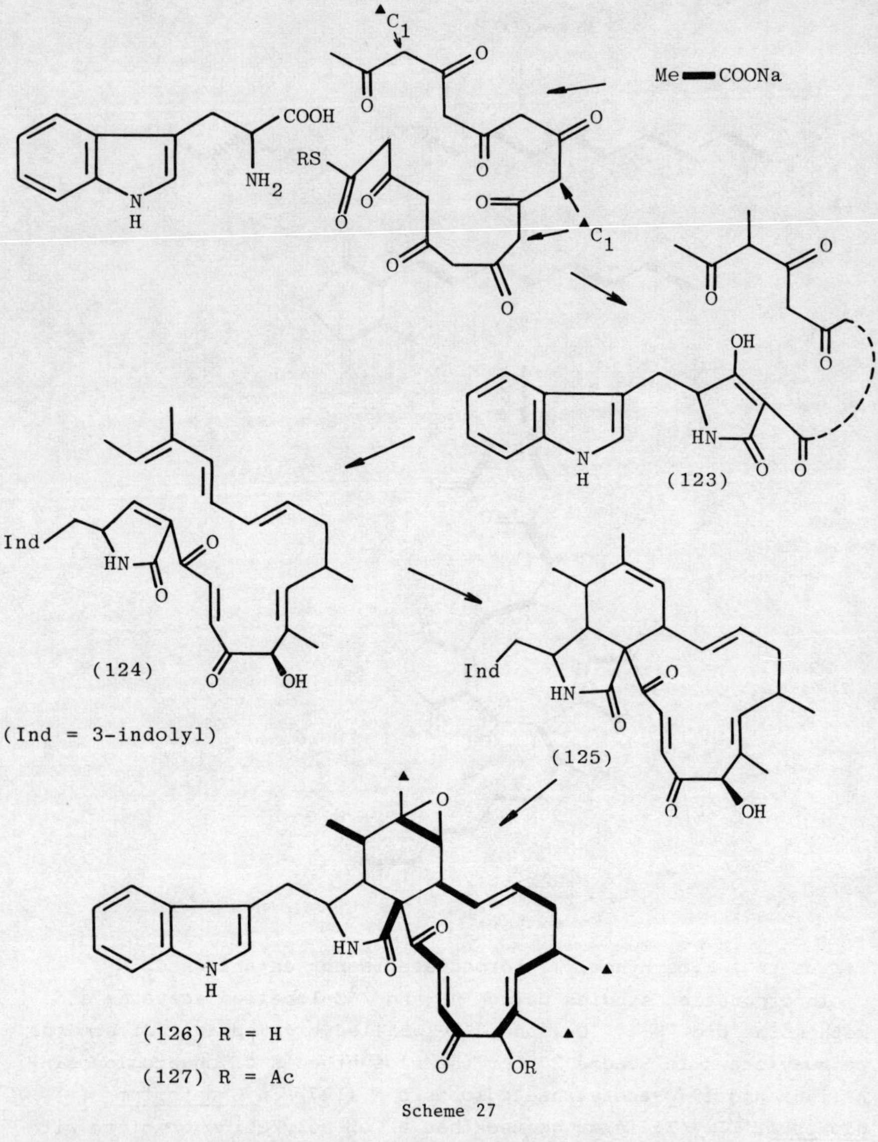

Scheme 27

chaetoglobosin J (125) before final acetylation to give (126).

A tetramic acid intermediate (128) has also been proposed in the biosynthesis of pseurotin A (129), a metabolite of Pseudorotium ovalis; Scheme 28. Feeding studies with ^{13}C-labelled propionate, acetate, methionine, and phenylalanine confirm the origin of all

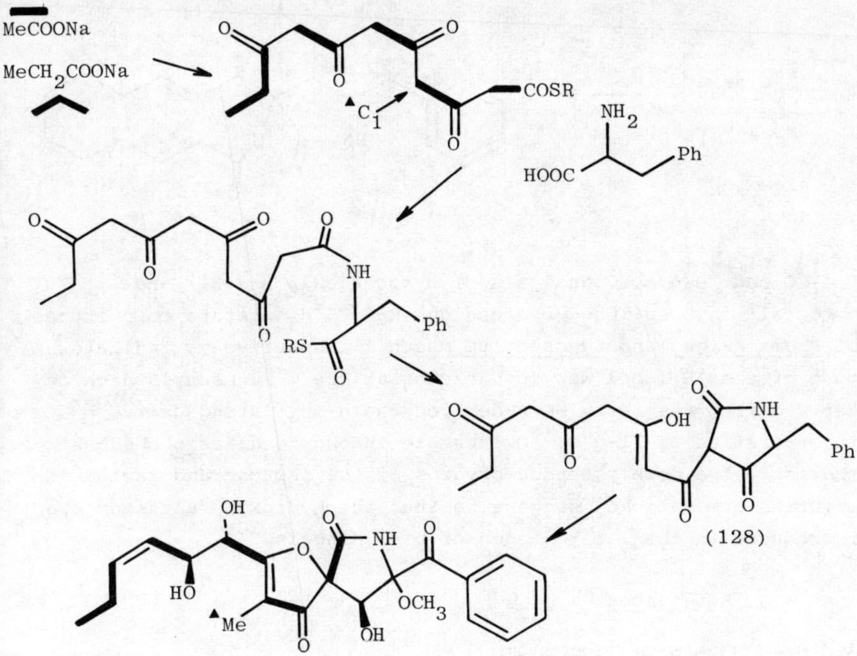

(129)

Scheme 28

the carbons. Incorporation of [2-^{13}C,^{15}N]phenylalanine shows that ^{15}N is retained, and so nitrogen-free intermediates can be excluded.[104]

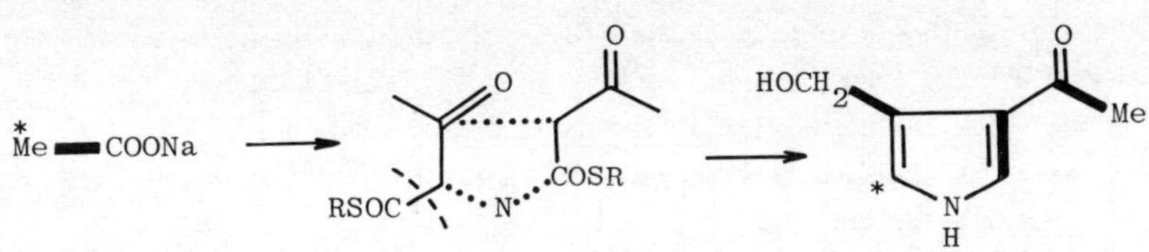

(130)

Scheme 29

Incorporation of [$^{13}C_2$]acetate into verrucarin E (130) by cultures of <u>Myrothecium verrucaria</u> indicates its formation by condensation of two molecules of acetoacetate as shown in Scheme 29.[107]

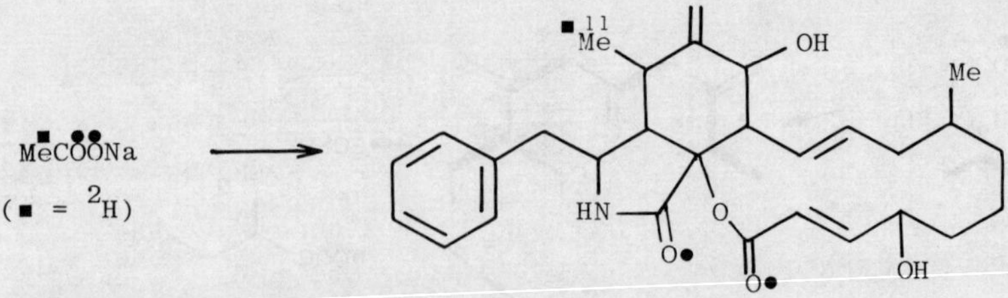

(131)

^{13}C and ^{2}H n.m.r. analysis of cytochalasin B (131) and cytochalasin D, enriched by feeding [2-^{13}C,^{2}H$_3$]acetate to cultures of Phoma exigua and Zygosporium masonii respectively, indicate most of the ^{2}H label was lost except at the C-11 methyl, which is derived from the acetyl-CoA-derived chain-initiating unit.[105] Incorporation of [1-^{13}C,^{18}O$_2$]acetate into cytochalasin B shows that only the doubly bonded oxygens of the lactone and lactam moieties are enriched, suggesting that the hydroxyl functions are introduced in the later stages of biosynthesis.[106]

References

1 T.J. Simpson, in 'Biosynthesis', vol. 6, ed. J.D. Bu'lock, Specialist Periodical Reports, The Chemical Society, London, 1977, p. 1.

2 M.J. Garson and J. Staunton, Chem. Soc. Rev., 1979, 8 , 539.

3 C. Abell and J. Staunton, J.C.S. Chem. Comm., 1981, 856.

4 J.C. Vederas, J. Amer. Chem. Soc., 1980, 102, 374.

5 'The Biosynthesis of Secondary Metabolites', R.B. Herbert, Chapman and Hall, London, 1981.

6 'Biosynthesis of Natural Products', P. Manitto, Ellis-Horwood, London, 1981.

7 'The Biosynthesis of Mycotoxins', ed. P.S. Steyn, Academic Press, London and New York, 1980.

8 B. Sedgwick and C. Morris, J.C.S. Chem. Comm., 1980, 96.

9 K. Saito, A. Kawaguchi, Y. Seyama, T. Yamakawa and S. Okuda, Eur. J. Biochem., 1981, 116, 581.

10 K. Saito, A. Kawaguchi, Y. Sayama, T. Yamakawa and S. Okuda, J. Biochem. (Tokyo), 1981, 90, 1697.

11 A. Kawaguchi, T. Yoshimura, K. Saito, Y. Seyama, T. Kasama, T. Yamakawa and S. Okuda, J. Biochem. (Tokyo), 1980, 88, 1.

12 A.G. McInnes, J.A. Walter and J.L.C. Wright, Tetrahedron Lett., 1979, 3245.

13 R.H. White, Biochemistry, 1980, 19, 9.

14 R.H. White, Biochemistry, 1980, 19, 15.

15 R.H. White, J. Amer. Chem. Soc., 1980, 102, 6605.

16 F. Lynen, Eur. J. Biochem., 1980, 112, 431.

17 J. Lari and R. Thomas, Tetrahedron, 1980, 36, 3305.

18 J. Sekiguchi and G.M. Gaucher, Biochemistry, 1979, 17, 1785; Can. J. Microbiol., 1979, 25, 881

19 J. Sekiguchi, G.M. Gaucher and Y. Yamada, Tetrahedron Lett., 1979, 41.

20 G.M. Gaucher, J. Food Protect., 1979, 42, 810.

21 A. Nakagawa, H. Ohno, K. Miyano, S. Omura, J. Org. Chem., 1980, 45, 3268.

22 A. Stoessl, C.H. Unwin and J.B. Stothers, Tetrahedron Lett., 1979, 2481.

23 A.E. de Jesus, P.S. Steyn, R. Vleggaar and P.L. Wessels, J.C.S. Perkin 1, 1980, 52.

24 D. Wiesleder and E. Lillehoj, Tetrahedron Lett., 1980, 21, 993.

25 J.S.E. Holker and T.J. Simpson, J.C.S. Perkin 1, 1981, 1397.

26 L. Colombo, C. Gennari, G. Severini Ricca, C. Scolastico and F. Aragozzini, J.C.S. Chem. Comm., 1981, 575.

27 R.G. Brereton, M.J. Garson and J. Staunton, J.C.S. Chem. Comm., 1980, 1165.

28 M.J. Garson and J. Staunton, J.C.S. Chem. Comm., 1981, 708.

29 A.A. Chalmers, C.P. Gorst-Allman, P.S. Steyn and R. Vleggaar, J.C.S. Perkin 1, 1979, 1481.

30 J. Barber, M.J. Garson and J. Staunton, J.C.S. Perkin 1, 1981, 2584.

31 U. Sankawa, H. Shimada, T. Sato, T. Kinoshita and K. Yamasaki, Chem. Pharm. Bull., 1981, 29, 3536.

32 E. Bardshiri and T.J. Simpson, unpublished results.

33 R.H. Carter, M.J. Garson and J. Staunton, J.C.S. Chem. Comm., 1979, 1097.

34 J. Barber and J. Staunton, J.C.S. Chem. Comm., 1979, 1098.

35 J. Barber and J. Staunton, J.C.S. Perkin 1, 1980, 2244.

36 J. Barber and J. Staunton, J.C.S. Chem. Comm., 1980, 552.

37 J. Barber, R.H. Carter, M.J. Garson and J. Staunton, J.C.S. Perkin 1, 1981, 2277.

38 J. Barber and J. Staunton, J.C.S. Chem. Comm., 1980, 1163.

39 L. Colombo, C. Gennari, C. Scolastico, F. Aragozzini and C. Merendi, J.C.S. Chem. Comm., 1980, 1132.

40 L. Colombo, C. Gennari, D. Potenza, C. Scolastico, F. Aragozzini and C. Merendi, J.C.S. Perkin 1, 1981, 2594.

41 U. Sankawa, Y. Ebizuka, H. Noguchi, Y. Ishikawa, S. Kitigawa, T. Kobayashi and H. Seto, Heterocycles, 1981, 16, 1115.

42 L. Colombo, C. Gennari, S. Ricca, C. Scolastico and F. Aragozzini, J.C.S. Perkin 1, 1980, 675.

43 L. Colombo, C. Gennari, C. Scolastico, F. Aragozzini and C. Merendi, J.C.S. Chem. Comm., 1979, 492; J.C.S. Perkin 1, 1980, 2549.

44 T.J. Simpson and D.J. Stenzel, J.C.S. Chem. Comm., 1981, 239.

45 R. Bentley, W.J. Banach, A.G. McInnes and J.A. Walter, Bioorg. Chem., 1981, 10, 399.

46 J.A. Gudgeon, J.S.E. Holker, T.J. Simpson and K Young, Bioorg. Chem., 1979, 8, 311.

47 F.M. Dean, R.A. Eade, R.A. Moubasher and A. Robertson, Nature, 1957, 179, 366; R. Thomas, Biochem. J., 1961, 78, 748.

48 S. Gatenbeck and K. Mosbach, Biochem. Biophys. Res. Commun., 1963, 11, 166.

49 I. Kurobane, C.R. Hutchinson and L.C. Vining, Tetrahedron Lett., 1981, 22, 493

50 I. Kurobane, L.C. Vining, A.G. McInnes and J.A. Walter, Can. J. Chem., 1980, 58, 1380.

51 I. Kurobane, L.C. Vining, A.G. McInnes, D.G. Smith and J.A. Walter, Can. J. Chem., 1981, 59, 422.

52 T.J. Simpson, J.C.S. Perkin 1, 1979, 1233.

53 A. Quick, R. Thomas and D.J. Williams, J.C.S. Chem. Comm., 1980, 1051.

54 Y. Sato, T. Oda, E. Miyata and H. Saito, FEBS Lett., 1979, 98, 271.

55 E. Bardshiri and T.J. Simpson, J.C.S. Chem. Comm., 1981, 195.

56 B. Franck and A. Stange, Liebigs Ann. Chem., 1981, 2106.

57 B. Franck, H. Backhaus and M. Rolf, Tetrahedron Lett., 1980, 21, 1185.

58 C.E. Snipes, C.-j. Chang and H.G. Floss, J. Amer. Chem. Soc., 1979, 101,701

59 A. Arnone, L. Camarada, R. Cardillo, G. Fronza, L. Merlini, R. Mondelli, G. Nasini and J. St. Pyrek, Helv. Chim. Acta, 1979, 62, 30.

60 C. Kitao, H. Tanaka, S. Minami and S. Omura, J. Antibiotics., 1980, 32, 711.

61 C.P. Gorst-Allman, B.A.M. Rudd, C.-j. Chang and H.G. Floss, J. Org. Chem., 1981, 46, 455.

62 A.J. Birch and T.J. Simpson, J.C.S. Perkin 1, 1979, 816.

63 K. Kakinuma, N. Imamura, N. Ikekawa, H. Tanaka, S. Minami and S. Omura, J. Amer. Chem. Soc., 1980, 102, 7493.

64 P.S. Steyn, R. Vleggaar and P.L. Wessels, J.C.S. Chem. Comm., 1979, 1041; J.C.S. Perkin 1, 1981, 1298.

65 A.E. de Jesus, C.P. Gorst-Allman, P.S. Steyn, R. Vleggaar, P.L. Wessels, C.C. Wan and D.P.H. Hsieh, J.C.S. Chem. Comm., 1980, 389.

66 P.S. Steyn, R. Vleggaar, P.L. Wessels and D.B.Scott, J.C.S. Perkin 1, 1979, 460.

67 M.S. Anderson and M.F. Dutton, Appl. Environ. Microbiol., 1980, 40, 706.

68 N.C. Wan and D.P.H. Hsieh, Appl. Environ. Microbiol., 1980, 39, 109.

69 L.O. Zamir and R. Ginsburg, J. Bacteriol., 1979, 138, 684.

70 L.O. Zamir and K.D. Hufford, Appl. Environ. Microbiol., 1981, 42, 168.

71 J.C. Vederas and T.T. Nakashima, J.C.S. Chem. Comm., 1980, 183.

72 C.T. Mabuni, L. Garlaschelli, R.A. Ellison and C.R. Hutchinson, J. Amer. Chem. Soc., 1979, 101, 707.

73 C.R. Hutchinson, I. Kurobane, C.T. Mabuni, R.W. Kumola, A.G. McInnes and J.A. Walter, J. Amer. Chem. Soc., 1981, 103, 3474.

74 C.R. Hutchinson, I. Kurobane, D.E. Cane, H. Hasler and A.G. McInnes, J. Amer. Chem. Soc., 1981, 103, 2477.

75 C.R. Hutchinson, M.M. Sherman, J.C. Vederas and T.T. Nakashima, J. Amer. Chem. Soc., 1981, 103, 5953.

76 C.R. Hutchinson, M.M. Sherman, A.G. McInnes, J.A. Walter and J.C. Vederas, J. Amer. Chem. Soc., 1981, 103, 5956.

77 D.E. Cane, T.-C. Liang and H. Hasler, J. Amer. Chem. Soc., 1981, 103, 5962.

78 D.E. Cane, H. Hasler and T.-C. Liang, J. Amer. Chem. Soc., 1981, 103, 5960.

79 T.S.S. Chen, C.-j . Chang and H.G. Floss, J. Org. Chem., 1981, 46, 2661.

80 T.S.S. Chen, C.-j. Chang and H.G. Floss, J. Amer. Chem. Soc., 1981, 103, 4565.

81 J.J. Kibby, I.A. McDonald and R.W. Rickards, J.C.S. Chem. Comm., 1980, 768.

82 M.G. Anderson, J.J. Kibby, R.W. Rickards and J.M. Rothschild, J.C.S. Chem. Comm., 1980, 1277; J.J. Kibby and R.W. Rickards, J. Antibiotics, 1981, 34, 58.

83 O. Ghisalba, H. Fuhrer, W.J. Richeter and S. Moss, J. Antibiotics, 1981, 34, 58.

84 O. Ghisalba and J. Nüesch, J. Antibiotics, 1981, 34, 64.

85 D.G.I. Kingston and M.X. Kolpak, J. Amer. Chem. Soc., 1980, 102, 5964.

86 C.J. Pearce, S.E. Ulrich and K.L. Rinehart, J. Amer. Chem. Soc., 1980, 102, 2510.

87 C.-M. Liu, T.H. Williams and R.G. Pitcher, J. Antibiotics, 1979, 32, 414.

88 L. Colombo, C. Bennari, D. Potenza, C. Scolastico and F. Aragozzini, J.C.S. Chem. Comm., 1979, 102.

89 D.L. Doerfler, L.A. Ernst and I.M. Campbell, J.C.S. Chem. Comm., 1980, 329.

90 A.J. Bartlett, J.S.E. Holker, E. O'Brien and T.J. Simpson, J.C.S. Chem. Comm., 1981, 1198.

91 T.J. Simpson, Tetrahedron Lett., 1981, 22, 3785.

92 T.J. Simpson and D.J. Stenzel, J.C.S. Chem. Comm., 1981, 1042.

93 C.R. McIntyre and T.J. Simpson, J.C.S. Chem. Comm., 1981, 1043.

94 R.O. Gould, T.J. Simpson and M.D. Walkinshaw, Tetrahedron Lett., 1981, 22, 1047.

95 N.J. McCorkindale, C.H. Calzadilla, S.A. Hutchinson, D.H. Kitson, G. Ferguson and I.M. Campbell, Tetrahedron, 1981, 37, 649.

96 R.J. Light and K. Hahlbrock, Z. Naturforsch., 1980, 35c, 717.

97 A. Stoessl and J.B. Stothers, Z. Naturforsch., 1979, 34c, 87.

98 M.K. Burns, J.M. Coffin, I. Kurobane and L.C. Vining, J.C.S. Chem. Comm., 1979, 426.

99 M.K. Burns, J.M. Coffin, I. Kurobane, L.C. Vining, A.G. McInnes, D.G. Smith and J.A. Walter, J.C.S. Perkin 1, 1981, 1411.

100 H. Shimada, H. Noguchi, Y. Iitaka and U. Sankawa, Heterocycles, 1981, 15, 1141.

101 P.W. Jeffs and D. McWilliams, J. Amer. Chem. Soc., 1981, 103, 6185.

102 S. Nieminen, T.G. Payne, P. Senn and C. Tamm, Helv. Chim. Acta, 1981, 64, 2162.

103 A. Probst and C. Tamm, Helv. Chim. Acta, 1981, 64, 2065.

104 P. Mohr and C. Tamm, Tetrahedron, 1981, 37, 201.

105 R. Wyss, C. Tamm and J.C. Vederas, Helv. Chim. Acta, 1980, 63, 1538.

106 J.C. Vederas, T.T. Nakashima and P. Diakus, Planta Med., 1980, 39, 201.

107 K.K. Chexal, C. Snipes and C. Tamm, Helv. Chim. Acta, 1980, 63, 761.

2

Phenolic Compounds Derived from Shikimate

BY P.M. DEWICK

1 Introduction

Biosynthetic research on shikimate-derived metabolites continues, although the comments of the previous reviewer about emphasis being placed on enzymology of the transformations are reinforced by results during the period of review, 1979-81. Studies in this area are now being developed mainly by biochemists, and the reasons for this are apparent when one examines biosynthetic methodology employed in some of the other areas reviewed in this Report. The majority of shikimate-derived metabolites are plant products, and plants do not readily lend themselves to experiments involving stable isotopes and n.m.r. methodology which can yield such valuable information. It is reassuring, therefore, to see reported a number of studies with stable isotopes in plant systems where disadvantageous conditions have to some extent been overcome. Areas previously neglected, e.g. stilbenes and lignans, are now attracting some attention, and results in other fields, e.g. isoflavonoids and quinones, show that simple feeding experiments can still yield useful data. Nevertheless, enzymology is progressing swiftly, and our understanding of the biotransformations grows rapidly each year.

Valuable review articles on a number of topics including plant phenolics,[1a] the shikimate pathway,[2a] phenolic acids,[3a] PAL,[3b] phenylpropanoids,[2b,2c,4] lignin,[1b,2d,3c,5] coumarins,[2e,3d,6] xanthones,[7,8] flavonoids,[2f,3e,5,9a] tannins,[3f] quinones,[3g,9b] and glycosylation[3h] have been published.

2 Phenols and Phenolic Acids

2.1 The Shikimate Pathway.- Immunological studies[10] have indicated that the three isoenzymes of 3-deoxy-D-*arabino*-heptulosonate 7-phosphate (DAHP) synthase from *Escherichia coli* have less structural similarity than might be expected for isoenzymes. Thus, only the phenylalanine-sensitive form complexed with antibodies. The tryptophan-sensitive isoenzyme from Neurospora crassa appears to exist as a tetramer (subunit has M 52000),[11] and kinetic examination[12] suggests that a rapid-equilibrium ordered mechanism is involved, rather than ping-pong as reported previously. In a Bacillus subtilis mutant, an artificially evolved bifunctional DAHP synthase-chorismate mutase has been reported.[13] This differs from the monofunctional synthase by being inhibited by chorismate in addition to prephenate, and is suggested to arise in the mutant from pre-existing monofunctional enzymes.

Chorismate mutase-prephenate dehydrogenase also exists as a bifunctional enzyme. Evidence from studies of inactivating conditions on the enzyme from Aerobacter aerogenes suggests that the two reactions do in fact occur at one active site, since both activities are affected simultaneously.[14] Several distinct aggregation states of the basic dimeric enzyme are reported for material from Escherichia coli.[15]

The conversion of shikimate into chorismate is severely inhibited by the herbicide N-(phosphonomethyl)glycine (glyphosate). Thus, in buckwheat (Fagopyrum esculentum), the incorporation of [^{14}C]shikimate into the aromatic amino acids was inhibited and shikimic acid accumulated in the tissue.[16] Accumulation of shikimate-derived flavonoids could be restored by administration of post-chorismate precursors, and similar results were observed with anthraquinones in Galium mollugo cells.[17] In tobacco cell cultures, free shikimic acid levels were increased over 300-fold by the administration of glyphosate.[18] Aerobacter aerogenes[17] and Escherichia coli[19] showed similar responses to the herbicide as did plant tissues, but in the latter organism, enzymes in addition to those involved in the shikimate/chorismate conversion, e.g. 3-dehydroquinate synthase, were shown to be affected.

Two pathways from prephenate (1) to tyrosine (4) are now clearly recognized (Scheme 1). In more recent studies of the "pretyrosine branchlet", the involvement of arogenate (pretyrosine) (2) has been demonstrated in coryneform bacteria Corynebacterium,[20] Brevibacterium[21] and plants (Vigna radiata and Zea mays).[22] Coryneform bacteria appear to lack the 4-hydroxyphenylpyruvate (3) pathway, though utilizing the analogous route for synthesis of phenylalanine.[23] However, enzymes of both branchlets to tyrosine were detected in V. radiata, prephenate dehydrogenase (PDH) and arogenate dehydrogenase (ADH) activities coeluting.[21] It is probable that a single protein (M 52000) catalyses both reactions, but only PDH was inhibited by L-tyrosine. ADH from Z. mays is inhibited by tyrosine, and requires NAD^+ as cofactor,[22] in contrast to the $NADP^+$ requirements of the Vigna ADH.

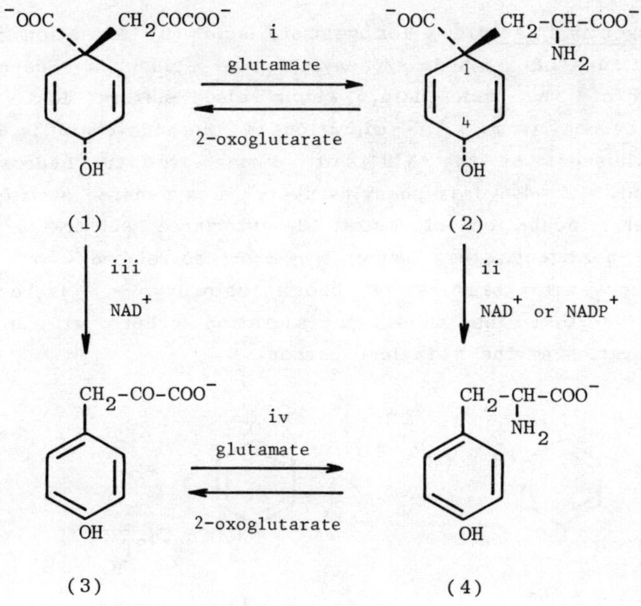

Enzymes: (i) prephenate aminotransferase;
(ii) arogenate dehydrogenase;
(iii) prephenate dehydrogenase;
(iv) 4-hydroxyphenylpyruvate aminotransferase.

Scheme 1

No PDH activity could be detected in corn (*Z. mays*).[24] The structure of arogenate has been confirmed as (2) although the configurations at C-1 and C-4 are assumed to be the same as in prephenate. The nomenclature pretyrosine has been abandoned in favour of arogenate in view of its conversion into phenylalanine by arogenate dehydratase.

The conversion of phenylalanine into tyrosine, particularly in animal tissues, has been known for many years. This transformation has now also been demonstrated in plant (*Portulaca grandiflora*) tissues,[25] although its relative importance as a route to tyrosine is not known. Crystalline phenylalanine hydroxylase has been obtained from *Chromobacterium violaceum*.[26]

2.2 Homogentisic Acid. - Homogentisic acid (5) formation involves hydroxylation of 4-hydroxyphenylpyruvate and a concomitant migration of the side-chain, which also suffers loss of the carboxyl carbon atom. The migration of the side-chain is in some ways reminiscent of the NIH shift during aromatic hydroxylation. In a study of 4-hydroxyphenylpyruvate dioxygenase purified from beef liver, incubation of chiral 3-deuteriated substrates yielded labelled homogentisates which were correlated by n.m.r. spectroscopy with samples of known chirality.[27] It could be demonstrated that the side-chain migration occurs with retention of configuration at the methylene carbon.

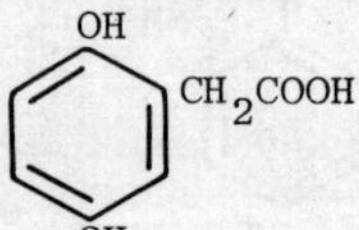

(5)

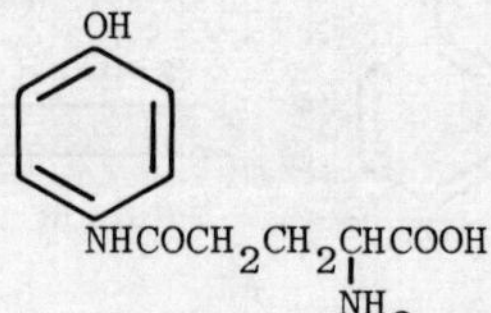

(6)

2.3 Brominated Derivatives of 4-Hydroxybenzaldehyde.- Chloroplast-containing fractions from the marine red alga *Odonthalia floccosa* have been shown to metabolize labelled L-tyrosine into 4-hydroxybenzaldehyde[28] and 3-bromo-4-hydroxybenzaldehyde.[29] In further studies, these two metabolites in ^{14}C-labelled form were shown to be modified further to 3,5-dibromo-4-hydroxybenzaldehyde, and other metabolites. It is suggested that these are likely precursors of a range of bromo-hydroxybenzyl alcohols found in this organism, previously regarded as artefacts of isolation.

2.4 Other Phenols.- 4-Hydroxybenzyl alcohol, although known as a plant metabolite has only just been reported from a bacterial source[30]. This appears to be a cleavage product produced during the biosynthesis of thiamine from tyrosine in *Escherichia coli*. The conversions of L- and D-phenylalanines into phenylacetate via phenylpyruvate in *Sorghum* leaf extracts are catalysed by two different transaminase enzyme systems.[31] Though the activities for the D- and L-forms were approximately equal in the crude particulate fraction, the proportions ranged from 1:1 to 1:3 in the supernatant.

N-(γ -L-glutamyl)-4-hydroxyaniline (6) is produced by the mushroom *Agaricus bisporus*, and the 4-hydroxyaniline fragment is derived from shikimate.[32] Feeding experiments followed by extensive degradations demonstrated that the amino function is introduced at the 4-position of shikimic acid. In this respect, there appears similarity with the route to 4-aminobenzoic acid, but the shikimate carboxyl must also be lost. The corresponding 3,4-dihydroxyaniline derivative is also produced from feedings of ^{14}C-labelled shikimic acid.[33]

3 Phenylpropanoids

3.1 Phenylalanine Ammonia Lyase (PAL).- PAL usually exhibits negative subunit cooperativity, i.e. the binding of the first molecule of substrate reduces the affinity for a second molecule. Extensive purification of PAL from soybean cell suspension cultures[34] resulted in an enzyme exhibiting negative cooperativity, but positive cooperativity after hydroxyapatite chromatography. A model for the cooperativity kinetics induced by D- and L- phenylalanine has been proposed[35] in which D- and L-isomers are assumed to pack into the same active sites, though in a mirror-image relationship.[36] D- and L-α-Aminooxy-β-phenylpropionic acids (AOPP) both act as superinhibitors, and mirror-image packing of these compounds into the active site is believed to mimic the conformation inferred for the transition state in the elimination. L-AOPP interferes strongly with phenylalanine metabolism. In many tissues, it is a competitive inhibitor of PAL and results in increased phenylalanine accumulation, e.g. in buckwheat,[37] though no increase in phenylalanine was observed in soybean.[38] PAL activity may actually be enhanced by the action of AOPP[39] but the synthesis of cinnamic acid derivatives is still markedly reduced. This has been correlated with decreased feedback inhibition of PAL synthesis by cinnamic acids.[39,40]

Changes in PAL activity resulting from illumination or treatment with phytoalexin-inducing elicitors have also been correlated with changes in the rate of synthesis of the enzyme, together with degradation or removal of the enzyme. Such conclusions were reached in studies in pea,[41] french bean,[42] and potato[43] tissues.

3.2 <u>Hydroxycinnamic Acids</u>.- Extended controversy over the biosynthetic origins of the C_3 side-chain in allylbenzenes, e.g. eugenol, has been satisfactorily resolved by double-labelling experiments during caffeic acid biosynthesis in *Ocimum basilicum*.[44] Whilst [3-^{14}C,G-^{3}H]phenylalanine suffered loss of about half of the ^{3}H label, as expected, [1-^{14}C,G-^{3}H] phenylalanine was incorporated into caffeic acid with little change in the ratio. The apparent ^{3}H retention was due to loss of some terminal ^{14}C because of exchange with a C_1-donor, affecting part of the caffeic acid produced in the plant. Both methionine and formate could act as sources of the carboxyl carbon in caffeic acid. All previous results are explicable if *p*-coumaric acid and/or caffeic acid, but not ferulic acid, can be involved in this exchange.

3.3 <u>Hydroxycinnamic Acid Esters</u>.- Enzymes catalysing the esterification of quinic acid with *p*-coumaroyl CoA and caffeoyl

CoA, giving p-coumaroylquinic and chlorogenic acids, have been isolated from two plant sources, Nicotiana alata and Stevia rebaudiana,[45] and partially purified. The hydroxycinnamoyl CoA:quinate hydroxycinnamoyl transferase activities in stored potatoes are markedly dependent on the temperature of storage.[46] Major activity present when tubers were stored at 0 °C was absent from those stored at 10 °C throughout the period after harvest. The related esterification of shikimic acid has been demonstrated with an enzyme from cultures of Cichorium endivia.[47] This enzyme shows absolute specificity for p-coumaroyl CoA and shikimic acid, forming the 5-O-p-coumaroyl ester, and its presence has been reported in a wide variety of plants and cultures. Aspergillus niger contains an apparent chlorogenic acid-specific esterase.[48] Though having a small activity towards isochlorogenic acid and ethyl cinnamate, all other esters tested were not hydrolysed.

The glycosylation of hydroxycinnamic acids can lead to glycosides, or sugar esters. In tomato fruits, a constant proportion of glucoside to glucose esters produced during various stages of enzyme purification suggested that both reactions are catalysed by the same enzyme.[49] The enzyme utilizes free hydroxycinnamic acids and UDP-glucose, not accepting hydroxycinnamoyl CoA esters/glucose-1-phosphate. Enzymes catalysing the formation of glucose esters of cinnamic acids have been reported in a range of plants,[50] that from Cestrum euanthes leaves producing 1-O-p-coumaroyl-β-D-glucose from p-coumaric acid and UDP-glucose. Although 1-sinapoylglucose may be formed in a similar manner by a cell-free system from Raphanus sativus seedlings,[51] this compound may also be produced during the hydrolysis of the choline ester sinapine,[52,53,54] which can act as a supply of choline. A further enzyme from R. sativus produces sinapoyl malate from sinapoyl glucose and L-malate[55], a reaction rather unusual in that most acylations involve acyl CoA derivatives.

Rosmarinic acid (7), the α-O-caffeoyl ester of 3,4-dihydroxyphenyllactic acid, was believed to arise by esterification of 3,4-dihydroxyphenyllactic acid produced from dopa. In Coleus blumei cell cultures,[56] no tyrosine-3-hydroxylase activity could be detected, and by the application of AOPP, tyrosine label accumulated in 4-hydroxyphenyllactic acid rather than dopa or 3,4-dihydroxyphenyllactic acid. These observations thus cast doubt on the intermediacy of dopa in the pathway to rosmarinic acid, and suggest the involvement of 4-hydroxyphenyllactic acid.

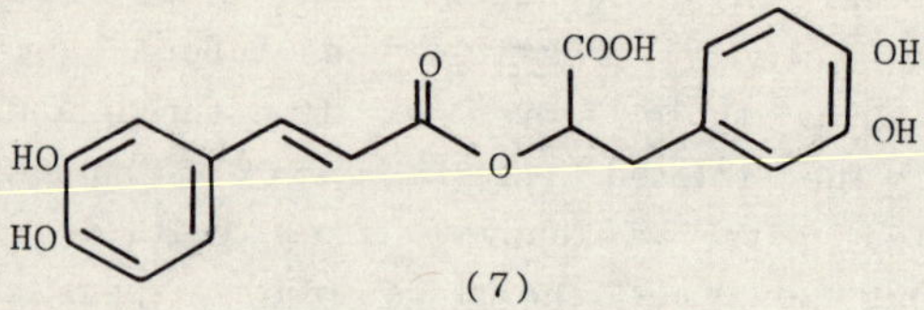

(7)

<u>3.4 Reduction Products of Cinnamic Acids</u>.- Fresh cinnamon (<u>Cinnamomum zeylanicum</u>) bark contains appreciable amounts of cinnamyl acetate, which is not present in the commercial fermented bark. Labelling studies[57] have demonstrated that the fermentative conditions result in the conversion of this ester into cinnamaldehyde, the major constituent of cinnamon oil. Results demonstrate an overall reductive sequence for the formation of cinnamaldehyde and cinnamyl acetate from cinnamic acid, coupled with hydrolytic and oxidative sequences transforming the ester to cinnamaldehyde.

Reduction of 3',4'-dimethoxycinnamyl alcohol (8) to the 3',4'-dimethoxyphenylpropanol (9) by <u>Saccharomyces cerevisiae</u> results in loss of one of the hydrogen atoms of the alcohol function, and thus proceeds via the intermediate aldehydes, involving alcohol dehydrogenases (Scheme 2).[58] Decarboxylation of 3,4-dimethoxycinnamic acid by <u>S. cerevisiae</u> occurs with retention of configuration at the side-chain double bond.[59] The enzyme shows specificity for the <u>E</u>-cinnamic acid. Reduction of <u>E</u>-cinnamic acid to phenylpropionic acid by two microorganisms has been investigated.[60] By analysing reduction products when the transformation was carried out in D_2O, it was established that in <u>Peptostreptococcus anaerobius</u> an <u>E</u> reduction to the <u>re</u> face had occurred, but in <u>Clostridium sporogenes,</u> an <u>E</u> reduction to the <u>si</u> face had taken place. The reductions were thus enzymically different.

Little is known about interconversions of allyl- and

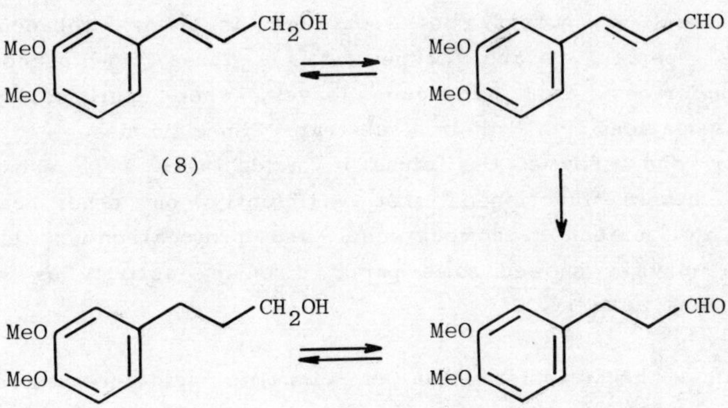

Scheme 2

propenyl-benzenes, but observations on the oil constituents of Croton aff. zehntneri[61] indicate that, in the bark, anethole (10) appears to be transformed into estragole (11). Light-induced transformation in the isolated oil gives the toxic *Z*-anethole instead.

(10) (11)

3.5 Methylation of Hydroxycinnamic Acids.- Preferential *meta* methylation of hydroxycinnamic acid substrates is usually observed, particularly since ferulic and sinapic acids are essential precursors for lignin biosynthesis. Enzymes methylating caffeic acid and 5-hydroxyferulic acid have been isolated from a variety of sources, including tobacco,[62,63] alfalfa,[64] poplar,[65] and aspen[66]. Three *o*-diphenol *O*-methyltransferases from tobacco leaves, when purified, showed marked variations in their substrate specificities.[62] Thus, whilst for one enzyme the cinnamic acids were best substrates, the two others functioned most efficiently on other catechol systems, e.g. catechol, homocatechol, and protocatechuic aldehyde. All three enzymes showed some *para* directing activity as well as *meta* directing activity.

3.6 Lignin.- The transformation of cinnamic acid precursors into cinnamyl alcohol lignin monomers proceeds through cinnamaldehyde intermediates, and the two steps are catalysed by cinnamoyl CoA reductases and cinnamyl alcohol dehydrogenase. The properties of these two enzymes from different sources, spruce (*Picea abies*) and soybean (*Glycine max*), have been compared.[67] There were pronounced differences in substrate specificities between enzymes from the two sources, reflecting the different lignin compositions of gymnosperms and dicotyledonous angiosperms. Thus, although feruloyl CoA was the preferred substrate for the reductase from both sources, sinapoyl CoA was only acceptable to the soybean reductase, and sinapaldehyde was similarly a substrate only for the soybean dehydrogenase. *p*-Coumaroyl CoA was a poor substrate for the reductase from both sources, again consistent with the low content of *p*-coumaryl units in these lignins. The cinnamyl alcohol dehydrogenase from soybean is known to be present as two isoenzymes. Extensive purification of isoenzyme 2 has shown it shares many similarities with alcohol dehydrogenase from horse, rat liver, and yeast,[68] and belongs to A-specific dehydrogenases, removing the *pro*-*R* hydrogen from coniferyl alcohol.

Enzyme activities (PAL, caffeate 3-O-methyltransferase, *p*-coumarate:CoA ligase, and cinnamyl alcohol dehydrogenase) in various types of reaction wood have been analysed[69] and correlated with the chemical nature of lignin in the wood types.

The incorporation of [^{14}C]phenylalanine into lignin in poplar xylem tissues was observed to be light-dependent,[70] syringyl residues only being labelled in the dark, but both guaiacyl and syringyl residues being labelled in the light. This effect was correlated with modification by light of membrane permeabilities, altering synthesis and transport or polymerisation of guaiacyl units.

β-Glucosidases from *Glycine max*[71] specific for coniferin and syringin, and from *Picea abies*[72] specific for coniferin, have been investigated. These are also believed to play some role in lignin biosynthesis, supplying the appropriate monomer units.

3.7 Lignans.- Feeding experiments and degradation studies[73] have established the specific incorporation of two phenylpropane units from phenylalanine and *p*-coumaric acid into the tumour-inhibitory aryltetralin lignan podophyllotoxin (12) in *Podophyllum hexandrum* (syn. *emodi*). Other studies[74] demonstrated the incorporation of phenylalanine, cinnamic acid, and ferulic acid into both podophyllotoxin and 4'-demethylpodophyllotoxin (13), and indicated that (12) could be obtained by both hydroxylation of desoxypodophyllotoxin (14) or reduction of podophyllotoxone (15).

O-Methyltransferases from *Thuja* species were able to convert 5-hydroxyferulic acid into sinapic acid, but did not catalyse the methylation of the lignan dihydroxythujaplicatin (16),[75] although (16) and its methyl ether (17) are commonly found in *Thuja*. This evidence was interpreted in favour of sinapic acid being a precursor of (17), rather than methylation occurring at the lignan level.

Two lignans having unusual *meta* hydroxylated aromatic groupings, isolated from mammalian urine, are now regarded as being produced by the action of intestinal bacteria, probably from dietary lignans.[76] Oxidative coupling of coniferyl alcohol with arctigenin, using horseradish peroxidase/H_2O_2, gave small yields of the natural C_6C_3 trimer lappaol B (18).[77] Of particular interest was the complete stereospecificity of the reaction, probably achieved because of the chirality present in arctigenin.

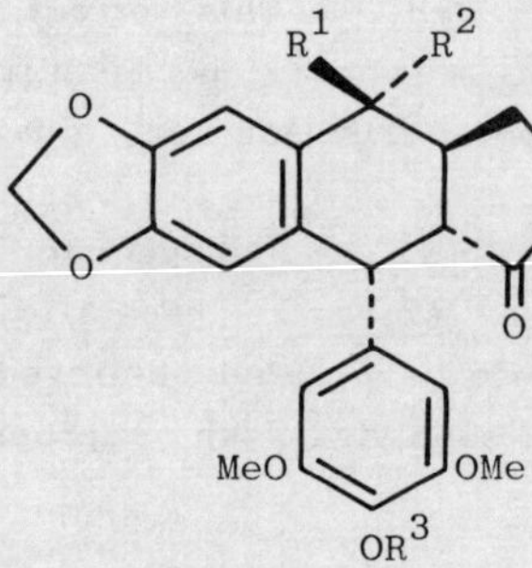

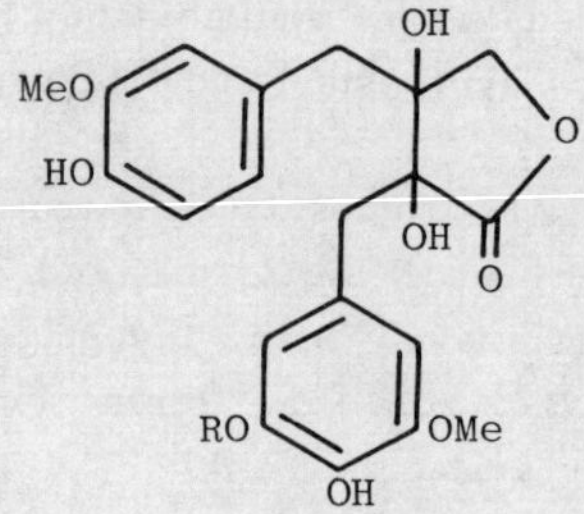

(12) $R^1 = H$, $R^2 = OH$, $R^3 = Me$

(13) $R^1 = R^3 = H$, $R^2 = OH$

(14) $R^1 = R^2 = H$, $R^3 = Me$

(15) $R^1R^2 = O$, $R^3 = Me$

(16) R = H

(17) R = Me

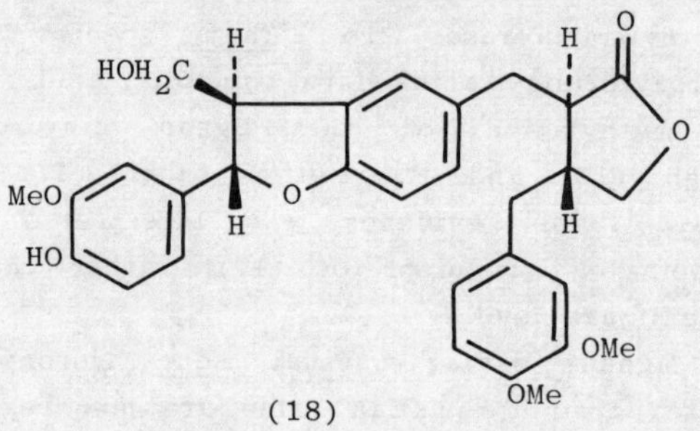

(18)

3.8 Coumarins.- The formation of linear O-alkylfuranocoumarins such as 5-methoxypsoralen (bergapten) (22) can proceed via hydroxylation and methylation of psoralen (20), or via an alternative pathway demonstrated in *Ficus carica* and *Ruta* graveolens.[78] In this sequence, marmesin (19) is hydroxylated to 5-hydroxymarmesin (21) prior to furan ring formation (Scheme 3). 5-Hydroxymarmesin was incorporated into bergapten in both plants, and trapping experiments demonstrated its formation from labelled marmesin.

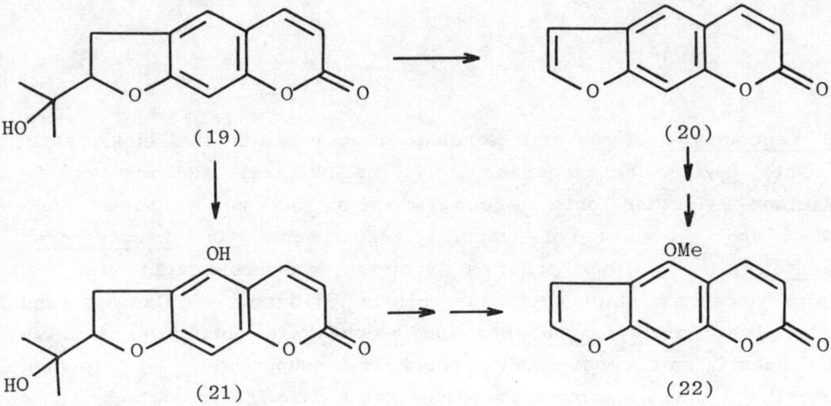

Scheme 3

Two O-methyltransferases from rue (Ruta graveolens) acting on the 5- or 8-hydroxyls of psoralen derivatives have been completely resolved.[79] The 5-O-methyltransferase acted on bergaptol (23) and 5-hydroxyxanthotoxin (24), the 8-O-methyltransferase on xanthotoxol (25) and 8-hydroxybergapten (26), and the evidence obtained suggested that only these two methylating enzymes were involved in production of the four products. A partially purified glucosyltransferase from tobacco cell cultures[80,81] showed strict positional specificity for the 7-hydroxy group of the dihydroxycoumarins daphnetin (7,8-dihydroxy) and esculetin (6,7-dihydroxy). The presence of 8-O-glucosyl and 6-O-glucosyl derivatives of these coumarins in tobacco suggests other enzymes present have high affinity for other positions.

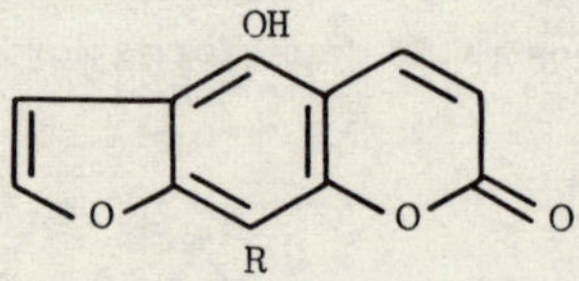

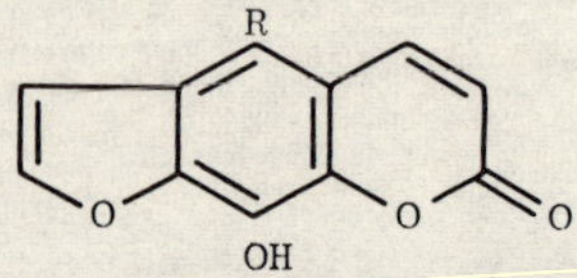

(23) R = H (25) R = H

(24) R = OMe (26) R = OMe

3.9 Leucodrin.- Of several schemes proposed for the biosynthesis of the C-glycoside leucodrin (27), a Michael addition of L-galactono-γ-lactone onto p-coumaric acid (Scheme 4) seems most likely from[82] results of feeding experiments in *Leucadendron argenteum*. Other schemes involving ascorbic acid are less likely since ascorbic acid was poorly utilized. Galactose and glucose were incorporated into the p-coumarate portion as well as the sugar unit, but this could be suppressed by feeding unlabelled phenylalanine at the same time, in which case incorporation became preferential into the carbohydrate part of the molecule.

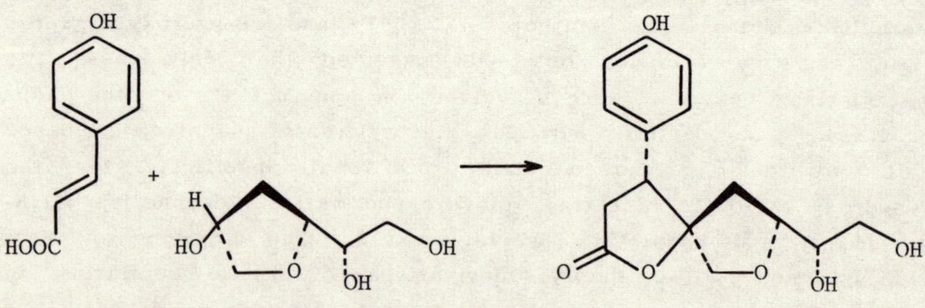

(27)

Scheme 4

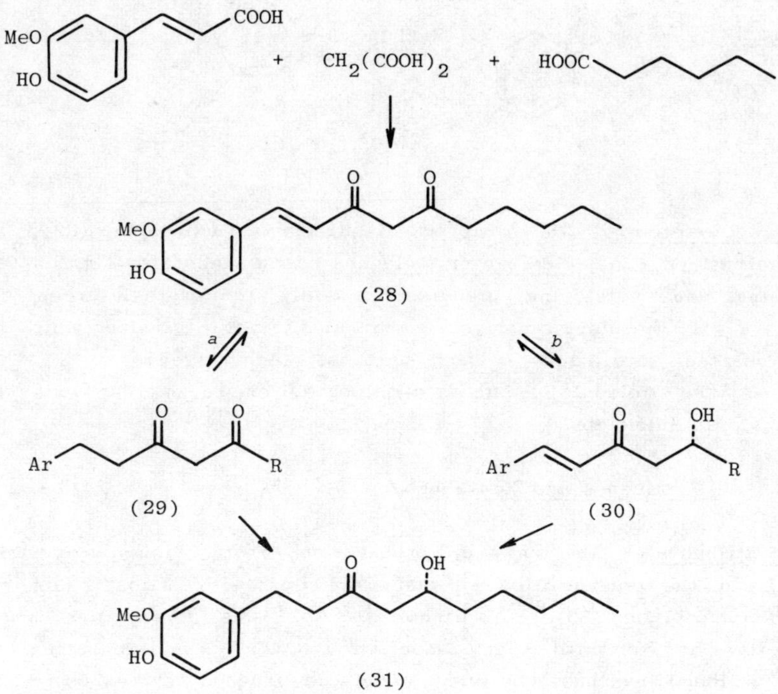

Scheme 5

3.10 [6]-Gingerol.- The biosynthesis of [6]-gingerol (31), a pungent principle of ginger (Zingiber officinale), involves an unusual biological Claisen reaction between ferulic acid, malonic acid, and hexanoic acid in which the malonate component loses both carboxyl groups. Double-labelling experiments with hexanoate and dihydroferulate[83] showed intact incorporation of the former, but dihydroferulate was incorporated via ferulate. In intact rhizomes and rhizome sections, dehydrogingerdione (28), gingerdione (29), and dehydrogingerol (30) were all incorporated, and thus both pathways a and b (Scheme 5) may operate, perhaps via a metabolic grid.

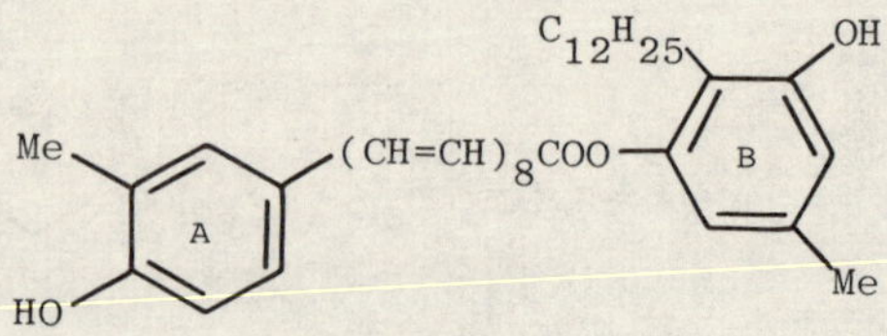

(32)

3.11 Flexirubin.- The bacterial pigment flexirubin (32) is biosynthesized in Flexibacter elegans from acetate, malonate, tyrosine, and methionine precursors. Only the methyl group on ring A is methionine-derived, and an intact tyrosine unit is incorporated into ring A and part of the polyene.[84,85] The rest of the molecule is acetate-malonate derived, perhaps via 3-dodecylorsellinic acid.

4 Stilbenes and Xanthones

4.1 Stilbenes.- An enzyme preparation from Pinus sylvestris catalysed the condensation of cinnamoyl CoA with malonyl CoA to yield pinosylvin (33).[86] Cinnamoyl CoA was five times more effective as substrate than p-coumaroyl CoA. The production of this stilbene synthase activity could be induced by exposure of the tissue to UV light. Stilbene synthase activities in leaves of several plants from the Vitaceae, catalysing the formation of resveratrol (34) from p-coumaroyl CoA, were also markedly increased (up to 100 x) by UV-irradiation.[87] PAL and CAH activities were induced concurrently.

Feeding experiments in rhubarb (Rheum rhaponticum) rhizomes[88] showed the incorporation of labelled acetate, phenylalanine, cinnamic acid, and p-coumaric acid into the stilbene glucoside rhaponticin (35). Isoferulic acid was a poor precursor, as was pinosylvin, but resveratrol was well-incorporated. High selectivity for substitution pattern was observed in membrane-bound enzyme preparations from rhubarb rhizomes which utilized p-coumaroyl CoA much better than cinnamoyl CoA, giving the corresponding stilbenes. At low pH's,

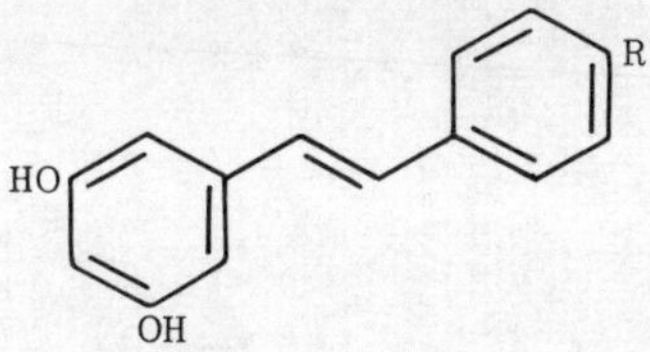

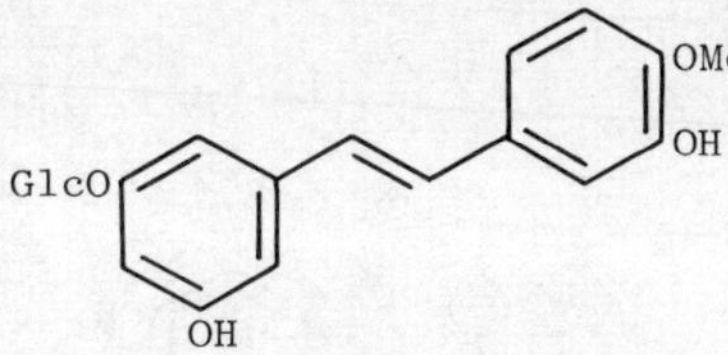

Pinosylvin (33) R = H

Resveratrol (34) R = OH

Rhaponticin (35)

caffeoyl CoA was also converted at a relatively low rate. Feruloyl CoA and isoferuloyl CoA were not acceptable as substrates. The results suggest the stilbene skeleton is synthesized from *p*-coumaroyl CoA, and that further substitution to give rhaponticin occurs at the stilbene level. *In vitro* methylation of astringin (36) to rhapontigenin and rhaponticin was demonstrated.

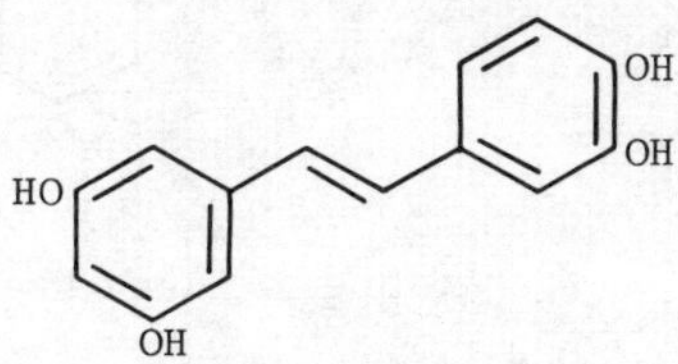

(36)

4.2 Xanthones.- The biosynthesis of the xanthone *C*-glucosides mangiferin (39) and isomangiferin (40) in *Anemarrhena asphodeloides* involved malonate chain-extension of a suitable cinnamic acid, carbon atoms C-1 and C-2 of which are incorporated into the phloroglucinol ring.[89,90] The benzophenone maclurin (38) was well-incorporated into mangiferin, whereas the aglycone of mangiferin was not a precursor, suggesting that *C*-glucosylation occurred at the benzophenone stage. Further studies[91] showed that *p*-coumaric acid was a much more efficient substrate than caffeic acid, though both were incorporated, and that the benzophenones iriflophenone (37) and maclurin were incorporated to similar extents. The sequence in Scheme 6 is proposed, although the results could be accommodated by a metabolic grid.

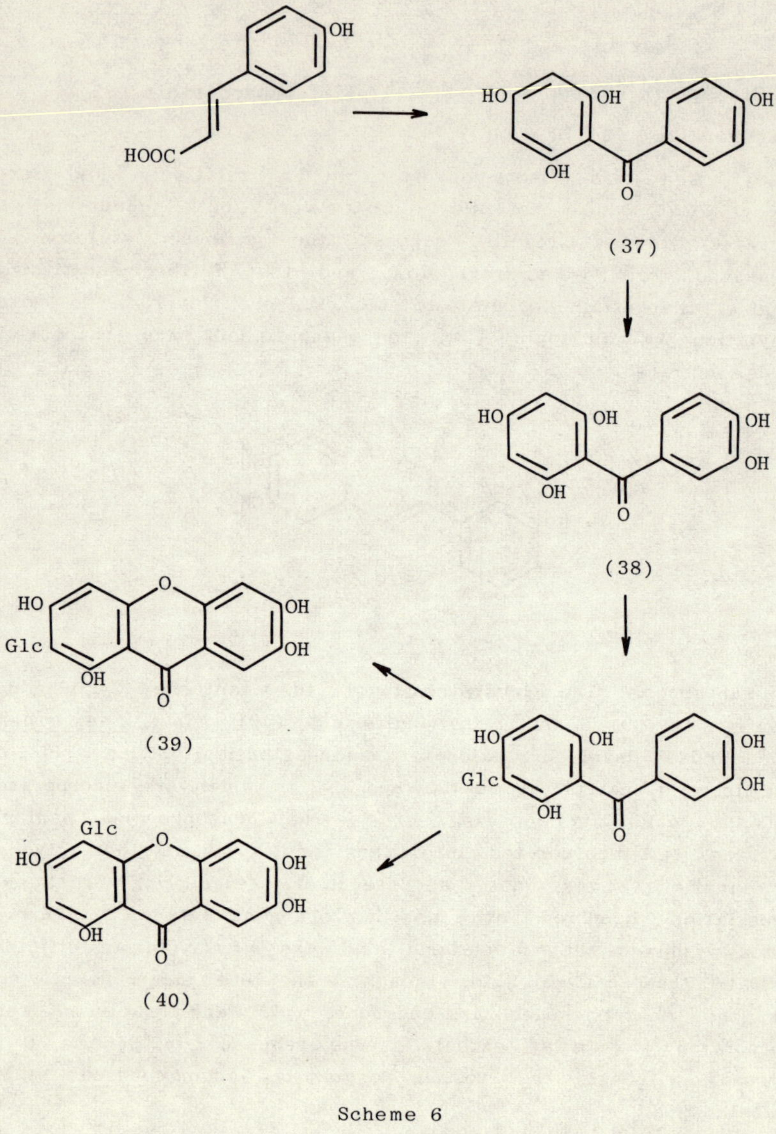
OH
HOOC
HO
OH
OH
OH
O
(37)
HO
OH
OH
OH
OH
O
(38)
HO
Glc
OH
OH
OH
O
O
(39)
HO
Glc
OH
OH
OH
O
Glc
HO
O
OH
OH
OH
O
(40)

Scheme 6

5 Flavonoids

5.1 Chalcone Synthase.- Enzyme preparations from parsley (*Petroselinum hortense*) giving the flavanone naringenin from *p*-coumaroyl CoA and malonyl CoA have been shown to be contaminated with chalcone isomerase,[92] and further purification has resolved the chalcone synthase activity. Enzymes from tulip anthers and *Cosmos* petals have also been purified and support the role of chalcones as intermediates.[93,94] These enzymes were very similar in properties, and would accept caffeoyl CoA or feruloyl CoA as substrates in addition to *p*-coumaroyl CoA. Lack of chalcone synthase activity in flowers of genotypes of *Matthiola incana* has been correlated with interruption of anthocyanin synthesis and production of white flowers.[95]

Further convincing evidence for the intermediacy of chalcones in flavonoid biosynthesis comes from a feeding experiment with sodium [1,2-$^{13}C_2$]acetate in *P. hortense* cell cultures.[96] Kaempferol (41) produced showed in its ^{13}C n.m.r. spectrum A-ring signals flanked by pairs of satellites, demonstrating free rotation of a symmetrical aromatic ring in

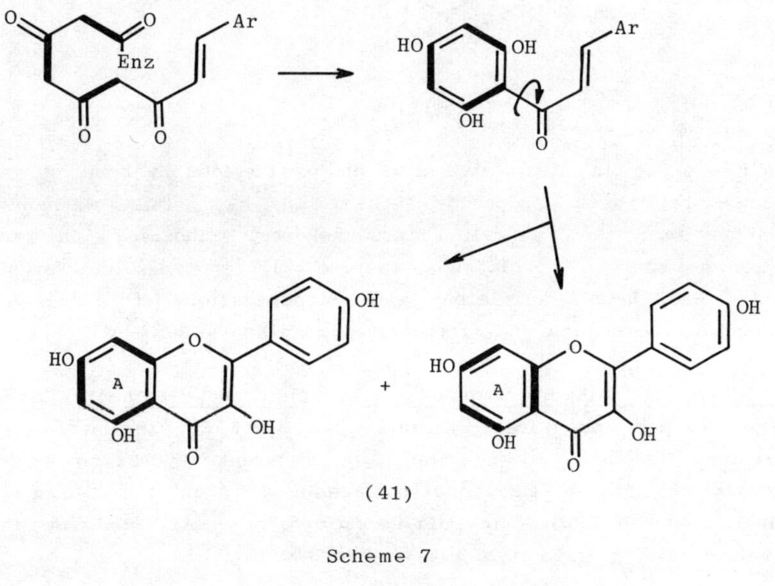

(41)

Scheme 7

the chalcone intermediate (Scheme 7). Similar results with the flavone apigenin were less certain, since chemical randomization could have occurred during the work-up.

5.2 Chalcone Isomerase.- From extensive studies of the pH profile of chalcone isomerase from soybean (Glycine max), a mechanism for the cyclization reaction involving a nucleophilic attack of an

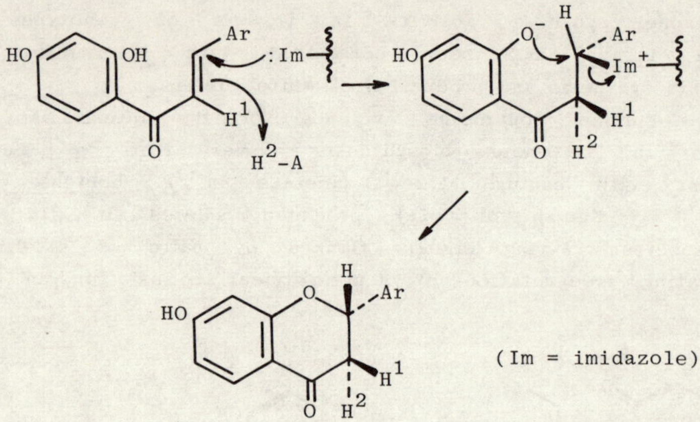

Scheme 8

imidazole group in the active site on to the double bond of the chalcone has been proposed (Scheme 8).[97] This mechanism accounts for the overall stereochemistry observed in the reaction. Lack of chalcone isomerase in genotypes of Petunia hybrida has been correlated with accumulation of naringenin chalcone, and deficiencies in flavonols and anthocyanins.[98]

5.3 Flavones, Dihydroflavonols, Flavonols, and Anthocyanidins.- Enzyme preparations from flowers of Matthiola incana[99] and Antirrhinum majus[100] contain two different hydroxylases for hydroxylation of the flavanone naringenin (42) in the 3-position (giving the dihydroflavonol dihydrokaempferol (44)) and in the 3'-position (giving eriodictyol (43)). The 3-hydroxylase requires 2-oxoglutarate, Fe^{2+}, and ascorbate as cofactors, and will also accept eriodictyol as substrate. The 3'-hydroxylase requires NADPH as cofactor, and also acts on dihydrokaempferol, giving dihydroquercetin (45) (Scheme 9). An enzyme from parsley cultures requiring the same cofactors as the 3-hydroxylase catalyses the conversion of naringenin into the corresponding flavone, dihydroflavonol, and flavonol.[101] The synthesis of the flavone and flavonol is suggested to proceed via dehydration of intermediate 2-hydroxy- and 2,3-dihydroxy-flavanones, formed by 2-hydroxylation of naringenin and dihydrokaempferol respectively. Flavone biosynthesis in Antirrhinum majus, however, in contrast to the parsley enzyme, is catalysed by a NADPH-dependent enzyme.[102] Both naringenin and eriodictyol were accepted as substrates.

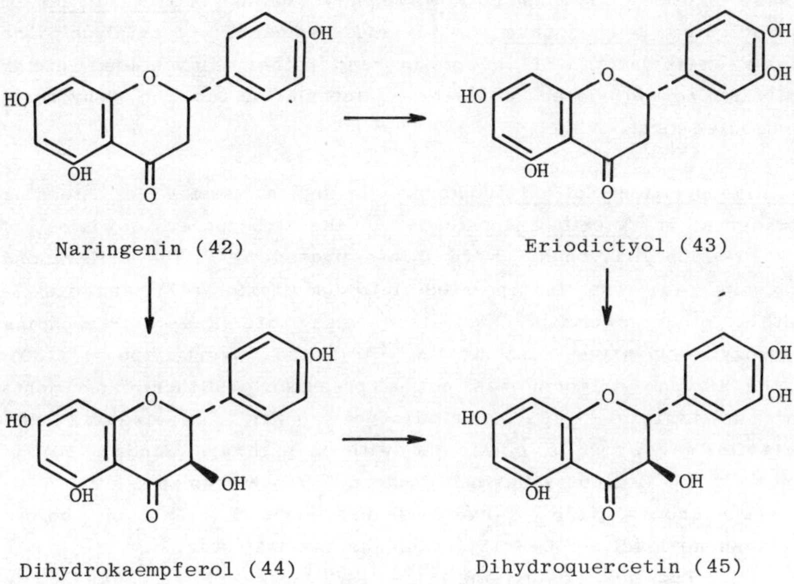

Scheme 9

Naringenin and dihydrokaempferol initiated anthocyanin synthesis when administered to acyanic flowers of genetically-defined lines of M. incana and A. majus,[103] emphasizing their role as biosynthetic intermediates. The participation of dihydroflavonols in anthocyanidin biosynthesis was again demonstrated in buckwheat hypocotyls treated with the PAL inhibitor α-aminooxyacetate.[104] Anthocyanin biosynthesis could be resumed by supplying a range of intermediates, including cinnamic acids, flavanones, and dihydroflavonols.

5.4 Methylation of Flavonoids.- The para O-methylation of apigenin (5,7,4'-trihydroxyflavone) to acacetin has been demonstrated in enzyme extracts from Robinia pseudoacacia.[105] This lacked activity towards p-coumaric acid, and supports earlier results that methylation in acacetin biosynthesis occurs at the C_{15} level. Para O-methylation of quercetin was the preferred reaction for enzymes from tobacco cell cultures, though not entirely to the exclusion of other substrates.[63] Cell-free extracts from a number of sources, including Citrus mitis,[106] Chrysosplenium americanum,[107] and spinach,[108] catalyse the stepwise methylation of quercetin and other flavonoids, giving eventually permethylated products. Methylation at the 3-hydroxyl was effected first.

5.5 Glycosylation of Flavonoids.- In a series of feeding experiments in Swertia japonica,[109] the biosynthetic origins of four C-glucosylflavones were investigated. The flavanone naringenin (42) was incorporated into isovitexin (47) and its 7-O-methyl ether swertisin (49), and also into the corresponding 3'-hydroxy derivatives isoorientin (48) and swertiajaponin (50), but the flavone apigenin was not a precursor. Other experiments with hemiphlein (46), eriodictyol (43), isovitexin, and isoorientin gave results consistent with a pathway (Scheme 10) in which the C-glucosylflavones were biosynthesized via C-glucosylflavanones, the 3'-hydroxylation occurring at or before the flavanone level. 7-O-Methylation was the last step.

Enzymes catalysing the synthesis of di- and tri-glycosides from monoglycosides have been reported in red campion (Silene dioica)[110] and pea (Pisum sativum).[111] The former enzyme catalyses the transfer of rhamnose from UDP-rhamnose to the 6-hydroxyl of glucose in anthocyanidin 3-O-glucosides

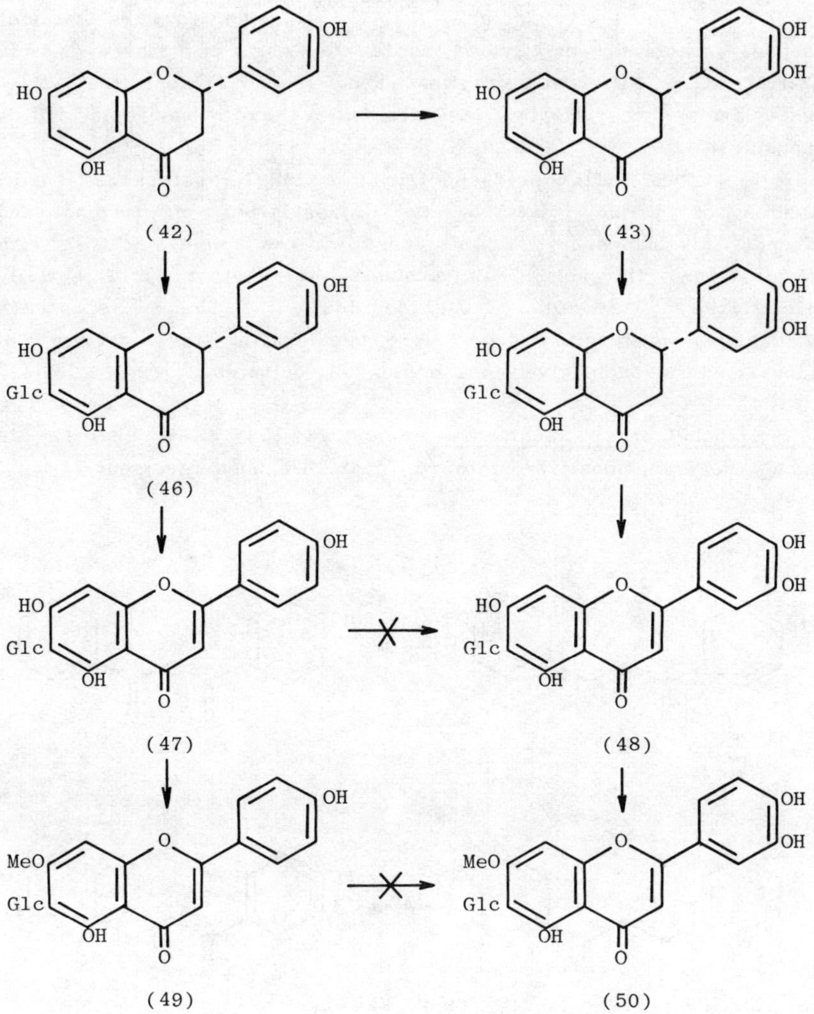
OH
HO
O
OH
O
(42)
OH
OH
HO
O
OH
O
(43)
OH
HO
Glc
O
OH
O
(46)
OH
OH
HO
Glc
O
OH
O
OH
HO
Glc
O
OH
O
(47)
OH
OH
HO
Glc
O
OH
O
(48)
OH
MeO
Glc
O
OH
O
(49)
OH
OH
MeO
Glc
O
OH
O
(50)

Scheme 10

(including those of cyanidin, pelargonidin, and delphinidin), the other the stepwise glucosylation of kaempferol and quercetin in the 3 position, giving triglucosides. 2"-*O*-Glycosylation of the flavone *C*-glucosides vitexin and isovitexin has also been reported in species of *Silene*.[112,113,114] The enzymes transfer glucose, arabinose, or xylose units to one or other of the substrates. Different enzymes are required for the 2"-*O*-glucosylation of vitexin (8-*C*-glucoside) and isovitexin (6-*C*-glucoside).

Two flavonoid-specific malonyltransferases from cultures of parsley[115] catalyse the malonylation of flavone and flavonol glycosides. One enzyme was most active with flavone and flavonol 7-*O*-glucosides, the other preferentially malonylated flavonol 3-*O*-glucosides. These substrate specificities were in good agreement with the pattern of malonylated flavonoid glycosides in the cell cultures.

5.6 Retrochalcone : Echinatin.- Echinatin (53) is a chalcone in which the functionality of the C_3 chain has been reversed.

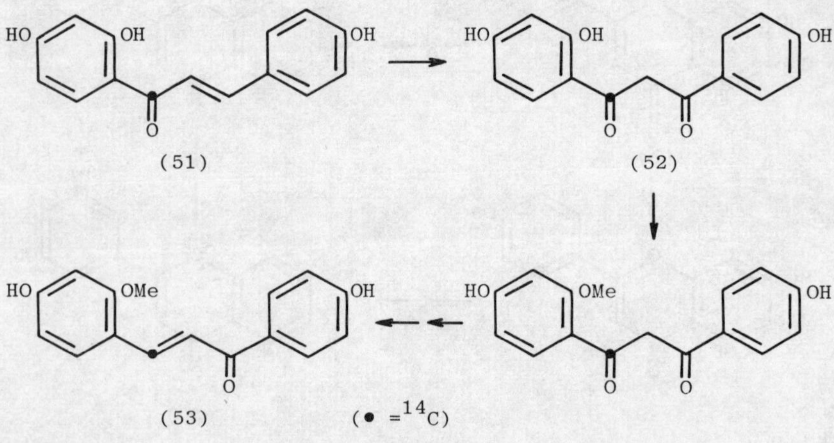

Scheme 11

Feeding experiments in *Glycyrrhiza echinata* cell suspension cultures[116] have demonstrated the incorporation of isoliquiritigenin (51) and the dibenzoylmethane licodione (52) into echinatin with retention of labelling pattern (Scheme 11). 7,4'-Dihydroxyflavone was a poor precursor, thus eliminating a possible flavone hydration mechanism. A specific O-methyltransferase from *G. echinata* catalysed the methylation of licodione,[117] thus supporting the pathway shown.

5.7 Chlorflavonin.- Chlorflavonin (54) is a novel flavonoid from *Aspergillus candidus*, and differs markedly from the plant flavonoids in its biosynthetic origin.[118] Although phenylalanine and cinnamic acid are utilized, they are not incorporated intact, but probably via benzoic acid. Salicylic acid was a poor precursor. The rest of the carbon skeleton is acetate-derived. This could be proven by ^{13}C n.m.r. studies after incorporation of sodium [1-^{13}C]acetate, and experiments with [1,2-$^{13}C_2$]acetate showed an intact unit for C-3/C-4, with the characteristic randomization of label in ring A, resulting from free rotation of a phloroglucinol-like ring. This means the substitution at C-8 must occur after the heterocyclic ring is formed. It is likely that chlorination is the last step of the sequence (Scheme 12).

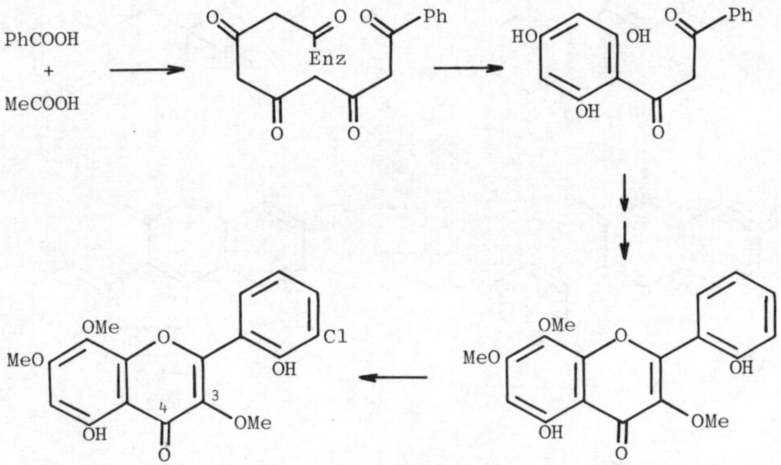

(54)

Scheme 12

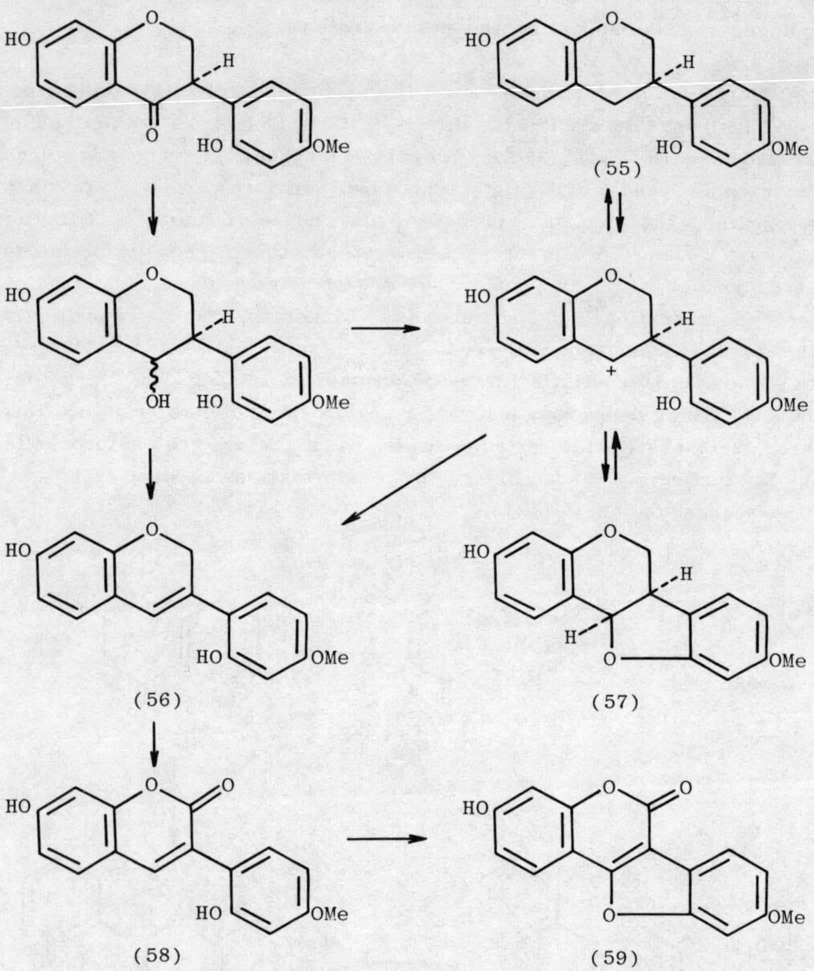

Scheme 13

5.8 Isoflavonoids.- Full details of the pathways leading to, and the interconversion of, pterocarpan and isoflavan phytoalexins in lucerne (*Medicago sativa*) seedlings have now been published.[119,120] Experiments with double-labelled medicarpin (57) and vestitol (55)[121] show that their interconversion cannot proceed via an intermediate isoflav-3-ene (56), and indeed, this compound was hardly incorporated into the phytoalexins. It was, however, a very efficient precursor of the coumestan (59), as was the arylcoumarin (58). A biosynthetic pathway to coumestans involving allylic oxidation of 2'-hydroxyisoflav-3-enes to 2'-hydroxy-3-arylcoumarins, followed by cyclization and further oxidation, is proposed. This differs from the previously accepted, but not proven, sequence involving allylic oxidation of pterocarpenes. Isoflav-3-enes could be derived from intermediates of the pterocarpan/isoflavan pathway as shown in Scheme 13.

The A-ring hydroxylation pattern of 7 or 5,7 is determined at the chalcone level, but this can be modified further (i.e. 7 → 6,7) at the isoflavone level, as is demonstrated by the incorporation of formononetin (60) and texasin (61) into

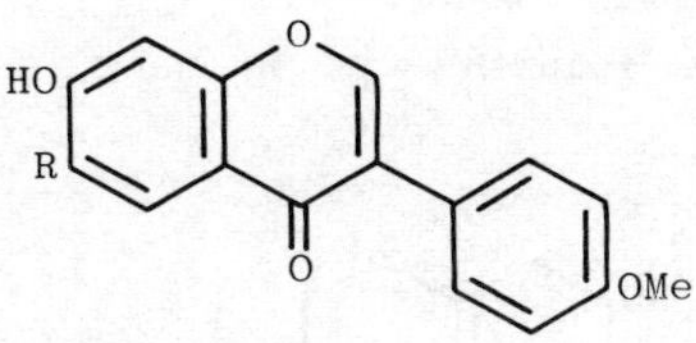

(60) R = H

(61) R = OH

(62) R = OMe

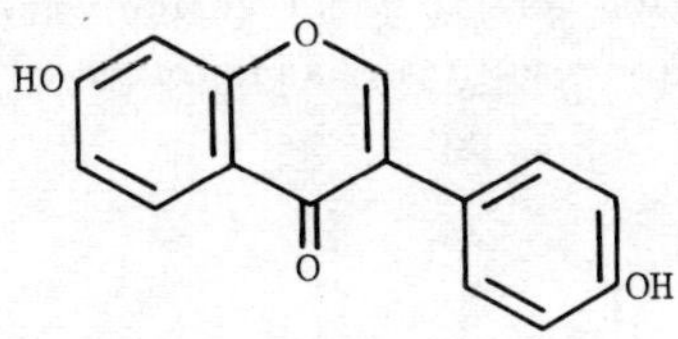

(63)

afrormosin (62) in sainfoin (*Onobrychis viciifolia*) seedlings.[122] As in many other studies, the stage at which methylation occurs in the B ring is not clear. Daidzein (63) and the appropriate methoxychalcone were poor precursors of both formononetin and afrormosin. The origin of the acetate-derived ring of the 6a-hydroxypterocarpan phytoalexin pisatin (64) has been investigated by feeding sodium [1,2-$^{13}C_2$]acetate to

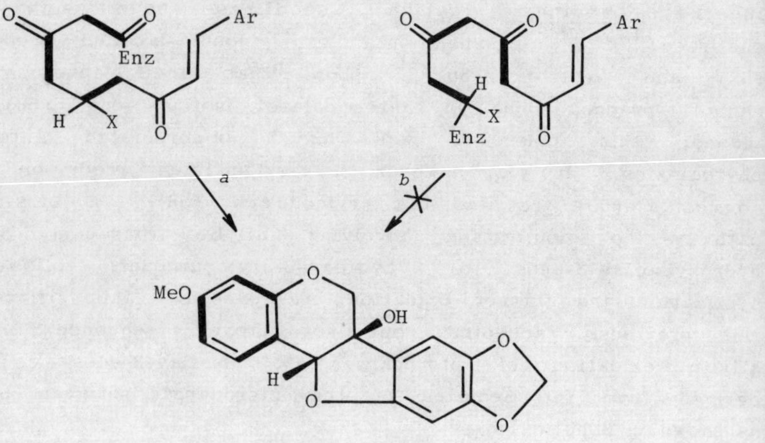

(64)

Scheme 14

$CuCl_2$-treated pea (Pisum sativum) pods.[123] Analysis of ^{13}C–^{13}C coupling in the ^{13}C n.m.r. spectrum showed that cyclization of the polyketide must have occurred as in a rather than b (Scheme 14), and proved that reduction of the 'lost'

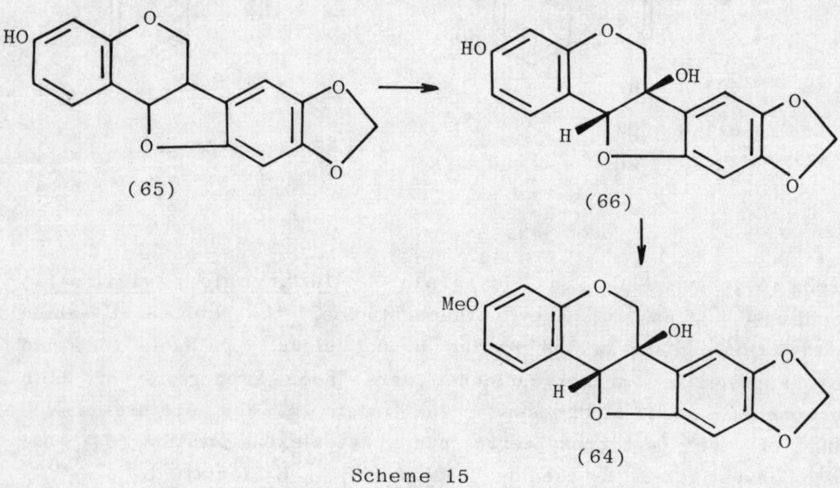

(65) (66) (64)

Scheme 15

oxygen function must have occurred before cyclization since no randomization of label was observed. Feeding experiments in pea pods and seedlings[124] have shown that the pathway to pisatin (64) probably involves 6a-hydroxylation of maackiain (65), then methylation (Scheme 15). Maackiain, 6a-hydroxymaackiain (66), and isoflavone precursors of maackiain were all efficiently incorporated into pisatin.

On the basis of earlier feeding experiments, a tentative pathway to the rotenoid amorphigenin (70) via (67), rot-2'-enonic acid (68), and rotenone (71) had been established (Scheme 16). In further experiments using seedlings of Amorpha fruticosa,[125] the presence of (68), (71), and dalpanol (72) has been established by isotope dilution. Rot-2'-enonic acid, but not rot-3'-enonic acid, was a good precursor of (70), (72), and 12a-hydroxyamorphigenin, the specific activities of the latter two compounds being some six times higher than that of (70).

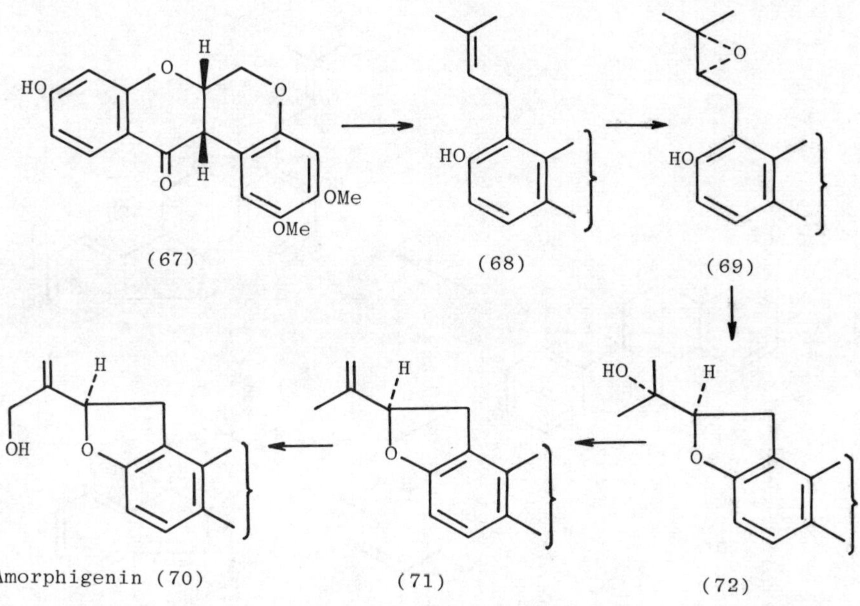

Scheme 16

This fits with the involvement of dalpanol as an intermediate, and establishes that the 12a-hydroxyrotenoid is not an artefact, but a true natural product. Degradation of the labelled rotenone showed that cyclization of the dimethylallyl side-chain was probably stereospecific, and thus the configuration of the postulated epoxide intermediate can be assigned as in (69). Neither of the 6'-epimers of amorphigenol was incorporated into amorphigenin, and thus allylic hydroxylation of rotenone must be involved, rather than hydroxylation of dalpanol followed by dehydration.[126]

From limited tracer experiments, such as with the rotenoids, it appears that dimethylallylation of isoflavonoids is a relatively late stage in biosynthesis. Evidence from enzyme studies now supports this hypothesis. A dimethylallyltransferase from elicitor-induced soybean (<u>Glycine max</u>) cotyledons[127,128] catalyses the formation of glyceollidins I (74) and II (75) from 3,6a,9-trihydroxypterocarpan (73) and dimethylallylpyrophosphate (Scheme 17), thus providing the likely key intermediates for

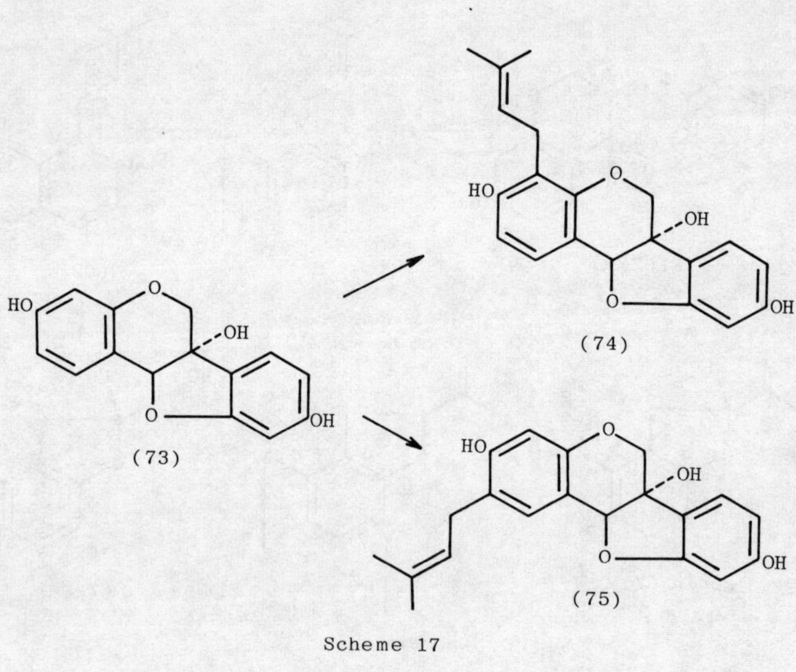

Scheme 17

soybean phytoalexin (glyceollins I-III) biosynthesis. The enzyme fraction appears to contain two different enzymes, catalysing dimethylallylation at the two aromatic sites.[129] Another prenyltransferase from Lupinus albus catalyses the 6-dimethylallylation of the isoflavones genistein and 2'-hydroxygenistein to give the antifungal isoflavones wighteone (76) and luteone (77) respectively.

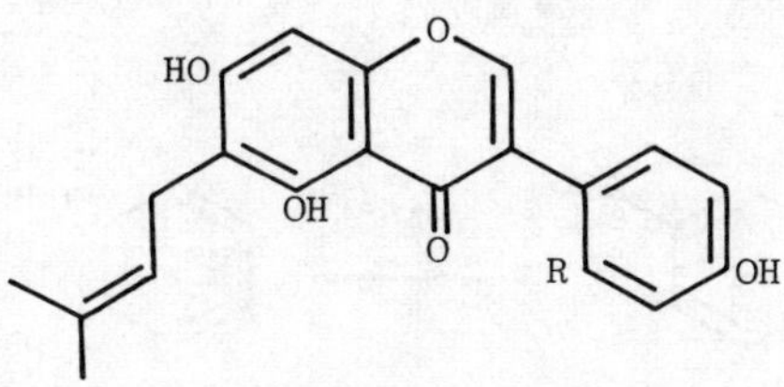

(76) R = H

(77) R = OH

A soluble UDP-glucose:isoflavone 7-O-glucosyltransferase from roots of chick pea (Cicer arietinum) catalysed the 7-O-glucosylation of 4'-methoxyisoflavones formononetin and biochanin A.[130] The corresponding 4'-hydroxyisoflavones daidzein and genistein were poor substrates, and a range of other isoflavones, isoflavanones, and flavonoids were not glucosylated. The same plant appears to contain β-glucosidases specific for isoflavone 7-O-glucosides.[131]

6 Quinones

6.1 o-Succinylbenzoic Acid.- o-Succinylbenzoic acid (78) is an intermediate in the biosynthetic pathway to a range of naphthoquinone and anthraquinone derivatives. This intermediate is accumulated by a menaquinone-deficient Escherichia coli mutant lacking dihydroxynaphthoate synthase, and labelling studies indicate that its biosynthesis from chorismate and L-glutamate required loss of C-1 from glutamate (Scheme 18).[132] A cell-free preparation utilized chorismate and 2-oxoglutarate, and required thiamine pyrophosphate for full activity,[133] supporting a mechanism in which the C_4 side-chain is derived from the thiamine pyrophosphate adduct of succinic semialdehyde.

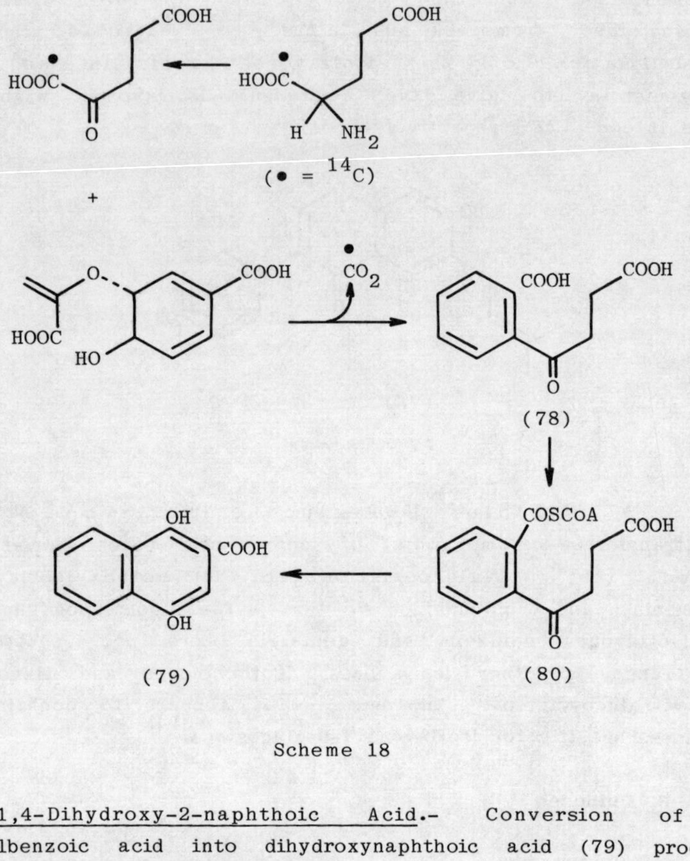

Scheme 18

6.2 1,4-Dihydroxy-2-naphthoic Acid.- Conversion of o-succinylbenzoic acid into dihydroxynaphthoic acid (79) proceeds via the unstable coenzyme A ester (80). The sequence (78) → (79) (Scheme 18) requires coenzyme A and ATP as cofactors, and has been demonstrated in Mycobacterium phlei[134] and Micrococcus luteus.[135] The intermediate ester has been isolated from cell-free preparations of M. phlei, and shown to be converted into (79) by an enzyme from the same organism.[136] Mutants of Bacillus subtilis lacking o-succinylbenzoyl CoA synthetase or dihydroxynaphthoate synthase had their growth requirements satisfied by complementation with the appropriate cell-free extracts from M. phlei.[137]

6.3 Anthraquinones.- The site of prenylation of dihydroxynaphthoic acid during biosynthesis of lucidin primeveroside (81) in *Galium mollugo* cell cultures has been investigated by ^{13}C n.m.r. spectroscopy.[138] Labelled *o*-succinylbenzoic acid (78) was transformed as shown in Scheme 19, demonstrating that prenylation occurs at C-3 of (78).

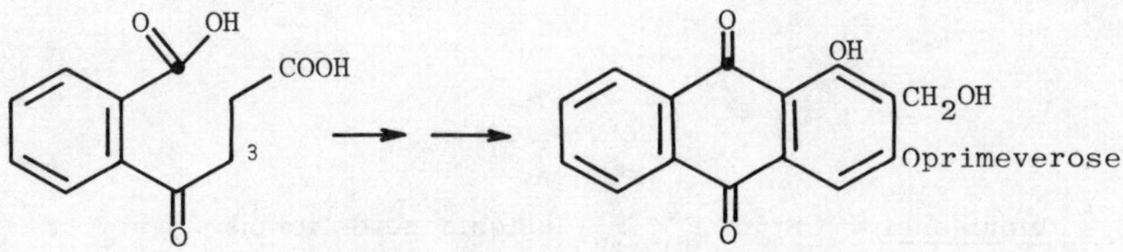

(78) (● = ^{13}C) (81)

Scheme 19

6.4 Naphthoquinones.- In contrast to the data obtained for lucidin primeveroside above, ^{14}C studies showed that *o*-succinylbenzoic acid was incorporated into phylloquinone (82) in maize (*Zea mays*) with retention of asymmetry, but the phytyl unit had been introduced at C-2 (Scheme 20),[139] in the same manner as observed with bacterial menaquinones. Prenylation at the same position is also a key reaction in the formation of a range of prenylnaphthoquinones and related compounds in *Catalpa ovata* (Scheme 21).[140,141] Dilution analyses with 2-prenyl-2-carboxy-4-oxotetralone (2-prenyl COT) (85) and catalponone (84) showed incorporation of (78) into only the (2*S*)- and (2*R*)-isomers of these compounds respectively. This showed that COT (83) (dihydroxynaphthoic acid in keto form) was prenylated stereospecifically, then decarboxylated with retention of configuration to give (2*R*)-catalponone. In agreement, this isomer was a better precursor than the (2*S*)-analogue of compounds including catalpalactone (86) and hydroxylapachone derivatives, e.g. (87). Whilst the major route to catalponone is as shown, there is evidence that the reduced 4-hydroxy derivatives of (83), (84), and (85) can function in the interconversions, though (83) is formed directly from (78).

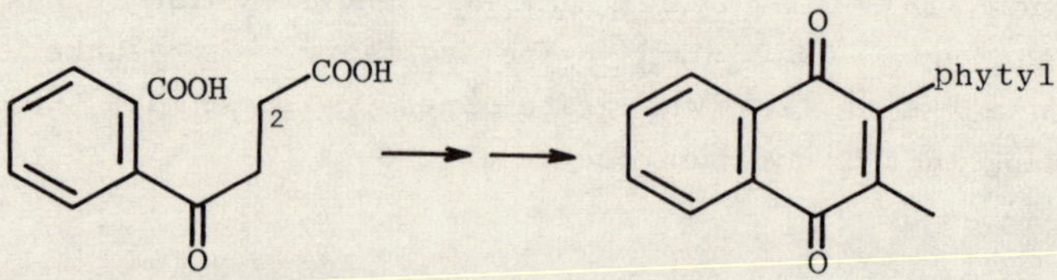

(78) (82)

Scheme 20

6.5 Ubiquinones.- Studies in ubiquinone-deficient mutants of *Saccharomyces cerevisiae*[142,143] have resulted in the isolation of two new compounds, 3,4-dihydroxy- and 3-methoxy-4-hydroxy-5-hexaprenylbenzoic acids (88) and (89), which can act as precursors of ubiquinone-6 (91). In prokaryotes, the 4-hydroxybenzoic acid derivative (90) is decarboxylated, prior to 3-hydroxylation and methylation during the sequence to ubiquinones. It is suggested that an alternative pathway via (88) and (89) operates in eukaryotes, in which hydroxylation and methylation precede decarboxylation.

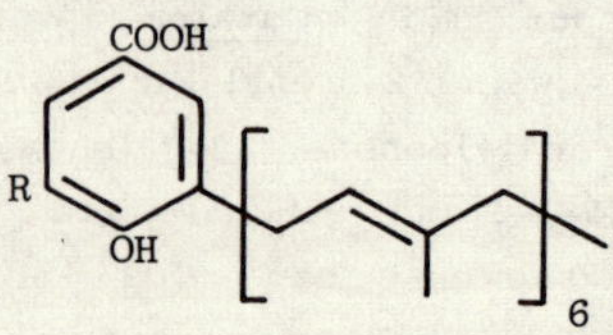

(88) R = OH

(89) R = OMe

(90) R = H

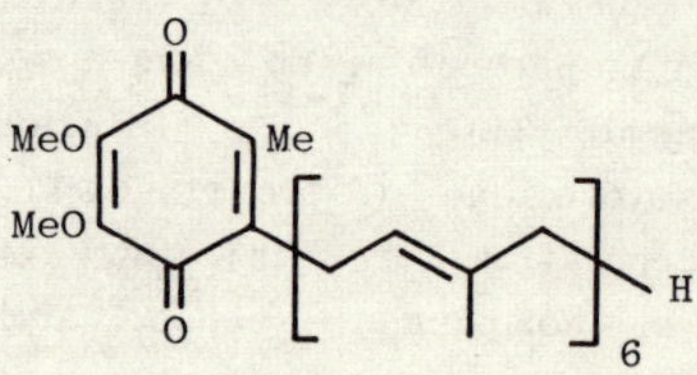

(91)

6.6 Plastoquinones and Tocopherols.- The incorporation of label from [Me-^{14}C]SAM into precursors in spinach chloroplasts suggests that the pathway from 2-methyl-6-phytylquinol (92) to α-tocopherol (94) involves firstly a methylation at the quinol stage, then ring cyclization to γ-tocopherol (93) followed by

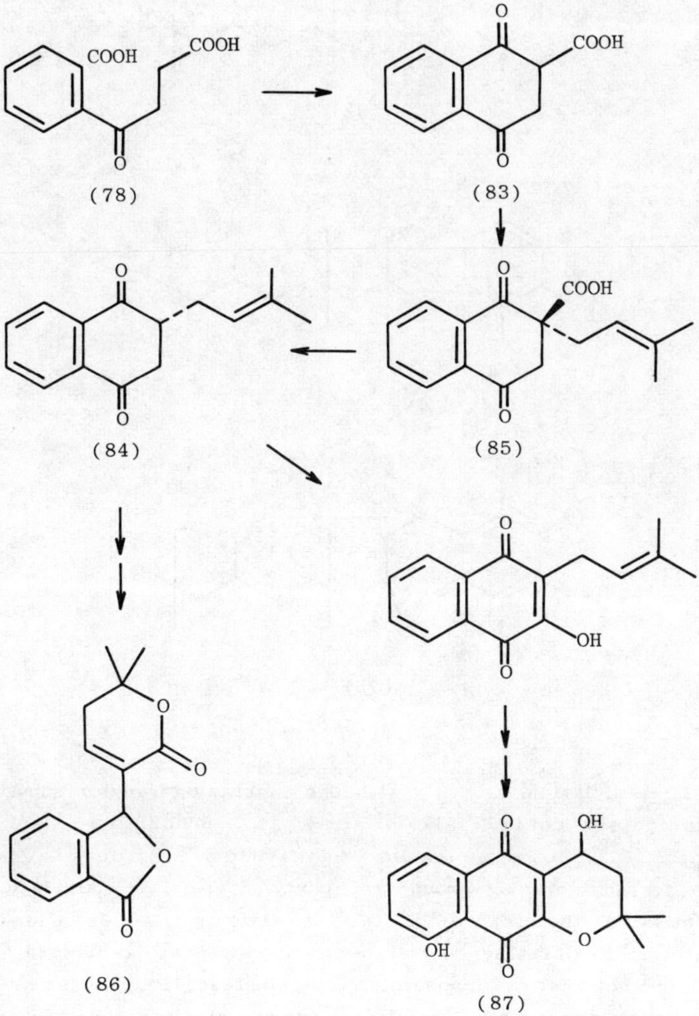

Scheme 21

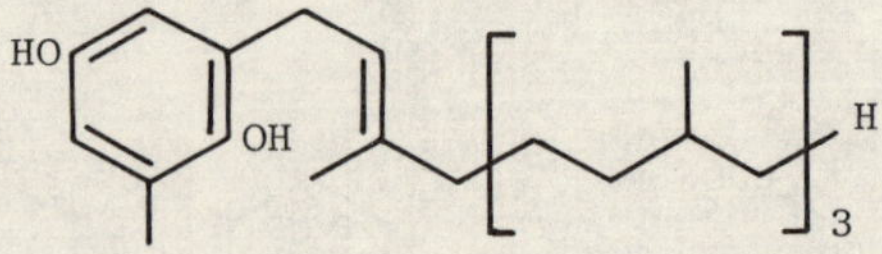

(92)

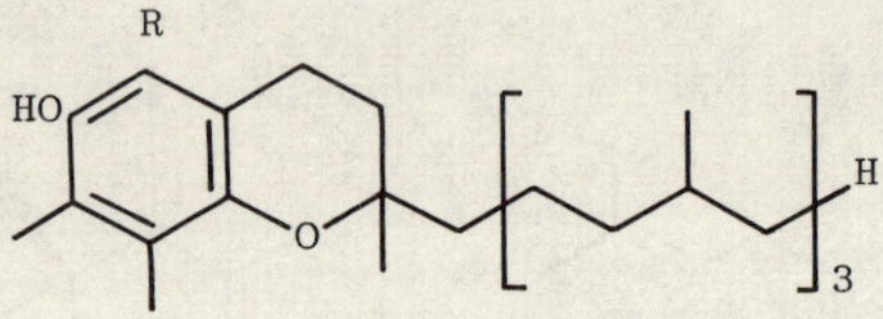

(93) R = H

(94) R = Me

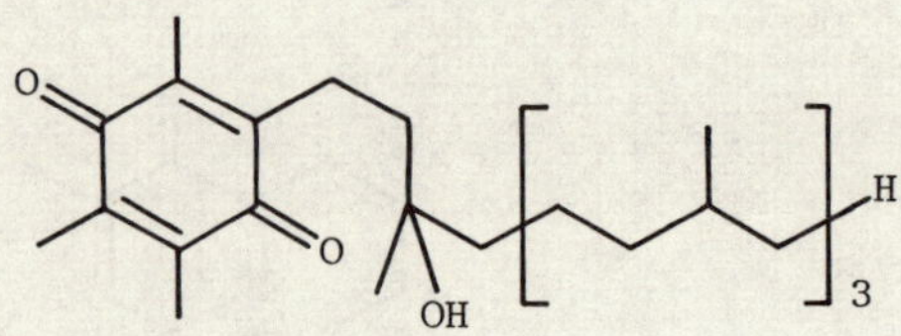

(95)

further methylation.[144] Although geranylgeranyl substituted methylquinols could also act as precursors of the tocopherols,[145] other results[146] favoured earlier reduction of the geranylgeranyl group to phytyl, prior to alkylation of homogentisate, the source of the aromatic ring. In illuminated lettuce chloroplasts,[147] label from homogentisate was channelled not into tocopherols, but instead into plastoquinones and phytylplastoquinones, resulting from oxidation of the quinol rather than the cyclization sequence. Although α-tocopherol and α-tocopherolquinone (95) are both present in rat tissues, labelling studies have shown that α-tocopherol is not the precursor of (95).[148]

References

1 'Encyclopaedia of Plant Physiology', New Series, eds. A. Pirson and M.H. Zimmermann, Springer-Verlag, Berlin, Heidelberg and New York. (a) J.B. Harborne, Vol.8, p.329, 1980; (b) T. Higuchi, Vol.13B, p.194, 1981.

2 'Recent Advances in Phytochemistry', eds. T. Swain, J.B. Harborne and C.F. van Sumere, Vol.12, Plenum Press, New York, 1979. (a) H.G. Floss, p.59; (b) K.R. Hanson and E.A. Havir, p.91; (c) M.H. Zenk, p.139; (d) G.G. Gross, p.177; (e) S.A. Brown, p.49; (f) H. Grisebach, p.221.

3 'The Biochemistry of Plants', eds. P.K. Stumpf and E.E. Conn, Vol.7, Academic Press, New York, 1981. (a) G.G. Gross, p.301; (b) K.R. Hanson and E.A. Havir, p.577; (c) H. Grisebach, p.457; (d) S.A. Brown, p.269; (e) K. Hahlbrock, p.425; (f) E. Haslam, p.527; (g) E. Leistner, p.403; (h) W. Hösel, p.725.

4 G.H.N. Towers and C. Wat, Planta Medica, 1979, 37, 97.

5 K. Hahlbrock and H. Grisebach, Annu. Rev. Plant Physiol., 1979, 30, 105.

6 S.A. Brown, Planta Medica, 1979, 36, 299.

7 M. Afzal and J.M. Al-Hassan, Heterocycles, 1980, 14, 1173.

8 M.U.S. Sultanbawa, Tetrahedron, 1980, 36, 1465.

9 'Pigments in Plants', ed. F. Czygan, 2nd ed. 1980. (a) H. Grisebach, p.187; (b) E. Leistner, p.352.

10 R.J. McCandliss and K.M. Herrmann, J. Biol. Chem., 1979, 254, 3761.

11 G.A. Nimmo and J.R. Coggins, Biochem. J., 1981, 197, 427.

12 G.A. Nimmo and J.R. Coggins, Biochem. J., 1981, 199, 657.

13 D.J. Llewellyn, A. Daday and G.D. Smith, J. Biol. Chem., 1980, 255, 2077.

14 E. Heyde, Biochemistry, 1979, 18, 2766.

15 D.J. Llewellyn and G.D. Smith, Biochemistry, 1979, 18, 4707.

16 H. Holländer and N. Amrhein, Plant Physiol., 1980, 66, 823.

17 N. Amrhein, B. Deus, P. Gehrke and H.C. Steinrücken, Plant Physiol., 1980, 66, 830.

18 J. Berlin and L. Witte, Z. Naturforsch., Teil. C, 1981, 36, 210.

19 U. Roisch and F. Lingens, Hoppe-Seyler's Z. Physiol. Chem., 1980, 361, 1049.

20 A.M. Fazel and R.A. Jensen, J. Bacteriol., 1979, 138, 805.

21 J.L. Rubin and R.A. Jensen, Plant Physiol., 1979, 64, 727.

22 G. Byng, R. Whitaker, C. Flick and R.A. Jensen, Phytochemistry, 1981, 20, 1289.

23 A.M. Fazel and R.A. Jensen, Arch. Biochem. Biophys., 1980, 200, 165.

24 L.O. Zamir, R.A. Jensen, B.H. Arison, A.W. Douglas, G. Albers-Schönberg and J.R. Bowen, J. Am. Chem. Soc., 1980, 102, 4499.

25 R. Endress, Plant Physiol., 1981, 68, 272.

26 H. Nakata, T. Yamauchi and H. Fujisawa, J. Biol. Chem., 1979, 254, 1829.

27 R. Leinberger, W.E. Hull, H. Simon and J. Rétey, Eur. J. Biochem., 1981, 117, 311.

28 S.L. Manley and D.J. Chapman, Plant Physiol., 1979, 64, 1032.

29 S.L. Manley and D.J. Chapman, Phytochemistry, 1980, 19, 1453.

30 R.H. White, Biochim. Biophys. Acta, 1979, 583, 55.

31 H.A. Stafford and L.L. Lewis, Plant Physiol., 1979, 64, 176.

32 H. Tsuji, N. Bando, T. Ogawa and K. Sasaoka, Biochim. Biophys. Acta, 1981, 677, 326.

33 H. Tsuji, N. Bando, T. Ogawa and K. Sasaoka, Agric. Biol. Chem., 1981, 45, 541.

34 E.A. Havir, Arch. Biochem. Biophys., 1981, 211, 556.

35 K.R. Hanson, Arch. Biochem. Biophys., 1981, 211, 564.

36 K.R. Hanson, Arch. Biochem. Biophys., 1981, 211, 575.

37 H. Holländer, H. Kiltz and N. Amrhein, Z. Naturforsch., Teil. C, 1979, 34, 1162.
38 E.A. Havir, Planta, 1981, 152, 124.
39 N. Amrhein and J. Gerhardt, Biochem. Biophys. Acta, 1979, 583, 434.
40 S.O. Duke, R.E. Hoagland and C.D. Elmore, Plant Physiol., 1980, 65, 17.
41 D.C. Loschke, L.A. Hadwiger, J. Schröder and K. Hahlbrock, Plant Physiol., 1981, 68, 680.
42 M.A. Lawton, R.A. Dixon and C.J. Lamb, Biochim. Biophys. Acta, 1980, 633, 162.
43 C.J. Lamb, T.K. Merritt and V.S. Butt, Biochim. Biophys. Acta, 1979, 582, 196.
44 L. Canonica, P. Gramatica, P. Manitto and D. Monti, J. Chem. Soc., Chem. Commun., 1979, 1073.
45 B. Ulbrich and M.H. Zenk, Phytochemistry, 1979, 18, 929.
46 M.J.C. Rhodes, L.S.C. Wooltorton and E.J. Lourenco, Phytochemistry, 1979, 18, 1125.
47 B. Ulbrich and M.H. Zenk, Phytochemistry, 1980, 19, 1625.
48 B. Schöbel and W. Pollmann, Z. Naturforsch., Teil. C, 1980, 35, 699.
49 A. Fleuriet, J.J. Macheix, R. Suen, and R.K. Ibrahim, Z. Naturforsch., Teil. C, 1980, 35, 967.
50 L. Nagels, M. Molderez and F. Parmentier, Phytochemistry, 1981, 20, 965.
51 D. Strack, Z. Naturforsch., Teil. C, 1980, 35, 204.
52 M. Bopp and W. Lüdicke, Z. Naturforsch., Teil. C, 1980, 35, 539.
53 D. Strack, G. Nurmann and G. Sachs, Z. Naturforsch., Teil. C, 1980, 35, 963.
54 D. Strack, Z. Naturforsch., Teil. C, 1981, 36, 215.
55 N. Tkotz and D. Strack, Z. Naturforsch., Teil. C, 1980, 35, 835.
56 B.E. Ellis, S. Remmen and G. Goeree, Planta, 1979, 147, 163.
57 J.D. Angmor, P.M. Dewick and W.C. Evans, Planta Medica, 1979, 35, 342.
58 P.Gramatica, B.M. Ranzi and P. Manitto, Bioorg. Chem., 1981, 10, 22.
59 P.Gramatica, B.M. Ranzi and P. Manitto, Bioorg. Chem., 1981, 10, 14.
60 H. Giesel, G. Machacek, J. Bayerl and H. Simon, FEBS Lett., 1981, 123, 107.
61 A.A. Craveiro and T.L.G. Lemos, J. Nat. Prod., 1980, 43, 634.
62 J. Collendavelloo, M. Legrand, P. Geoffrey, J. Barthelemy and B. Fritig, Phytochemistry, 1981, 20, 611.
63 Y. Tsang and R.K. Ibrahim, Z. Naturforsch., Teil. C, 1979, 34, 46.
64 C.P. Vance and J.W. Bryan, Phytochemistry, 1981, 20, 41.
65 Y. Charrière-Ladreix, Phytochemistry, 1979, 18, 43.
66 H. Kuroda, M. Shimada and T. Higuchi, Phytochemistry, 1981, 20, 2635.
67 T. Lüderitz and H. Grisebach, Eur. J. Biochem., 1981, 119, 115.
68 D. Wyrambik and H. Grisebach, Eur. J. Biochem., 1979, 97, 503.
69 H. Kutsuki and T. Higuchi, Planta, 1981, 152, 365.
70 C. Grand, R. Ranjeva, A.M. Boudet and G. Alibert, Planta, 1979, 146, 281.
71 W. Hösel and R. Todenhagen, Phytochemistry, 1980, 19, 1349.
72 S. Marcinowski, H. Falk, D.K. Hammer, B. Hoyer and H. Grisebach, Planta, 1979, 144, 161.
73 D.C. Ayres, A. Farrow and B.G. Carpenter, J. Chem. Soc., Perkin Trans. 1, 1981, 2134.
74 D.E. Jackson and P.M. Dewick, J. Pharm. Pharmacol., 1981, 33, 18P.
75 H. Kutsuki, M. Shimada and T. Higuchi, Mokuzai Gakkaishi, 1981, 27, 39.
76 M. Axelson and K.D.R. Setchell, FEBS Lett., 1981, 123, 337.
77 A. Ichihara, H. Kawagishi, T. Sasagawa and S. Sakamura, Agric. Biol. Chem., 1981, 45, 2163.

78 G. Caporale, G. Innocenti, A. Guitto, P. Rodighiero and F. Dall'Acqua, Phytochemistry, 1981, 20, 1283.
79 S.K. Sharma, J.M. Garrett and S.A. Brown, Z. Naturforsch., Teil. C, 1979, 34, 387.
80 R.K. Ibrahim, Phytochemistry, 1980, 19, 2459.
81 R.K. Ibrahim and B. Boulay, Plant Sci. Lett., 1980, 18, 177.
82 C.G. Glennie and G.W. Perold, Phytochemistry, 1980, 19, 1463.
83 P. Denniff, I. MacLeod and D.A. Whiting, J. Chem. Soc., Perkin Trans. 1, 1980, 2637.
84 E. Fautz and H. Reichenbach, Phytochemistry, 1979, 18, 957.
85 H. Achenbach, A. Böttger, W. Kohl, E. Fautz and H. Reichenbach, Phytochemistry, 1979, 18, 961.
86 A. Schoeppner and H. Kindl, FEBS Lett., 1979, 108, 349.
87 K. Fritzemeier and H. Kindl, Planta, 1981, 151, 48.
88 N. Rupprich, H. Hildebrand and H. Kindl, Arch. Biochem. Biophys., 1980, 200, 72.
89 M. Fujita and T. Inoue, Chem. Pharm. Bull., 1980, 28, 2476.
90 M. Fujita and T. Inoue, Chem. Pharm. Bull., 1980, 28, 2482.
91 M. Fujita and T. Inoue, Phytochemistry, 1981, 20, 2183.
92 W. Heller and K. Hahlbrock, Arch. Biochem. Biophys., 1980, 200, 617.
93 R. Sütfeld and R. Wiermann, Arch. Biochem. Biophys., 1980, 201, 64.
94 R. Sütfeld and R. Wiermann, Z. Naturforsch., Teil. C, 1981, 36, 30.
95 R. Spribille and G. Forkmann, Z. Naturforsch., Teil. C, 1981, 36, 619.
96 R.J. Light and K. Hahlbrock, Z. Naturforsch., Teil. C, 1980, 35, 717.
97 M.J. Boland and E. Wong, Bioorg. Chem., 1979, 8, 1.
98 G. Forkmann and B. Kuhn, Planta, 1979, 144, 189.
99 G. Forkmann, W. Heller and H. Grisebach, Z. Naturforsch., Teil. C, 1980, 35, 691.
100 G. Forkmann and G. Stotz, Z. Naturforsch., Teil. C, 1981, 36, 411.
101 L. Britsch, W. Heller and H. Grisebach, Z. Naturforsch., Teil. C, 1981, 36, 742.
102 G. Stotz and G. Forkmann, Z. Naturforsch., Teil. C, 1981, 36 411.
103 G. Forkmann, Phytochemistry, 1979, 18, 1973.
104 N. Amrhein, Phytochemistry, 1979, 18, 585.
105 G. Kuroki and J.E. Poulton, Z. Naturforsch., Teil. C, 1981, 36, 916.
106 G. Brunet and R.K. Ibrahim, Phytochemistry, 1980, 19, 741.
107 F.W. Collins, V. De Luca, R.K. Ibrahim, B. Voirin, and M. Jay, Z. Naturforsch., Teil. C, 1981, 36, 730.
108 Y. Charrière-Ladreix, R. Douce and J. Joyard, FEBS Lett., 1981, 133, 55.
109 M. Fujita and T. Inoue, Yakugaku Zasshi, 1979, 99, 165.
110 J. Kamsteeg, J. van Brederode and G. van Nigtevecht, Z. Naturforsch., Teil. C, 1980, 35, 249.
111 J.L. Shute, P.S. Jourdan and R.L. Mansell, Z. Naturforsch., Teil. C, 1979, 34, 738.
112 R. Heinsbroek, J. van Brederode, G. van Nigtevecht and J. Kamsteeg, Phytochemistry, 1979, 18, 935.
113 R. Heinsbroek, J. van Brederode, G. van Nigtevecht, J. Maas, J. Kamsteeg, E. Besson and J. Chopin, Phytochemistry, 1980, 19, 1935.
114 J. van Brederode and R. Kamps-Heinsbroek, Z. Naturforsch., Teil. C, 1981, 36, 484.
115 U. Matern., J.R.M. Potts and K. Hahlbrock, Arch. Biochem. Biophys., 1981, 208, 233.
116 S. Ayabe and T. Furuya, Tetrahedron Lett., 1981, 22, 2097.
117 S. Ayabe, T. Yoshikawa, M. Kobayashi and T. Furuya, Phytochemistry, 1980, 19, 2331.

118 M.K. Burns, J.M. Coffin, I. Kurobane, L.C. Vining, A.G. McInnes, D.G. Smith and J.A. Walter, J. Chem. Soc., Perkin Trans. 1, 1981, 1411.
119 P.M. Dewick and M. Martin, Phytochemistry, 1979, 18, 591.
120 P.M. Dewick and M. Martin, Phytochemistry, 1979, 18, 597.
121 M. Martin and P.M. Dewick, Phytochemistry, 1980, 19, 2341.
122 H.A.M. Al-Ani and P.M. Dewick, Phytochemistry, 1980, 19, 2337.
123 A. Stoessl and J.B. Stothers, Z. Naturforsch., Teil. C, 1979, 34, 87.
124 P.M. Dewick and S.W. Banks, Planta Medica, 1980, 39, 287.
125 L. Crombie, I. Holden, G.W. Kilbee and D.A. Whiting, J. Chem. Soc., Chem. Commun., 1979, 1143.
126 L. Crombie, I. Holden, G.W. Kilbee and D.A. Whiting, J. Chem. Soc., Chem. Commun., 1979, 1144.
127 U. Zähringer, J. Ebel, L.J. Mulheirn, R.L. Lyne and H. Grisebach, FEBS Lett., 1979, 101, 90.
128 U. Zähringer, E. Schaller, and H. Grisebach, Z. Naturforsch., Teil. C, 1981, 36, 234.
129 G. Schröder, U. Zähringer, W. Heller, J. Ebel and H. Grisebach, Arch. Biochem. Biophys., 1979, 194, 635.
130 J. Köster, and W. Barz, Arch. Biochem. Biophys., 1981, 212, 98.
131 G. Burmeister and W. Hösel, Planta, 1981, 152, 578.
132 R. Meganathan and R. Bentley, Biochemistry, 1981, 20, 5336.
133 R. Meganathan, J. Biol. Chem., 1981, 256, 9386.
134 R. Meganathan and R. Bentley, J. Bacteriol., 1979, 140, 92.
135 R. Meganathan, T. Folger and R. Bentley, Biochemistry, 1980, 19, 785.
136 L. Heide and E. Leistner, FEBS Lett., 1981, 128, 201.
137 R. Meganathan, R. Bentley and H. Taber, J. Bacteriol., 1981, 145, 328.
138 K. Inoue, Y. Shiobara, H. Nayeshiro, H. Inouye, G. Wilson and M.H. Zenk, J. Chem. Soc., Chem. Commun., 1979, 957.
139 K.G. Hutson and D.R. Threlfall, Phytochemistry, 1980, 19, 535.
140 H. Inouye, S. Ueda, K. Inoue and Y. Shiobara, Phytochemistry, 1981, 20, 1707.
141 K. Inoue, S. Ueda, Y. Shiobara, I. Kimura and H. Inouye, J. Chem. Soc., Perkin Trans. 1, 1981, 1246.
142 R.R. Goewert, C.J. Sippel and R.E. Olson, Biochemistry, 1981, 20, 4217.
143 R.R. Goewert, C.J. Sippel, M.F. Grimm and R.E. Olsen, Biochemistry, 1981, 20, 5611.
144 J. Soll and G. Schultz, Phytochemistry, 1980, 19, 215.
145 J. Soll and G. Schultz, Biochem. Biophys. Res. Commun., 1979, 91, 715.
146 J. Soll, M. Kemmerling and G. Schultz, Arch. Biochem. Biophys., 1980, 204, 544.
147 K.G. Hutson and D.R. Threlfall, Biochim. Biophys. Acta, 1980, 632, 630.
148 P.E. Hughes and S.B. Tove, J. Biol. Chem., 1980, 255, 7095.

3

The Biosynthesis of C_5—C_{20} Terpenoid Compounds

BY J.R. HANSON

1 Introduction

This chapter covers the three-year period between the completion of the last report and December, 1981. It follows the pattern of the earlier reports[1] with sections on each class of terpenoid. During the period under review there has been an increasing amount of information which has been obtained from stable isotope studies. In particular there have been a number of applications of deuterium n.m.r. in terpenoid biosynthesis as well as a large number of carbon-13 n.m.r. studies. The latter have been effective in defining the way in which the prenyl units are assembled and in unravelling rearrangements.

A number of books and reviews[2-5] covering aspects of terpenoid biosynthesis have appeared including one devoted solely to the biosynthesis of isoprenoid compounds which contains a series of extensive and valuable surveys of the area. A review[6] of chiral methyl group methodology includes various terpenoid examples as does a review[7] of deuterium n.m.r. in biosynthesis.

A further preparation of [3'-^{13}C]-mevalonic acid has been described.[8]

The prenyl transferase reactions have continued to attract attention and progress has been reviewed.[9] The partial purification and properties of prenyl transferases from microbiological, plant and animal sources have been described.[10,11] It has been suggested that there are two or more closely related prenyl transferases that are stereoselective for E and Z products. Homogeneous preparations of a dimeric farnesyl pyrophosphate synthetase of a molecular weight of about 85 000 daltons have been obtained. Some prenyl transferases will form C_{15} or larger chains without the apparent accumulation of intermediate prenylogues. The substrate specificity of prenyl transferase is not very stringent and thus the use of analogues has afforded some insight into its mode of action. Results obtained with fluorinated analogues have been used to support a stepwise ionization-condensation-elimination mechanism for the 1'-4 condensation step.[9,12-14] This was discussed in the previous report. Evidence has also been presented[15] for a rigid geranyl cation - pyrophosphate anion pair in the chain-elongation step.

An extensive review[16] discusses the stereochemistry of the displacement of allylic pyrophosphates in the subsequent metabolism and cyclization of prenyl pyrophosphates. Aspects of this work will be described later.

2 Hemiterpenoids

A number of developments in this area are recorded elsewhere in these reports. 5-Hydroxymarmesin (1) plays[17] a significant role in the biosynthesis of the linear O-alkylfuranocoumarin 5-methoxypsoralen (2).

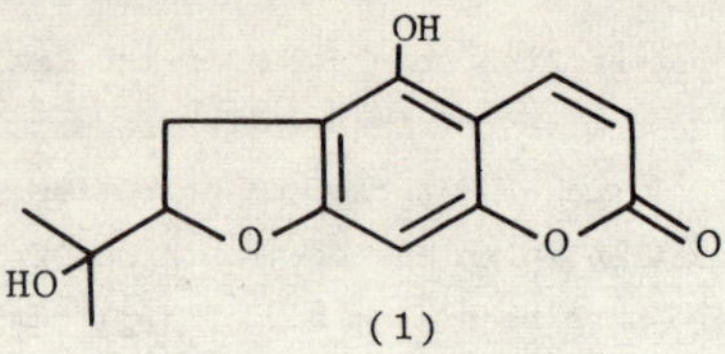

(1)

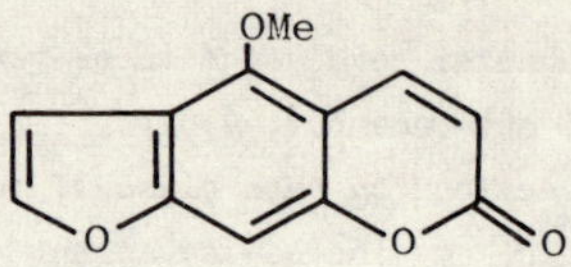

(2)

3 Monoterpenoids

A number of reviews of monoterpenoid biosynthesis have appeared.[2, 18 - 20] In contrast to the absence of definitive evidence for the E - Z isomerization of the allylic pyrophosphates, further evidence has been presented[21] for a redox mechanism in the interconversion of geraniol (3) and nerol (4) in *Rosa damascena*. The isomerization of geraniol to nerol involves the loss of the pro-(1S)-hydrogen atom whereas the reverse isomerization involves the loss of the pro-(1R)-hydrogen atom. Cell-free extracts that mediated the reaction were shown to be $NADP^+$ - NADPH dependent, which suggested that an aldehyde was involved. Degradation of geraniol, biosynthesized by *Rosa dilecta* from mevalonate, has shown[22] that C-10 is derived exclusively from the C-2 of mevalonate. However, the geminal dimethyl groups of pulegone are scrambled[23] in its biosynthesis from [10-^{14}C] geraniol by *Mentha* species. Further evidence has been presented[24,25] to show that the dimethylallyl pyrophosphate pool is at least partially amino-acid-derived in plants.

The preferential participation of linaloyl pyrophosphate rather than neryl pyrophosphate has been demonstrated[26] in the biosynthesis of some cyclic monoterpenoids such as α-terpineol and carvone in *Mentha spicata*. The oxidation of limonene to form carvone involves[27] a shift of the endocyclic double bond in (5) to form (6), i.e. (+)-limonene and (-)-carvone are biogenetically related. It has been suggested that they are probably formed

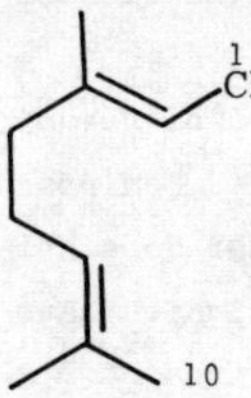

(3)

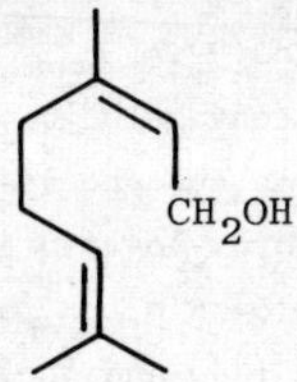

(4)

(5)

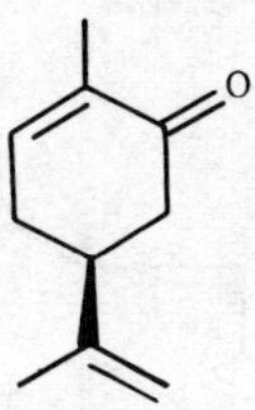

(6)

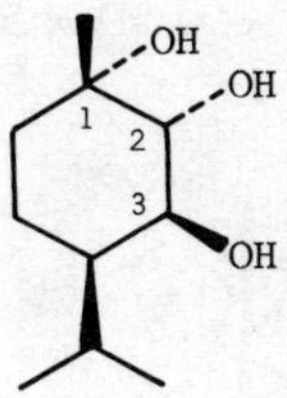

(7)

on a divergent pathway from a common intermediate. Degradation showed that the exocyclic double bond of carvone was not formed regiospecifically. The biosynthesis of 1α,2α,3β-trihydroxy-p-menthane by Fusicoccum amygdali has been reported.[28] Measurement of isotope ratios from [5-^{3}H,2-^{14}C] —mevalonate showed that four tritium atoms are retained - one more than the hydroxylation pattern would suggest. It was proposed that the menth-2-en-1-ol precursor was generated from an α-terpinyl cation through a 1,3-hydride shift. Trans cleavage of an epoxide would afford the 1α,2α,3β-trihydroxy-p-menthane (7).

Time course studies with $^{14}CO_2$ in Thymus species suggest[29] that γ-terpinene was converted via p-cymene into thymol. Measurement of ^{3}H:^{14}C ratios in carene (9) biosynthesized from various mevalonates showed[30] that the carene

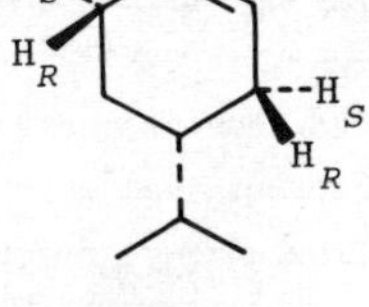

(8)

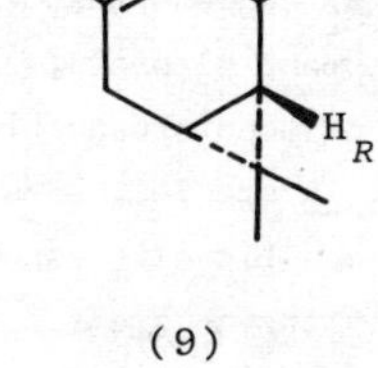

(9)

skeleton is constructed from its presumed monocyclic precursor with migration of the double bond and an unexpected 1,2-shift of a proton to the site of the original double bond, i.e. (8) to (9).

A full paper describing the conversion of 10-oxoderivatives of geraniol and nerol into secologanin in Catharanthus species has appeared.[31] Geniposide may be easily converted[32] into 10-hydroxyloganin although the latter does not appear to be a precursor of secologanin. Deuterium n.m.r. studies have shown that the biosynthesis of lamiide (11) and ipolamiide (12) utilizes[33,34] 8-*epi*-deoxyloganin (10) rather than deoxyloganin itself. Compounds of the 8-*epi*-iridodial series are involved in this sequence.

The role of (−)-ipsdienol, derived from myrcene, in the biosynthesis of ipsenol (13) in Ips beetles has been investigated.[35]

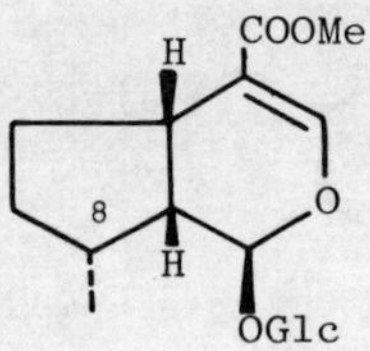

(10)

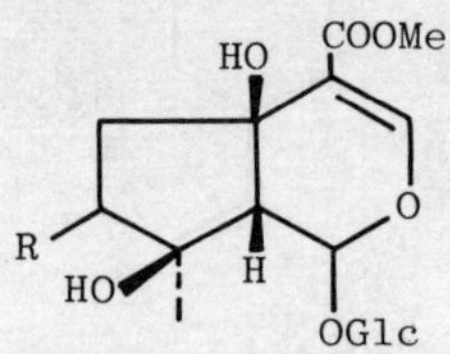

(11) R = OH

(12) R = H

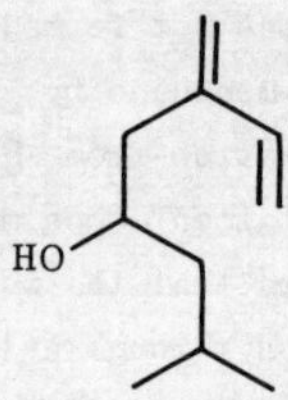

(13)

The production of linalool and geraniol by hydrolytic breakdown of bound forms in disrupted tea shoots has been described.[36] Geranyl β-D-glucoside is formed, transported and metabolized when apples were injected with geraniol with results[37] that suggest that glycosides may be storage and transportable forms of the monoterpene alcohols.

The enzymology of monoterpene biosynthesis has been extensively studied[38-41] particularly in Salvia species. Systems have been obtained which mediate the formation of mono- and bicyclic monoterpenoids from geranyl pyrophosphate. Not unexpectedly the key enzyme systems are at their most active during maximum leaf expansion. The substrate specificity of the monoterpenol dehydrogenases from Foeniculum vulgare and Tanacetum vulgare have been examined. A number of non-enzymic model studies, which are outside the scope of this report, have been presented on monoterpenoid biosynthesis, in particular with regard to the solvolysis of pyrophosphates.

4 Sesquiterpenoids

A comprehensive review[42] of terpenoid metabolites of the Basidiomycetes has sections on the biosynthesis of these compounds. Many of the sesquiterpenoids found in these fungi can in theory be derived by various cyclizations of humulene. Conformational calculations on humulene have been made[43] with a view to predicting the role of stable conformers in its biosynthetic reactions.

The enzymatic conversion of farnesyl pyrophosphate to nerolidyl pyrophosphate and the cyclization of the latter to cyclonerodiol (15) has been studied.[44,45] Labelling experiments have established that the conversion of farnesyl pyrophosphate to nerolidyl pyrophosphate (14) takes place by a net suprafacial process and the subsequent cyclization of nerolidyl pyrophosphate involves a trans addition of water across the vinyl and central double bond to form cyclonerodiol (15). Oxygen-18 experiments lead to an ion-pair mechanism for the isomerization of farnesyl pyrophosphate to nerolidyl pyrophosphate.

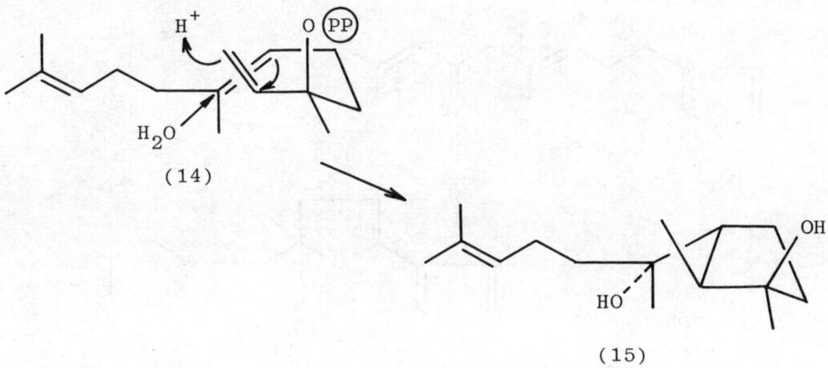

An important role has been proposed[46] for nerolidyl pyrophosphate in the formation of the hydrocarbon trichodiene (16). However, contrary to previous reports, this cyclization does not involve the loss of a proton from C-1 of farnesyl pyrophosphate. The trichoverrin diesters (17) may be the precursors of the macrocyclic trichothecene esters in Myrothecium verrucaria.[47]

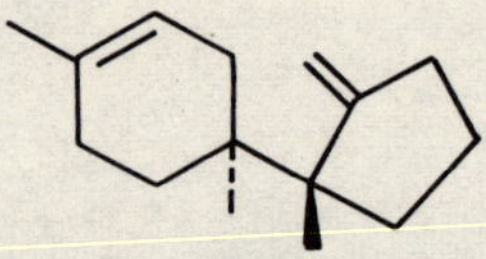

(16)

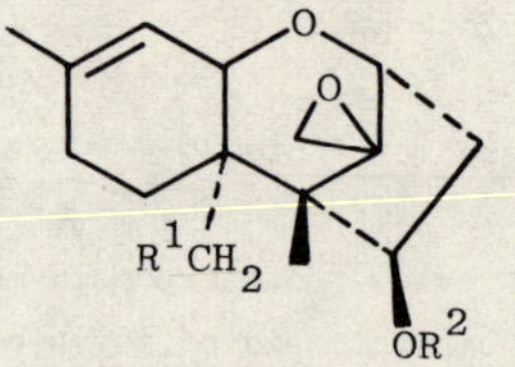

(17) $R^1 = OCOCH{=}CH(CH_3)CH_2CH_2OH$,
$R^2 = COCH{=}CH{-}CH{=}CHCH(OH){-}CH(OH)CH_3$

9-Oxofarnesol and 6-oxodendrolasin (18) are precursors of ipomeamarone (19) which is a stress metabolite of the sweet potato, Ipomoea batatas.[48] Confirmation has been provided by a deuterium n.m.r. study for the hydride shift (C-5 - C-4) which accompanies the C-10 - C-5 methyl group migration in the biosynthesis of the antifungal sesquiterpenoid capsidiol (20).[49]

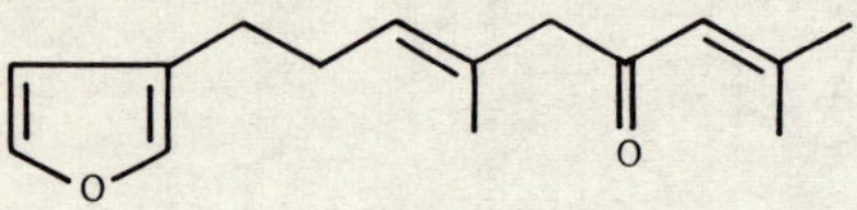

(18)

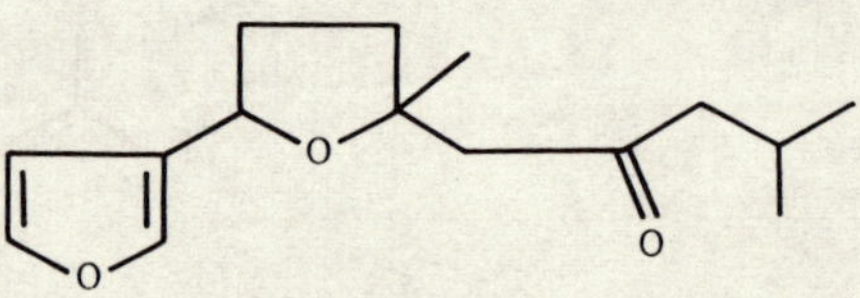

(19)

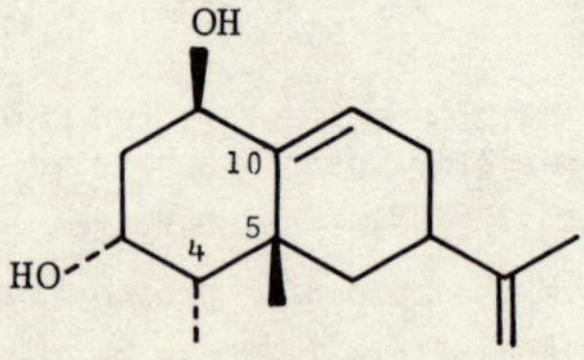

(20)

Two independent studies have been reported[50,51] on the biosynthesis of the PR-toxin (22) in Penicillium roqueforti. The results of [1,2-$^{13}C_2$]-acetate experiments show that the biosynthesis of the eremophilane skeleton in (22) follows the scheme proposed by Robinson involving a 1,2-methyl group shift from a eudesmane skeleton (21). In both cases some induced couplings were observed which probably arise from multiple labelling of a farnesyl pyrophosphate precursor.

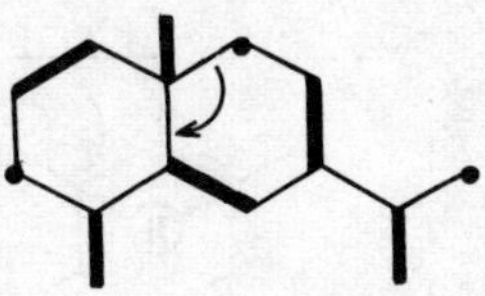

(21)

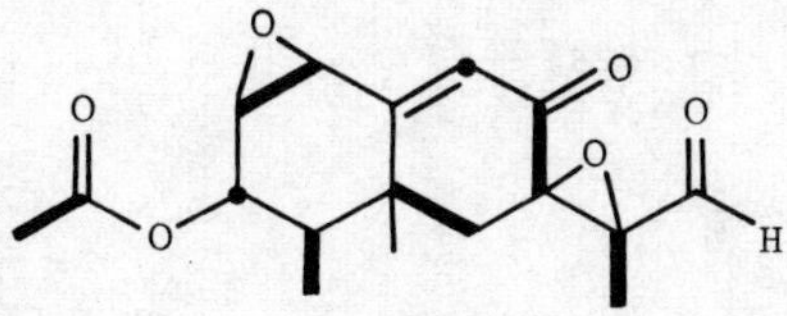

(22)

The labelling pattern of dihydrobotrydial (24) derived from [1-^{13}C]-, [1,2-$^{13}C_2$]-acetate and [4,5-$^{13}C_2$]-mevalonate by Botrytis cinerea has shown[52] that farnesyl pyrophosphate is folded as in (23). The initial cyclization is similar to that which leads to caryophyllene. Examination of the deuterium: carbon-13 n.m.r. coupling patterns in material biosynthesized from [4-$^{2}H_2$,4-^{13}C]-mevalonate revealed[53] that a 1,3-hydride shift occurred during the cyclization. This provides another example of the observation that secondary methyl groups often mark the terminus of hydrogen rearrangements in the biosynthesis of terpenoid compounds. The incorporation of oxygen-18 from oxygen-18-labelled water showed[54] that the C-9 carbocation was discharged by hydration. The labelling pattern from [pro-2(R)- and [pro-5(R)-^{3}H] mevalonates established[55] that the formation of the hemi-acetal ring occurs with the retention of these atoms at C-15 and C-10 respectively. This provided stereochemical information on the formation of the hemi-acetal ring which probably takes place through the cleavage of a trans-glycol to afford the dialdehyde, botrydial. N.m.r. studies using material biosynthesized from [2-$^{2}H_3$]-acetate showed that the reduction of the aldehyde at C-15 to form dihydrobotrydial proceeded with the re-stereospecificity which is typical of a microbial dehydrogenase.

The labelling pattern of fomannosin (26), biosynthesized from [1,2-$^{13}C_2$]-acetate, supports[56] a pathway involving the intervention of humulene (25). However, during studies on its biosynthesis from [5,5-$^{2}H_2$]-mevalonate, the presence of label was established at C-5 and C-10 but not at C-12, where it

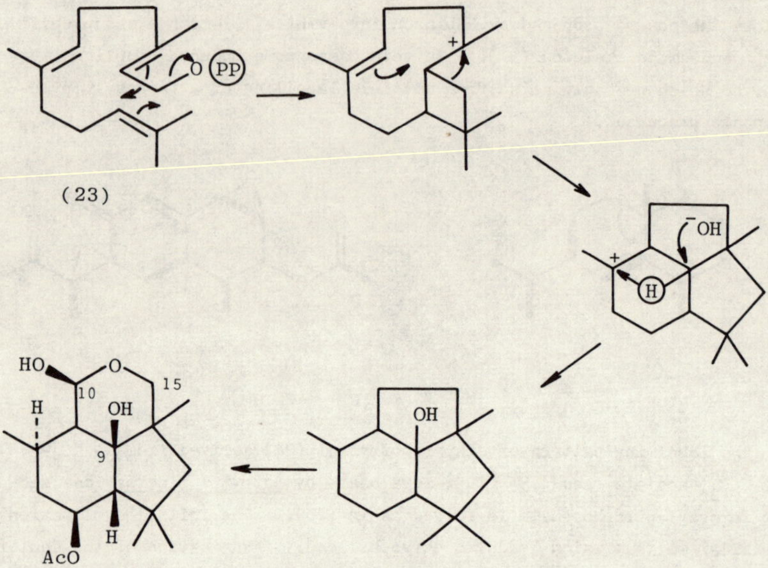

(23)

(24)

might have been found had a hydrogen shift occurred during its biosynthesis. It has now been established that a similar hydrogen shift, at one time proposed in the illudin series, does not occur. The amount of deuterium which was present at C-5 in the fomannosin suggested an isotope effect in the formation of the double bond at that centre.

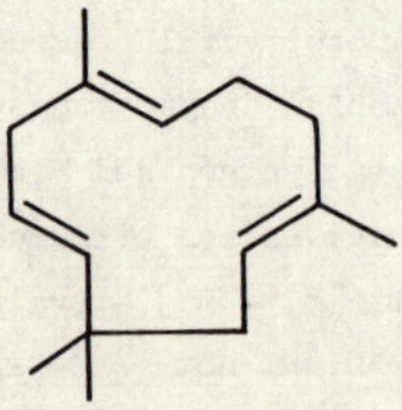

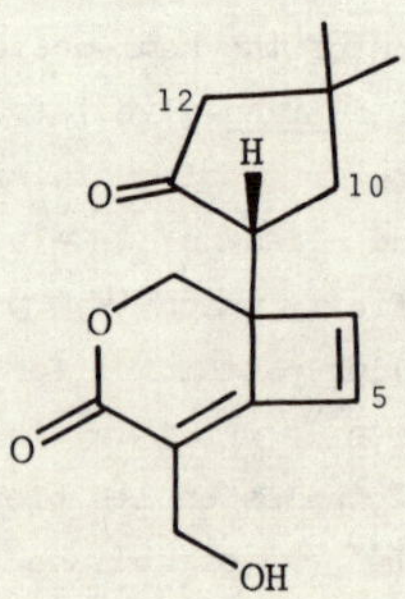

(25) (26)

The manner of folding of farnesyl pyrophosphate (27) to form the skeleton of pentalenolactone (28) has been established[57] by an interesting and potentially useful technique using [1,2-$^{13}C_2$]-acetate generated *in situ* from [U-$^{13}C_6$] glucose. The isoprene units of alliacolide (29) have been defined using [1,2-$^{13}C_2$]-acetate.[58,59] However, this did not establish the way in which farnesyl pyrophosphate is folded to generate this skeleton. The sites of labelling of alliacolide by [2-$^{2}H_3$]acetate and by [2-$^{2}H_2$]-, [4-$^{2}H_2$]-, and [5-$^{2}H_2$]-mevalonates have been determined. The occurrence of hydrogen rearrangements to C-1 and from C-5 to C-6 during the biosynthesis were demonstrated by deuterium n.m.r. studies. The labelling pattern of abscisic acid (30), biosynthesized from [1,2-$^{13}C_2$] acetate by *Cercospora rosicola*, has been reported.[60] The biosyntheses of a number of meroterpenoids which comprise a sesquiterpenoid and a polyketide fragment have been investigated[61 - 63] and are reported in the polyketide chapter.

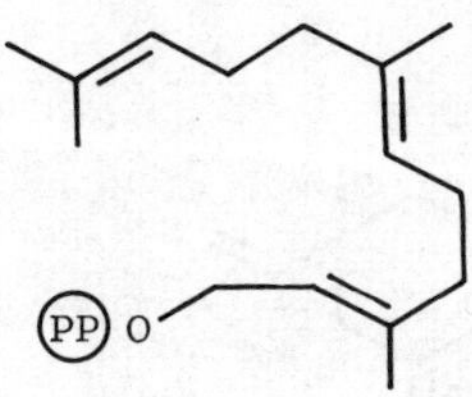

(27)

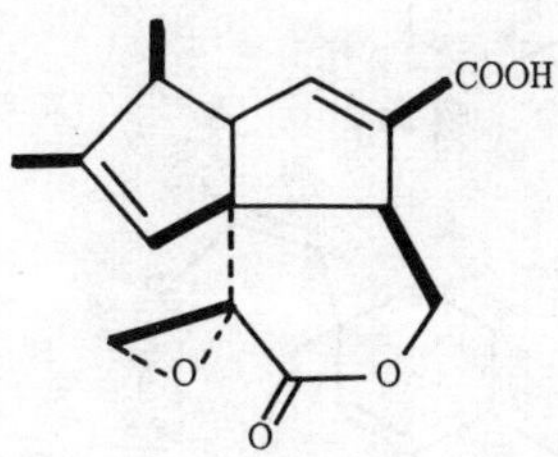

(28)

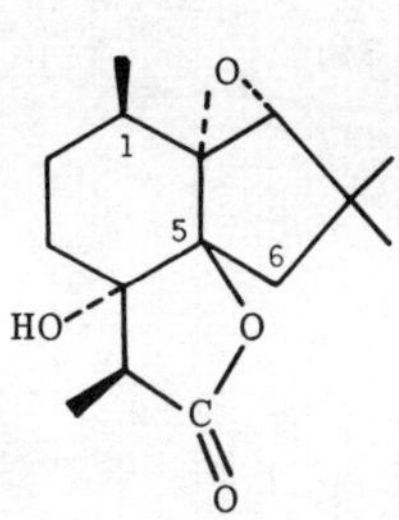

(29)

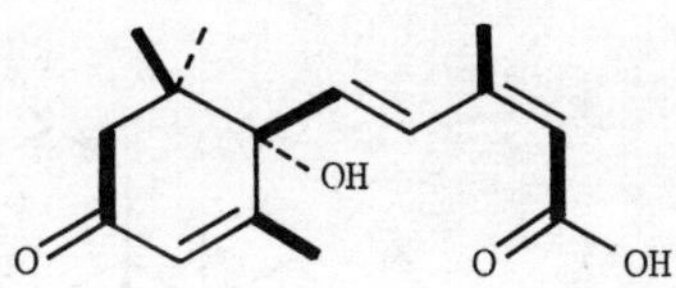

(30)

5 Diterpenoids

The stereochemistry of the S_N' cyclization in the biosynthesis of ent-sanadaracopimaradiene (33) has been examined[64] with an enzyme system which was prepared from *Ricinus communis*. Incubation of [1(*S*)-1-^{2}H]geranylgeranyl pyrophosphate (31) produced (*E*)-[16-^{2}H]-*ent*-sandaracopimaradiene (33), showing that the S_N' cyclization of the intermediate copalyl pyrophosphate (32) occurs with the *anti* stereochemistry. Similar results have been obtained[65] in the stereochemistry of the allylic displacement of a pyrophosphate in the biosynthesis of virescenol B (34). It has also been suggested[16,66] that the cyclization of copalyl pyrophosphate to *ent*-kaurene (35) follows the same *anti* pathway.

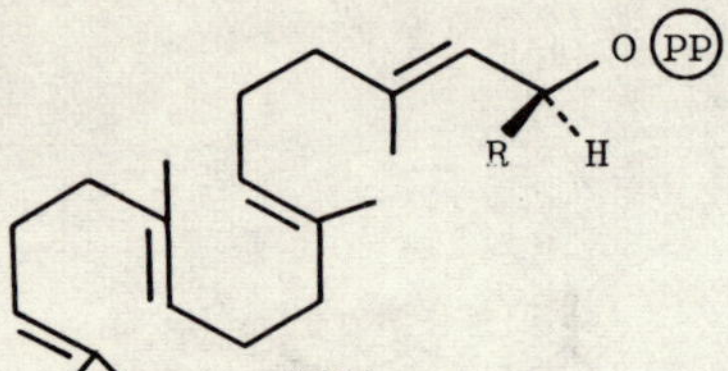

(31)

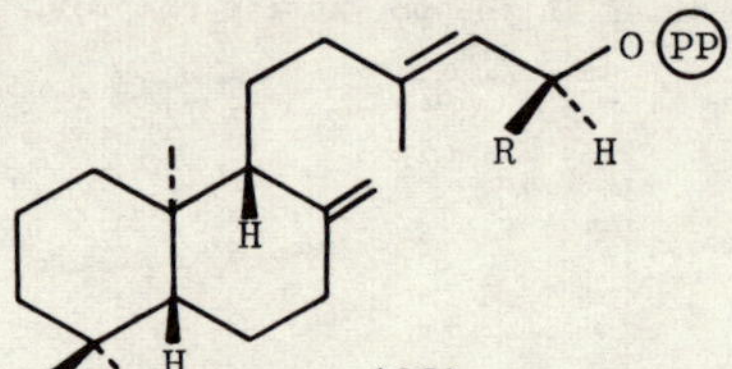

(32)

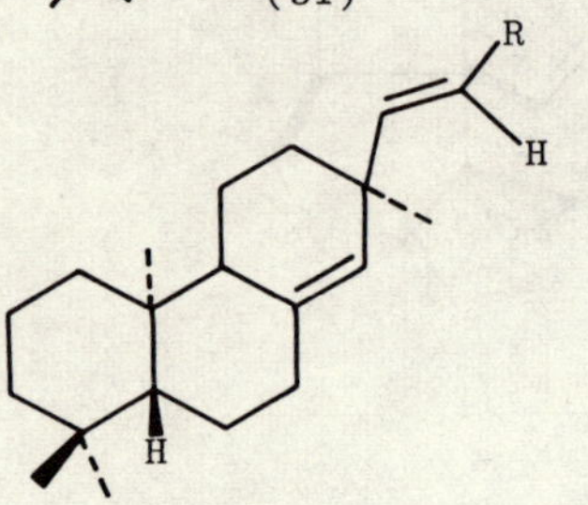

(33)

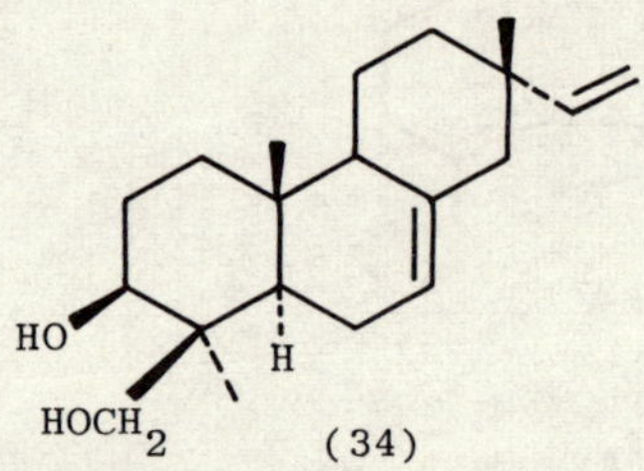

(34)

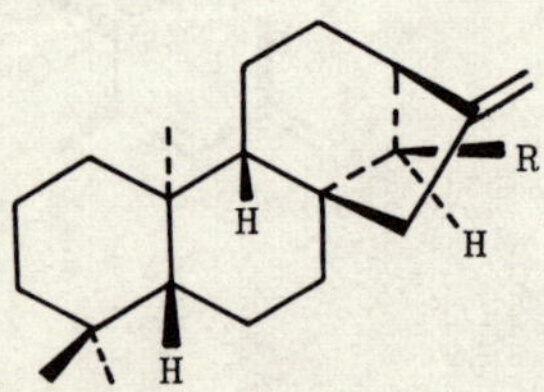

(35)

A carbon-13 n.m.r. study of the formation of ring D of ent-kaurene has been reported[67] confirming previous biosynthetic conclusions. A comparison of ent-kaur-16-ene and ent-kaur-15-ene biosynthesis in cell-free systems from etiolated shoots of normal and dwarf-d-5 maize has established[68] that the latter is formed by the dwarf maize system. An enzyme preparation from Phaseolus coccineus that mediates the biosynthesis of kaurenol from mevalonate has been described.[69] The biosynthesis of the bitter principles enmein and oridonin from some 7- and 15- mono and dioxygenated kaurenoids has been reported.[70]

ent-Kaur-6,16-dien-19-oic acid (36) rather than ent-7α-hydroxykaurenoic acid (37) has been shown[71,72] to be a key intermediate in the formation of the kaurenolide lactones (e.g. 38) in Gibberella fujikuroi. The formation of the trans-diaxial oxygen functions on ring B may arise through the opening of an epoxide ring. The relationship between some of the kaurenolides and the seco-ring B metabolites of Gibberella fujikuroi has been described.[73]

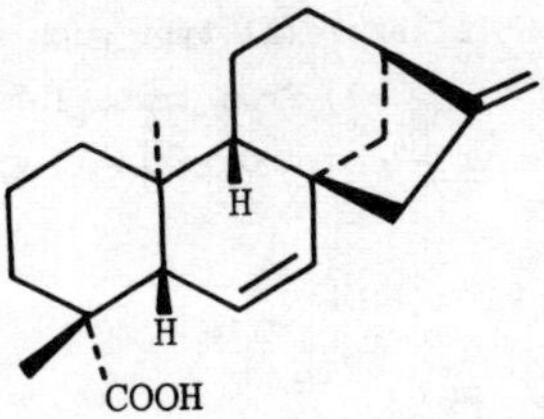

(36)

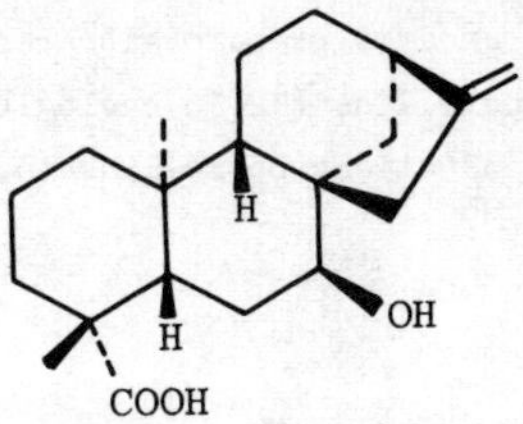

(37)

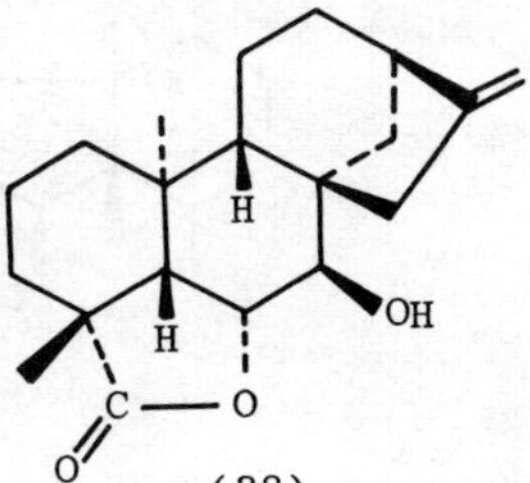

(38)

The formation of the gibberellin plant hormones and their metabolism has been reviewed on a number of occasions.[74 - 77] Hitherto Gibberella fujikuroi has been the sole fungal source of these compounds but recently a second source, Sphaceloma manihoticola, has been described.[78,79] Attention has been directed at the metabolism of the gibberellins in various higher plants. Thus metabolism of gibberellin A_{29} (39) by Pisum sativum affords[80 - 82] the

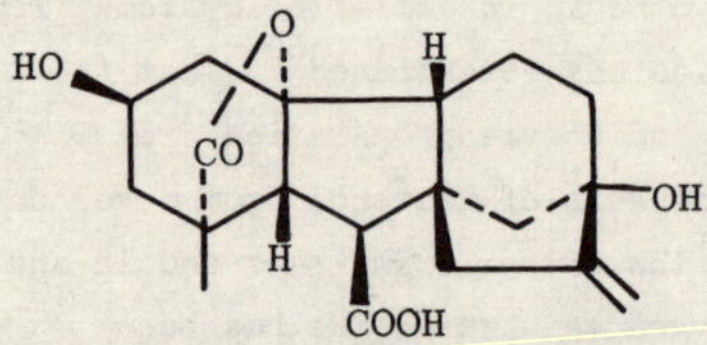

(39)

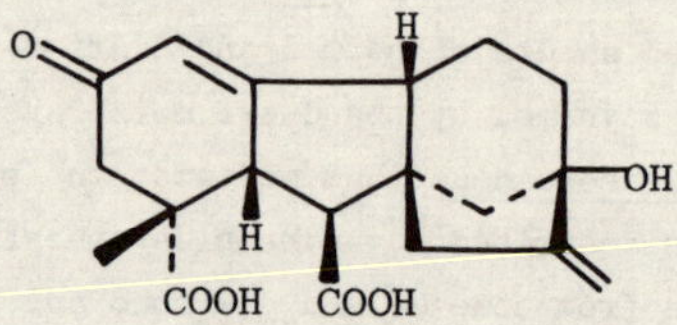

(40)

catabolite (40). Hydroxylation at C-2 and conjugation also appear to be deactivation processes. The use of gas chromatography coupled to mass spectrometry has enabled the changes in the gibberellin content of a plant to be monitored at various stages in its life cycle.[82,83]

Parts of the gibberellin pathway in Gibberella fujikuroi have a relatively low substrate specificity and can accept analogues of their normal substrates. This has been used to prepare gibberellins of a novel skeletal type such as the trachylobagibberellins (12,16-cyclogibberellins) (e.g. 42) from trachylobanic acid (41)[84] and atisagibberellins (e.g. 44) from ent-7α-hydroxyatis-16-en-19-oic acid (43).[85]

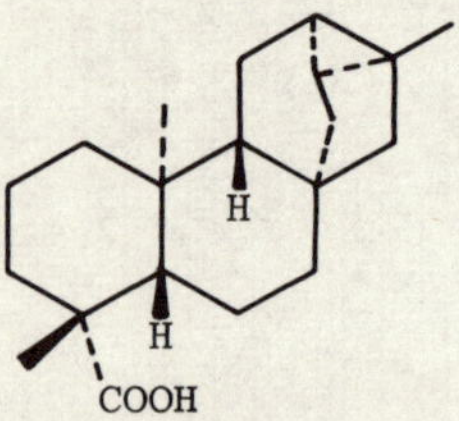

(41)

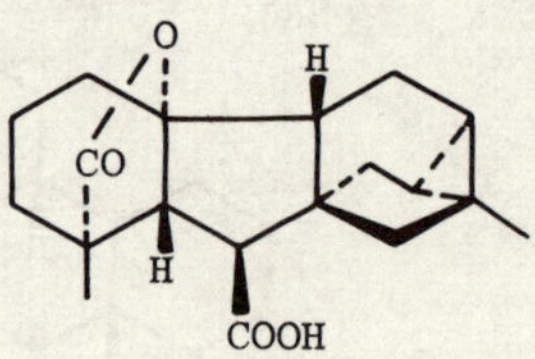

(42)

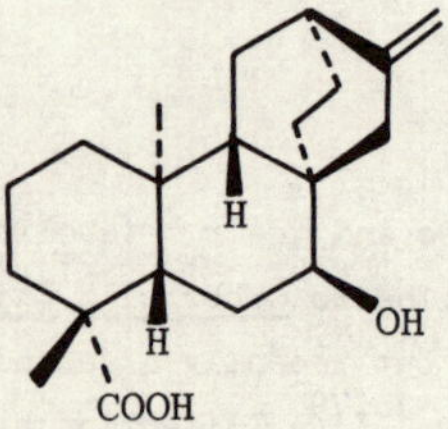

(43)

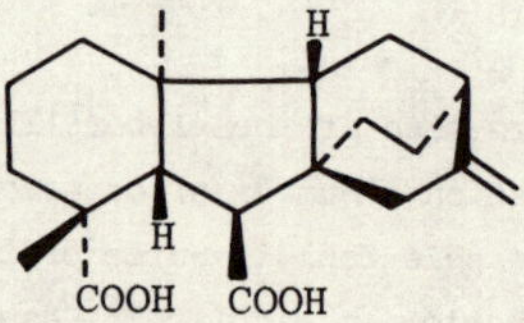

(44)

In studies on the microbiological transformation of ent-3β-hydroxy-kaurenes, the inhibitory effect of a 3α-hydroxyl group on microbiological oxidation at C-19 was established.[86] The effect of the kaurenoid 7,15 and 18-hydroxyl groups on microbiological transformation along the gibberellin pathway has also been examined.[87] An 18-substituent appears to exert an inhibitory effect on transformations involving the 6β-position. The microbiological production of 12- and 15-hydroxygibberellins has also been reported[88-90] from the corresponding kaurenes whilst the metabolism of steviol by Gibberella fujikuroi has again been examined.[91] In several of these studies Gibberella fujikuroi was grown in the presence of plant growth retardants (e.g. CCC and AMO 1618) which block certain steps in kaurene formation. This prevents the production of the endogenous metabolites and facilitates the detection of abnormal metabolites. The microbiological production of a series of fluorogibberellins, using fluorinated analogues of gibberellin precursors, has been successfully demonstrated[92-97] with the synthesis of 15-fluorogibberellins A_4 and A_{14} and 16,16-difluoro-17-nor-gibberellin A_7.

The inhibition of gibberellin biosynthesis provides a means of developing novel plant growth regulators. As might be expected, the effect of exogenous gibberellic acid on the wild strain of Gibberella fujikuroi is to suppress its own biosynthesis.[98] A number of fluorogibberellins (e.g. 45) have been prepared as possible regulators of gibberellin biosynthesis.[99] ent-Kauran-16,17-epoxide binds competitively to proteins that also bind ent-kaurene and consequently the epoxide blocks the biosynthesis of gibberellic acid.[100] A number of mimics of ent-7α-hydroxykaur-16-en-19-oic acid (e.g. the B-nor-alcohol, 46) have been shown to block the ring contraction stage in gibberellin biosynthesis and to act as plant growth regulators.[101] The effects of 1-n-decylimidazole and a series of amine inhibitors of gibberellin biosynthesis have been reported.[102,103]

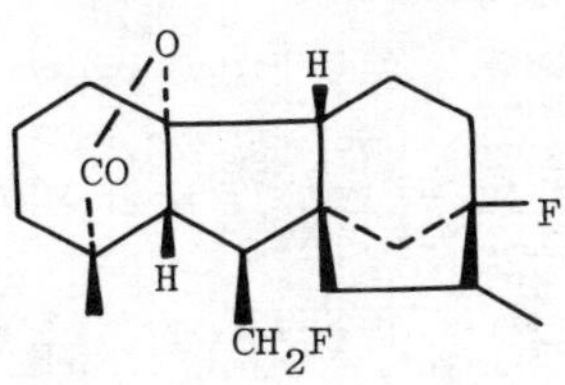

(45)

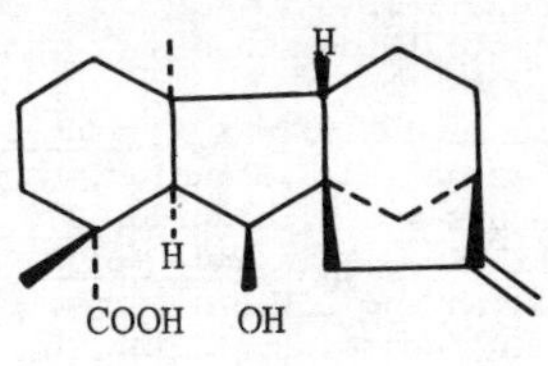

(46)

The use of deuterium as a tracer with carbon-13 n.m.r. spectroscopy has been employed in detecting some hydride shifts in the cyclization of geranylgeranyl pyrophosphate in fusicoccin biosynthesis.[104] The results of feeding various mevalonates to Fusicoccum amygdali and their incorporation into fusicoccin have been reported.[105]

The feeding of [1-^{13}C]-, [2-^{13}C]- and [1,2-$^{13}C_2$]-acetate to Cyathus earlei has shown[106] that geranylgeranyl pyrophosphate is folded as in (47) to form the cyathin skeleton of 11-O-acetylcyathatriol (48). [1,2-$^{13}C_2$] -Acetate studies have also been reported on the terpenoid moieties of the cochlioquinones[107] and the penitrems.[108]

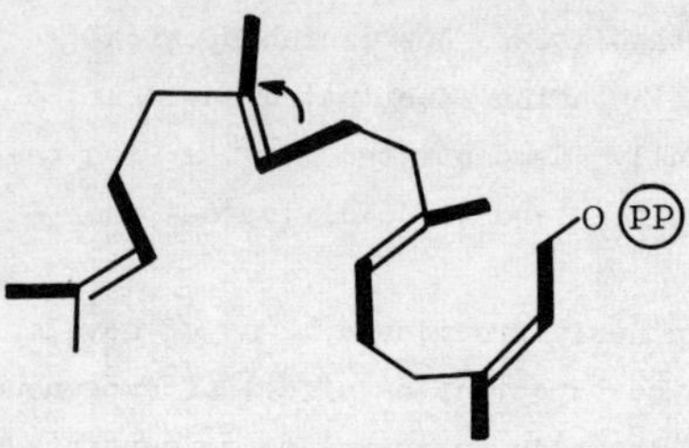

(47)

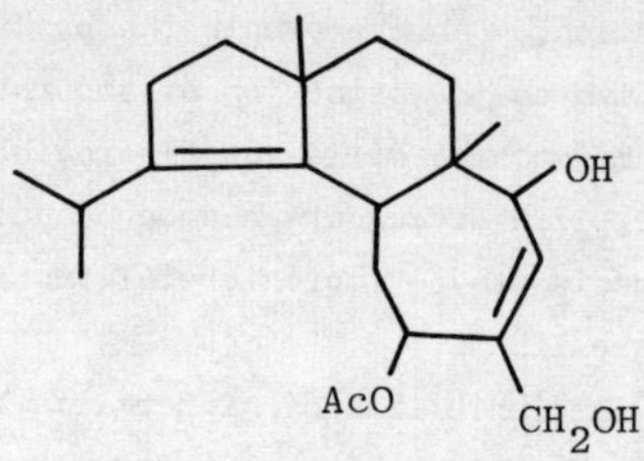

(48)

References

1 J.R.Hanson , in 'Biosynthesis' vol.6 ed. J.D.Bu'Lock, Specialist Periodical Reports,The Chemical Society, London, 1980, p. 76
2 'Biosynthesis of Isoprenoid Compounds' ed. J.W.Porter and S.L.Spurgeon, John Wiley, New York, 1981.
3 D.V.Banthorpe and B.V.Charlwood,in 'Encyclopedia of Plant Physiology' ed. E.A.Bell and B.V.Charlwood, Springer,Berlin, 1980,vol.8 p. 185.
4 R.B.Herbert,'Biosynthesis of Secondary Metabolites', Chapman and Hall, London, 1981.
5 P.Manitto,'Biosynthesis of Natural Products', Ellis Horwood,Chichester, 1981.
6 H.Floss and M.D.Tsai,'Advances in Enzymology, 1979, 50, 243.
7 M.J.Garson and J.Staunton, Chemical Society Reviews, 1979, 8, 539.
8 A.I.Scott and K.Shishido, J.C.S.Chem.Comm.,1980, 400.
9 H.C.Rilling, Pure and Appl.Chem.,1979,51,697.
10 B.E.Allan and D.V.Banthorpe, Phytochemistry, 1981,20,35.
11 M.de la Fuente, L.M.Perez, U.Hashagen, L.Chayet, C.Rojas, G.Portilla and O.Cori, Phytochemistry, 1981, 20, 1551.
12 T.Koyama, A.Saito, K.Ogura and S.Seto, J.Amer.Chem.Soc., 1980, 102, 3614.
13 C.D.Poulter, E.A.Mash, J.Argyle, O.J.Muscio and H.C.Rilling, J.Amer.Chem. Soc., 1979, 101, 6761.

14 C.D.Poulter, P.L.Wiggins and A.T.Le, J.Amer.Chem.Soc., 1981, 103, 3926.
15 E.A.Mash, G.M.Gurria and C.D.Poulter, J.Amer.Chem.Soc., 1981, 103, 3927.
16 D.E.Cane, Tetrahedron, 1980, 36, 1109.
17 G.Caporale, G.Innocenti, A.Guiotto, P.Rodighiero and F. Dal'Acqua, Phytochemistry, 1981, 20, 1283.
18 B.V.Charlwood and D.V.Banthorpe, Progress in Phytochemistry, 1978, 5, 65.
19 W.D.Loomis and R.Croteau, in 'Biochemistry of Plants', ed. P.K.Stumpf, Academic Press, New York, 1980, vol.4, p.363.
20 R.Croteau, Perfum.Flav. Int., 1980, 5, 35.
21 D.V.Banthorpe and I.Poots, Phytochemistry, 1979, 18, 1297.
22 A.Akhila and D.V.Banthorpe, Phytochemistry, 1980, 19, 1429.
23 A.Akhila and D.V.Banthorpe, Phytochemistry, 1980, 19, 277.
24 K.Tange, T.Hirata and T.Suga, Chemistry Letters, 1979, 269.
25 K.Tange, H.Okita, Y.Nakao, T.Hirata and T.Suga, Chemistry Letters, 1981, 777.
26 T.Suga, T.Shishibori and H.Morinaka, J.C.S.Chem.Comm., 1980, 167.
27 A.Akhila, D.V.Banthorpe and M.G.Rowan, Phytochemistry, 1980, 19, 1433.
28 G.Randazzo, A.Evidente, A.Boccalatte and C.Rossi, Phytochemistry, 1980, 20, 2177.
29 A.J.Poulose and R.Croteau, Arch.Biochem.Biophys., 1978, 177, 307; 191, 400.
30 A.Akhila and D.V.Banthorpe, Phytochemistry, 1980, 19, 1691.
31 A.R.Battersby, M.Thompson, K.H.Glusenkamp and L.F.Tietze, Chem.Ber., 1981, 114, 3430.
32 K.Inoue, Y.Takeda, T.Tanahashi and H.Inouye, Chem.and Pharm.Bull. 1981, 29, 970.
33 S.Damtoft, S.R.Jensen and B.J.Nielsen, J.C.S.Chem.Comm., 1980, 42.
34 S.Damtoft, J.C.S.Chem.Comm., 1981, 228
35 R.H.Fish, L.E.Browne, D.L.Wood and L.B.Hendry, Tetrahedron Letters, 1979, 1465.
36 T.Takeo, Phytochemistry, 1981, 20, 2145.
37 R.B.H.Wills and F.M.Scriven, Phytochemistry, 1979, 18, 785.
38 R.Croteau and F.Karp, Arch.Biochem.Biophys., 1979, 198, 523.
39 R.Croteau and N.M.Felton, Phytochemistry, 1980, 19, 1343.
40 R.Croteau, N.M.Felton and R.C.Ronald, Arch.Biochem.Biophys., 1980, 200, 524 and 534.
41 R.Croteau and N.M.Felton, Arch.Biochem.Biophys., 1981, 207, 460.
42 W.A.Ayer and L.M.Browne, Tetrahedron, 1981, 37, 2199.
43 H.Shirahama, E.Osawa and T.Matsumoto, J.Amer.Chem.Soc., 1980, 102, 3208.
44 D.E.Cane and R.Iyengar, J.Amer.Chem.Soc., 1979, 101, 3385.
45 D.E.Cane, R.Iyengar and M.S.Shiao, J.Amer.Chem.Soc., 1981, 103, 914.
46 D.E.Cane, S.Swanson and P.P.N.Murthy, J.Amer.Chem.Soc., 1981, 103, 2136.
47 B.B.Jarvis, G.Pavanasasivam, C.E.Holmlund, T.DeSilva, G.P.Stahly and E.P.Mazzola, J.Amer.Chem.Soc., 1981, 103, 472.
48 L.T.Burka, L.J.Felice and S.W.Jackson, Phytochemistry, 1981, 20, 647.
49 Y.Hoyano, A.Stoessl and J.B.Stothers, Canad.J.Chem., 1980, 58, 1894.
50 A.A.Chalmers, A.E.De Jesus, C.P.Gorst-Allman and P.S.Steyn, J.C.S. Perkin Trans. 1, 1981, 2899.
51 S.Moreau, A.Lablache-Combier and J.Biguet, Phytochemistry, 1981, 20, 2339.
52 A.P.W.Bradshaw, J.R.Hanson and R.Nyfeler, J.C.S.Perkin Trans.1, 1981, 1469.
53 A.P.W.Bradshaw, J.R.Hanson, R.Nyfeler and I.H.Sadler, J.C.S.Chem.Comm., 1981, 649.
54 A.P.W.Bradshaw, J.R.Hanson, and I.H.Sadler, J.C.S.Chem.Comm., 1981, 1169.
55 A.P.W.Bradshaw and J.R.Hanson, J.C.S.Chem.Comm., 1979, 924.
56 D.E.Cane and R.B.Nachbar, Tetrahedron Letters, 1980, 437.
57 D.E.Cane, T.Rossi and J.P.Pachlatko, Tetrahedron Letters, 1979, 3639.
58 A.P.W.Bradshaw, J.R.Hanson and I.H.Sadler, J.C.S.Chem.Comm., 1981, 631.
59 J.R.Hanson, Pure and Appl.Chem., 1981, 53, 1155.
60 R.D.Bennett, S.M.Norman and V.P.Maier, Phytochemistry, 1981, 20, 2343.
61 T.J.Simpson and D.J.Stenzel, J.C.S.Chem.Comm., 1981, 1042.

62 C.R.McIntyre and T.J.Simpson, J.C.S.Chem.Comm.,1981, 1043.
63 A.J.Bartlett, J.S.E.Holker, E.O'Brien and T.J.Simpson, J.C.S.Chem.Comm., 1981, 1198.
64 K.A.Drengler and R.M.Coates, J.C.S.Chem.Comm.,1980, 856.
65 D.E.Cane, H.Hasler, J.Materna, N.Cagnoli-Bellavita, P.Ceccherelli, G.F.Madruzza and J.Polonsky, J.C.S.Chem.Commun.,1981, 280.
66 R.M.Coates and P.L.Cavender, J.Amer.Chem.Soc., 1980, 102, 6358.
67 K.Honda, T.Shishibori and T.Suga, J.Chem.Research(S), 1980, 218.
68 P.Hedden and B.O.Phinney, Phytochemistry, 1979, 18, 1475.
69 N.Ceccarelli, R.Lorenzi and A.Alpi, Phytochemistry, 1979,18, 1657
70 T.Fujita, S.Takao and E.Fujita, J.C.S.Perkin Trans.1, 1979, 2468.
71 P.Hedden and J.E.Grabe, Phytochemistry, 1981, 20, 1011.
72 M.H.Beale, J.R.Bearder, G.H.Down, M.Hutchinson,J.MacMillan and B.O.Phinney, Phytochemistry, 1982, 21, 1279.
73 J.R.Hanson and F.Y.Sarah, J.C.S.Perkin Trans.1, 1979, 3151.
74 J.MacMillan, Pure and Appl.Chem., 1978, 50, 995.
75 J.E.Graebe and H.J.Ropers in 'Phytohormones' Elsevier, Amsterdam, 1978, vol.1, p.107.
76 B.O.Phinney, American Chemical Society Symposium Series, 1979, 111, 57.
77 'Gibberellins, Chemistry, Physiology and Use' ed.J.R.Lenton, British Plant Growth Regulator Group Monograph no. 5, 1980.
78 W.Rademacher and J.Graebe, Biochem,Biophys.Res.Comm.,1980,91, 35.
79 R.S.Zeigler,L.E.Powell and H.D.Thurston, Phytopathology, 1980, 70, 589.
80 V.M.Sponsel and J.MacMillan, Planta,1978, 144, 69.
81 V.M.Sponsel and J.MacMillan, Planta, 1980, 150, 46.
82 V.M.Sponsel in 'Gibberellins, Chemistry, Physiology and Use' British Plant Growth Regulator Group Monograph no.5' 1980 p.49
83 S.Kurogochi, N.Murofushi, Y.Ota and N.Takahashi, Planta, 1979, 146, 185.
84 J.R.Bearder, J.MacMillan,A.Matsuo and B.O.Phinney, J.C.S.Chem.Comm., 1979, 649.
85 J.R.Hanson,F.Y.Sarah, B.M.Fraga and M.G.Hernandez, Phytochemistry,1979, 18, 1875.
86 B.M.Fraga, A.G.Gonzalez, J.R.Hanson and M.G.Hernandez, Phytochemistry, 1981, 20, 57.
87 B.M.Fraga, J.R.Hanson,M.G.Hernandez and F.Y.Sarah, Phytochemistry, 1980, 19, 1087.
88 K.Wada and H.Yamashita, Agric. and Biol.Chem.(Japan), 1980, 44, 2249.
89 K.Wada, T.Imai, and H.Yamashita, Agric.and Biol.Chem.(Japan),1981,45,1833.
90 K.Wada,TImai, and K.Shibata,Agric.and Biol.Chem.(Japan), 1979,43,1157.
91 N.Murofushi, S.Nagura and N.Takahashi, Agric. and Biol.Chem. (Japan), 1979, 43, 1159.
92 B.E.Cross and A.Erasmuson,J.C.S.Chem.Comm.,1978, 1013.
93 B.E.Cross and A.Erasmuson,J.C.S.Perkin Trans.1, 1981, 1918.
94 B.E.Cross and P.Filippone,J.C.S.Chem.Comm.,1980, 1097.
95 B.E.Cross and P.Filippone, J.Chem.Research, 1981, 166.
96 B.E.Cross and A.Erasmuson, J.C.S.Perkin Trans. 1, 1981, 1918.
97 B.E.Cross and P.Filippone, J.C.S.Chem.Comm., 1980, 1097.
98 B.Dockerill and J.R.Hanson, Phytochemistry, 1981, 20, 2679.
99 K.Boulton and B.E.Cross, J.C.S.Perkin Trans. 1, 1981, 427.
100 J.R.Hanson, C.L.Willis and K.P.Parry, Phytochemistry, 1980, 19, 2323.
101 J.R.Hanson, K.P.Parry and C.L.Willis, J.C.S.Chem.Comm., 1981 285.
102 K.Wada and T.Imai, Agric. and Biol.Chem.(Japan), 1980, 44, 2511.
103 L.C.Echols, V.P.Maier, S.M.Poling and P.R.Sterling, Phytochemistry, 1981, 20, 433
104 A.Banerji, R.Hunter, G.Mellows,Kim-y-Sim, and D.H.R.Barton, J.C.S.Chem. Comm., 1978, 843.
105 G.Randazzo, A.Evidente, R.Capasso, F.Colantuoni, L.Tuttobello, and A.Ballio, Gazz. Chim.Ital.,,1979, 109, 101.

106 W.A.Ayer, S.P.Lee and T.T.Nakashima, Canad.J.Chem., 1979, 57, 3338.
107 L.Canonica, M.G.Beretta, L.Colombo, C.Gennan, B.M.Ranzi and C.Scholastico, J.C.S.Perkin Trans. 1, 1980, 2686.
108 A.E.De Jesus, P.S.Steyn, F.R.van Heerden, R.Vleggaar and P.L.Wessels, J.C.S.Chem.Comm., 1981, 289.

4

The Biosynthesis of Alkaloids

BY E. LEETE

1 Introduction

This chapter is organized as in previous volumes of 'Biosynthesis' and I have attempted to review critically all work on the biosynthesis of alkaloids which has been published between January 1979 and December 1982. The alkaloids are the most heterogeneous class of secondary natural product, both from a structural and biosynthetic standpoint. It is hard to reach agreement on which nitrogen-containing compounds should be called alkaloids. Pelletier addresses this problem in some depth in a new series of monographs (of which he is the editor) concerned with the chemical and biological aspects of alkaloids[1]. I am rather fond of one definition: 'An alkaloid is like my wife, I can recognize her when I see her, but I can't define her'. In a more serious vein, Pelletier proposes the following: 'An alkaloid is a cyclic organic compound containing nitrogen in a negative oxidation state which is of limited distribution among living organisms' . I am in fair agreement with this proposal, although this chapter contains information on compounds which would be excluded by this defination, such as spermine and spermidine (acyclic), and aristolochic acid (a nitro compound in which the oxidation state of nitrogen is +3). In the past I have included information on the origin of biotin[2,3],nicotinic acid[4,5] pyridoxal[6,7] , and thiamine[8-15], in the chapter on alkaloid biosynthesis. With reluctance, since space has been limited, they have been omitted from the present chapter, except for the above cited references. They would, of course, be excluded by Pelletier's definition, since these components of coenzymes are presumably found in all species. Most of the work involving the use of isotopically labelled substrates is recorded in a Table in which the species studied are arranged alphabetically . Alkaloids whose biosynthesis or catabolism is reported in this Table are indexed before this Table, citing the species in which they were studied.

Unlike other classes of natural products such as the flavonoids, novel biosynthetic pathways continue to be discovered for the alkaloids, and some of these are discussed in more detail in Section 2. In the last few years considerable progress has been made in the isolation of enzymes in varying degrees of purity, which catalyze some of the steps involved in the formation of alkaloids, and this work is summarized in Section 3. Often it has been discovered that a cell-suspension culture is a better source of enzymes than the differentiated whole plant .

Many reviews and monographs relating to various aspects of alkaloid biosynthesis have appeared in the last four years. Herbert has written an excellent account of the whole subject in a book[16] and recent comprehensive texts[17,18] . He also continues to write the annual surveys in Specialist Periodical Reports entitled 'The Alkaloids'[19-22]. A text by D. R.Dalton[23] discusses the biosynthesis of representative members of different classes of alkaloids . Another general text[24] includes alkaloid biosynthesis. A major opus by G. A. Cordell[25] reviews the biosynthesis of all the major classes of alkaloids, along with their chemistry and pharmacology. T. Robinson has written a well-referenced book on the biochemistry of the alkaloids[26]. Another monograph[27] contains reviews of many of the substances I claim as alkaloids: ergot alkaloids[28], pyrazines, fumitremorgin, paspaline and slaframine[29], cytochalasins[30], gliotoxin and related dioxopiperazines[31], cyclopiazonic acid[32]. The proceedings of a symposium on the indole alkaloids (April 1979) have been published[33] and this monograph contains reviews on the biosynthesis of the ergot alkaloids[34], camptothecin, perloline and vinblastine[35], and the heteroyohimbine alkaloids[36]. Volume 8 of the Encyclopedia of Plant Physiology[37] contains surveys on the biosynthesis of alkaloids which are classified according to their origin (ornithine, lysine, phenylalanine, tyrosine, tryptophan, nicotinic acid etc.). Volume 7 of a comprehensive treatise[38] is concerned with secondary plant products, and the chapter on alkaloids is devoted to a review on the enzymology of their metabolism[39].Volume IV of the series of monographs entitled 'Antibiotics'[40] is devoted to biosynthesis, and includes chapters on the ansamycins[41], β-lactam antibiotics[42], pyrrolo[1,4]benzodiazepines[43] (see also ref. 59), mitomycins[44], and nucleoside antibiotics[45]. Specialized reviews have appeared on the following subjects: Special topics in alkaloid biosynthesis[46], biosynthesis of the *Cephalotaxus* alkaloids[47], Enzymology of indole alkaloid production[48], biosynthesis of pactamycin, berninamycin, and streptolydigin[49], biosynthesis of cyclopenin and viridicatin[50], biosynthesis of isoquinoline alkaloids[51], biosynthesis of isoquinoline alkaloids including colchicine and aristolochic acid[52], biosynthesis of penicillin and related β-lactam antibiotics[53-55], production of indole alkaloids in cell-suspension cultures[56], biosynthesis of the tropane alkaloids[57], biosynthesis and metabolism of the tobacco alkaloids[58], biosynthesis of the ergot alkaloids[60], use of ^{13}C n.m.r. to study the biosynthesis of alkaloids[61].

2 Some Highlights of 1979-1982

2.1 Cocaine Biosynthesis.- The first definitive work on the biosynthesis of the tropane moiety of this alkaloid has appeared[62-64]. The results obtained are consistent with speculations[57] that it is formed by the same general route which has long been established for the origin of hyoscyamine (16) in *Datura* species and *Atropa belladonna*. Extensive tracer experiments extending over 25 years have

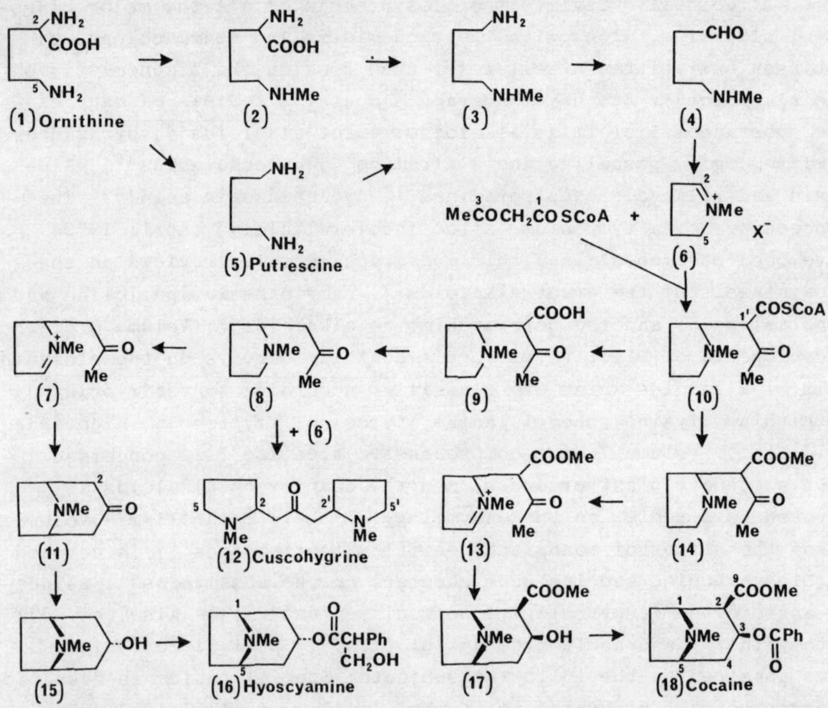

Scheme 1 *Biosynthesis of Cocaine and Related Alkaloids*

established that ornithine is incorporated into the pyrrolidine ring of its tropane moiety unsymmetrically. Thus [2-^{14}C]ornithine affords hyoscyamine which is labelled only at the C-1 bridgehead carbon. The proposed intermediates in this sequence are illustrat- in Scheme 1.

It was suggested that ornithine (1) is incorporated via δ-*N*-methylornithine (2), and this amino-acid was shown to be a direct precursor of hyoscyamine[65,66] in *Datura stramonium* and *Atropa belladonna*. This amino-acid has been shown to be an authentic natural product since it was isolated in radioactive form after feeding [5-^{14}C]ornithine or [5-^{3}H]ornithine to *A. belladonna*[67]. Decarboxylation then yields *N*-methylputrescine (3), an established precursor of the tropane ring of hyoscyamine and scopolamine[68-70]. Oxidation of *N*-methylputrescine yields 4-methylaminobutanal (4), which was detected in *Datura* plants which had been fed [2-^{14}C]ornithine[71]. The cyclized form of this amino-aldehyde is the *N*-methyl-Δ^1-pyrrolinium salt (6). Condensation of this iminium salt with acetoacetyl coenzyme A yields the coenzyme A ester of hygrine-1'-carboxylic acid (10). Hydrolysis of this ester and decarboxylation of the resultant β-keto-acid (9) affords hygrine (8), which has been shown to be a precursor of tropine (15)[72–74], which is plausibly formed via dehydrohygrine (7) and tropinone (11). Cocaine (18) was thus considered to arise by transesterification of (10) to afford the methyl ester (14), which could cyclize via the dehydro derivative (13) to yield 2-carbomethoxytropinone . Reduction of this ketone yields ecgonine (17), which affords cocaine on esterification with benzoic acid. Even though this biogenetic scheme for cocaine seemed very reasonable, feeding experiments with the obvious precursors were equivocal. At a meeting in 1963, it was reported[75] that radioactive alkaloids were obtained when Formosan coca trees (*Erythroxylon novogranatense*) were injected with solutions of [3'-^{14}C]phenylalanine, sodium [1-^{14}C]acetate, or [*methyl*-^{14}C]methionine. However, no additional information on this work has been published and one assumes that the incorporation of radioactivity was not significant. The benzoyl moiety of cocaine is derived from phenylalanine by cleavage of the side chain[76]. In a dissertation[77] the incorporation of labelled ornithine and putrescine into cocaine was reported; however, since no subsequent publications have appeared, one surmises that incorporations were low, making it impossible to determine the location of radioactive atoms by degradations. Rather strange results were obtained[72] on feeding [1-^{14}C]acetate to *Erythroxylon coca*. About 60% of the activity of the cocaine was apparently located in the ester methyl group. One suspects that the cocaine was not radiochemically pure since the carboxyl group of acetic acid is not normally the source of one-carbon units. Furthermore, if this had occurred, one would also have expected to find a significant amount of label on the *N*-methyl group of cocaine. The carboxyl group of the benzoic acid moiety contained 30% of the activity. Only 8.7% of the activity was present in the ecgonine . This was distributed: 33% on the carboxyl group, 36% at C-3, and 22% at C-1, 5,6,7, and the *N*-methyl group. It was thus claimed, in my opinion somewhat optimistically, that the acetic acid was incorporated into ecognine in agreement with the previously discussed biogenetic scheme for cocaine.

After many unsuccessful experiments, a significant incorporation of [5-^{14}C]ornithine into cocaine was achieved[62,63]. The tracer in aqueous solution containing a little Tween 80 was applied to the leaves of the *coca* plant by means of an artists paint brush. Other methods of feeding (hydroponics, wick in stems, immersion of excised stems in tracer) all failed to yield specifically labelled cocaine from its putative precursors. A systematic degradation of the labelled cocaine established that all the activity was located at the bridgehead carbons : C-1 and C-5, and was equally divided between these positions. If the [5-^{14}C]ornithine had been incorporated via δ-*N*-methylornithine according to Scheme 1, all the activity should have been at C-5. It is thus proposed that the ornithine is incorporated via free putrescine (5), a symmetrical intermediate, which would afford the pyrrolinium salt (6) with equal labelling at C-2 and C-5. Radioactive cuscohygrine (12) was also obtained from the *coca* plants which had been fed [5-^{14}C]ornithine. It has been shown previously[72-74] that this alkaloid is formed by reaction of hygrine with second molecule of the iminium salt (6). Degradation of the labelled cuscohygrine indicated that about 25% of its total activity was present at the C-5 and C-5' positions, a result which is consistent with the pyrrolidine rings of this alkaloid being labelled symmetrically at the positions α- to the nitrogen. In tobacco, ornithine is converted to the pyrrolinium salt (6) via free putrescine[78]. It is thus not too surprising that the tropane ring system of cocaine should arise by a slightly different pathway from that previously established in *Datura* species.

Having obtained a successful incorporation of ornithine into cocaine by the leaf-painting technique, the same method was used for administering [1-^{14}C]acetate and [4-^{3}H]phenylalanine[64]. These two potential precursors were fed at the same time, the phenylalanine affording the benzoyl moiety in which almost all the tritium (96%) was located in the *para* position. Most of the ^{14}C was found at C-3 (48%) and C-9 (38%). The somewhat higher activity at the former position is consistent with the formation of [1,3-^{14}C]-acetoacetate in which C-3 will have a higher activity than C-1 since it has often been observed that the "starter" acetate unit in polyketides has a higher level of activity than subsequent units which are incorporated via malonate. A small, but significant (4.4%), amount of activity was found at C-1 and C-5 of cocaine. This result is consistent with the incorporation of [1-^{14}C]acetate, via the Krebs cycle, affording [1,5-^{14}C]-α-ketoglutarate, [1,5-^{14}C]-ornithine, and then [1,4-^{14}C]putrescine.

2.2 Biosynthesis of Retronecine.- Early work[79,80] on the biosynthesis of this compound, which is the basic moiety (necine) of several pyrrolizidine alkaloids (monocrotaline, retrorsine, senecionine), was consistent with the hypothesis of Robert Robinson[81], that this ring system is derived from two C_4 units derived from ornithine. It was found by Geissman[80] that C-9 of retronecine (33) contained 25% of the total activity of the base when either [2-^{14}C]- , [5-^{14}C]ornithine or [1,4-^{14}C]putrescine were fed to a *Senecio* species. It was proposed that the bicyclic system is formed by an intramolecular Mannich reaction, involving the dialdehyde-amine (26) as illustrated in Scheme 2. Indeed this reaction, yielding the pyrrolizidine-1-aldehyde (27), has been achieved *in vitro* under physiological conditions[82,83]. This dialdehyde could arise from homospermidine (25), which is formed from two molecules of putrescine, presumably via the Schiff base (21) and 4-aminobutanal (22)[84]. By this biosynthetic pathway the expected location of ^{14}C in retronecine derived from [2-^{14}C]ornithine will be C-3,5,8, and 9. Doubt was raised concerning this pathway by a publication of Warren[85]. Labelled retronecine was obtained from [2-^{14}C]ornithine. A Kuhn-Roth oxidation of retronecanol (32) derived from this retronecine yielded acetic acid having 26% of the activity of the original alkaloid, the bulk of the activity being in the methyl group of the acetic acid. This result is in agreement with the previous work of Geissman. Subsequent steps in the degradation involved the conversion of (32) to heliotridane methiodide (31), which was subjected to a Hofmann degradation to yield 2-allyl-1,3-dimethylpyrrolidine (30). Hydrogenation, methylation, and a second Hofmann reaction afforded 4-dimethylamino-3-methyl-1-hexene (29). The formaldehyde formed on ozonolysis of (29) and (30) was found

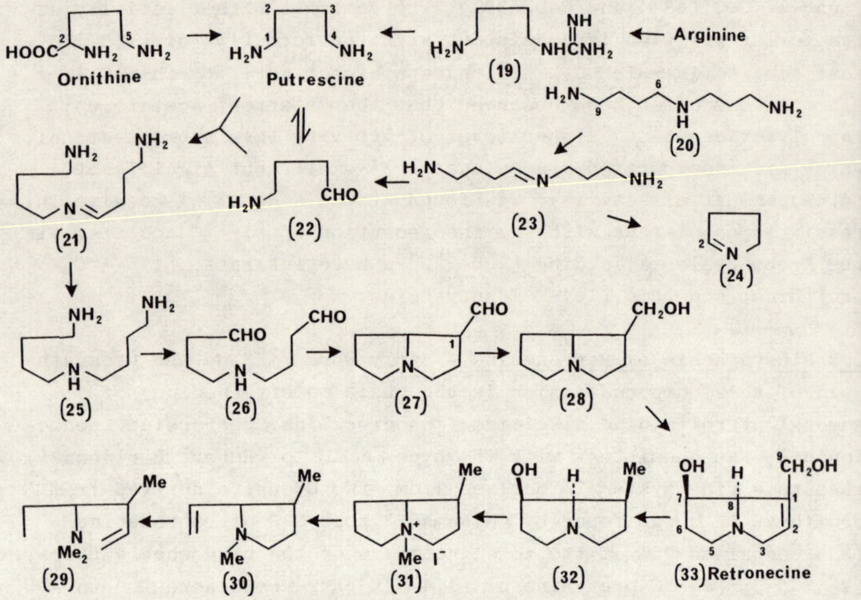

Scheme 2 *Biosynthesis of Retronecine*

to be inactive, apparently indicating that C-3 and C-5 of retronecine were devoid of activity. It is hard to evaluate the validity of this degradation since no yields of the degradation products were given. The heliotridane methiodide had a low specific activity (1.16 x 10^4 dpm/mmole). Thus if there had been 25% of the activity at C-3 and C-5 the resultant formaldehyde-dimedone derivative would only have had 9 dpm/mg. above background. I am inclined to believe that the inactive formaldehyde which was obtained in this degradation originated from another source, perhaps from oxidation of the *N*-methyl groups of (29) and (30). Albeit, these ambiguities have lead others[86-92] to reinvestigate the biosynthesis of retronecine, and they have obtained results which have confirmed and extended Geissman's initial observations. More sophisticated degradations of retronecine derived from [5-^{14}C]ornithine or [1,4-^{14}C]putrescine established that C-5 and C-8 contained 25% of the activity. The same pattern of labelling was obtained in retronecine derived from spermidine (20) or spermine which were labelled with ^{14}C at the 1,4-positions of their C_4 units. It is proposed that these polyamines are serving as a source of 4-aminobutanal, since it is known that they are oxidized to Δ^1-pyrroline (24) in higher plants[89]. Later feeding experiments with ^{13}C-labelled putrescine [90,92], determining the location of the isotope with n.m.r., established unequivocally the symmetrical labelling of the two halves of the pyrrolizidine nucleus. Thus retronecine derived from [1,4-^{14}C, 2,3-^{13}C]putrescine (0.5% specific incorporation) exhibited satellites due to contiguous ^{13}C atoms, at C-1,2 and C-6,7, in its ^{13}C n.m.r. spectrum. Complementary work has been carried out by Spenser[91]. Retronecine derived from [1-^{13}C, 1-^{15}N]-putrescine (6.6% specific incorporation) had enhanced ^{13}C signals in its ^{13}C n.m.r. spectrum at the peaks due to C-3,5,8, and 9. Furthermore, the signals at C-3 and C-5 exhibited satellites ($^1J_{3,N}$= 4.1 Hz, $^1J_{5,N}$= 3.8 Hz) due to contiguous ^{13}C and ^{15}N atoms. The existence of satellites on both sides of the nitrogen eliminates an alternate route to the pyrrolizidine ring system which is illustrated in Scheme 3. Here it is proposed that the ring system

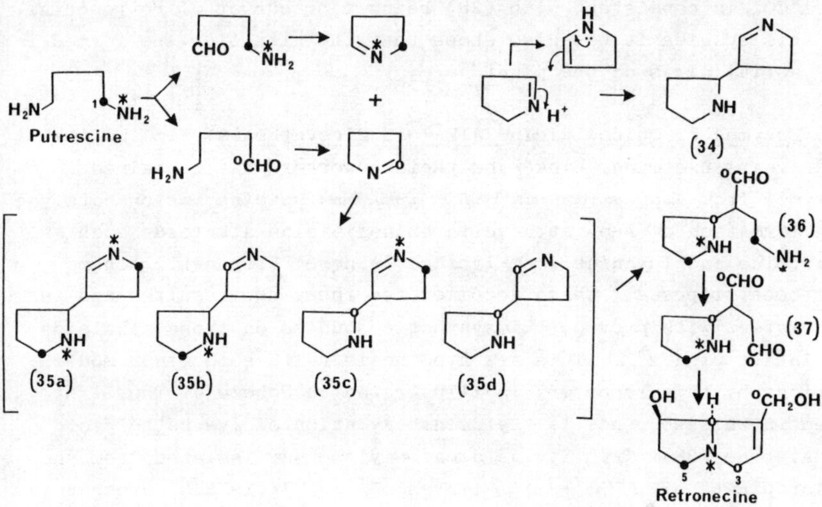

Scheme 3 *Hypothetical Biosynthesis of Retronecine via a Dimer*

is derived from a dimer (34) of Δ^1-pyrroline. The fate of ^{13}C and of ^{15}N is illustrated in this scheme. When the ^{13}C remains contiguous with ^{15}N it is represented with a heavy dot, otherwise it is represented with an open circle. The Δ^1-dimer will thus be a mixture of the four labelled species : (35a, b, c, and d), together with dimers resulting from condensations between unlabelled Δ^1-pyrroline in the plant and enriched material derived from the labelled putrescine. Ring opening of the Δ^1-pyrroline ring of the dimer affords (36), which is then oxidized to (37). Unexceptional steps then afford retronecine in which only the bond between C-5 and the nitrogen maintains its integrity throughout this sequence. On the other hand, the retronecine derived from a symmetrical intermediate such as (26) in Scheme 2 will have some molecules with a contiguous ^{13}C-3,^{15}N , and an equal number with contiguous ^{13}C-5, ^{15}N. An identical experiment was carried out by Khan and Robins,[92] with the same results. These authors also obtained labelled homospermidine (25) from a *Senecio isatideus* plant which had been fed [5-^{14}C]ornithine, thus indicating that this compound is a normal intermediate in retronecine biosynthesis. Specifically labelled homospermidine[88,92] also served as a precursor of retronecine. Arginine is also an efficient precursor of retronecine[93], and it could be serving as a source of putrescine via agmatine (19). The higher level of activity found in trachelanthamidine (28) compared with retronecine, after exposure of *Heliotropium spathulatum* plants to $^{14}CO_2$, is consistent with (28) being a precursor of retronecine. In this species it was also shown that the alkaloids are formed in the aerial parts of the plant[94].

2.3 Enzymes of Quinolizidine Alkaloid Biosynthesis.- In the past four years Hartmann, Wink, and their coworkers have obtained enzymes from *Lupinus* (especially *L.polyphyllus*) species which catalyse the formation of the tetracyclic quinolizidine alkaloids such as sparteine and lupanine from lysine. A novel biogenetic scheme has been proposed[95] which accommodates these new results, and is consistent with previous biosynthetic studies on these alkaloids in intact plants[96]. This new hypothesis (with some minor modifications by this reporter) is illustrated in Scheme 4. The first step in this sequence is the decarboxylation of lysine to yield cadaverine (38). This lysine decarboxylase was isolated from the chloroplasts of *L. polyphyllus* leaves[97-102]. It is also present in parts of the plant, eg. the roots, which do not produce alkaloids[103]. There was a direct correlation between the level of lysine decarboxylase and the chlorophyll content of the leaves[102]. This enzyme was not unique to species which produce the quinolizidine alkaloids; indeed it was found in all plant species examined. However, the amount found was up to 10^3 times smaller than the decarboxylase involved in the formation of lysine from diaminopimelic acid[99]. The lysine decarboxylase from *L. polyphyllus* had a maximum activity at pH 8.0, and was not inhibited by the ultimate alkaloids,

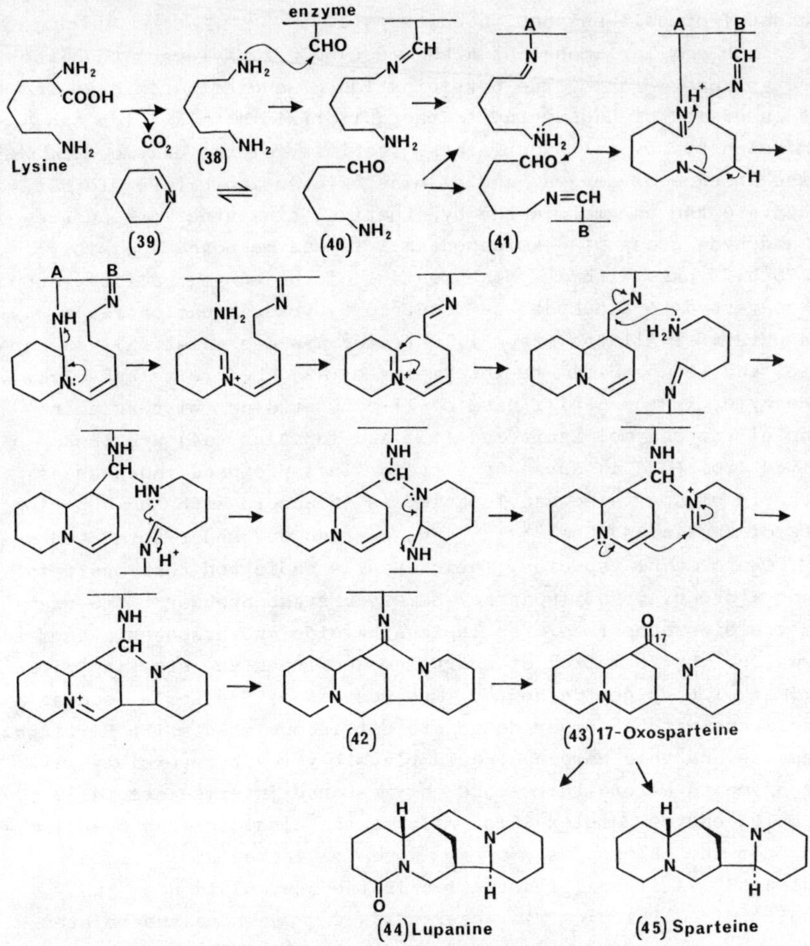

Scheme 4 *Biosynthesis of the Quinolizidine Alkaloids via Enzyme-Bound Intermediates*

17-oxosparteine and sparteine. The next step in the biosynthetic sequence is the transamination of cadaverine with pyruvate to yield alanine and 5-aminopentanal (40). It is proposed that this compound is enzyme-bound (41) , since none of the self-condensation products of this amino-aldehyde (the open-chain form of Δ^1-piperideine (39)) such as tetrahydroanabasine or tripiperideine could be detected in the incubation mixture. The enzyme is called 17-oxosparteine synthase since no free intermediates between cadaverine and 17-oxosparteine could be isolated. The crude enzyme was first obtained from cell-suspension cultures of *L. polyphyllus*[104] which produce a much smaller amount of alkaloid (~500 times lower) than the intact plants[105,106]. The participation of a diamine oxidase in the conversion of cadaverine to the quinolizidine alkaloids was eliminated by several experiments. Activity of the enzyme *increased* in the absence of oxygen, and diamine oxidase inhibitors did not inactivate the enzyme. In the hypothetical biosynthetic sequence the aldehyde group of 5-aminopentanal is enzyme-bound, possibly as a Schiff base with a lysine residue of the enzyme. Amino-groups are converted to aldehydes as required by transamination reactions with pyruvate. All the steps illustrated are mechanistically acceptable, and the penultimate intermediate (42), before release from the enzyme, is the Schiff base of 17-oxosparteine with an amino-group of the enzyme. Sparteine (45) and lupanine (44) are then derived from (43) in subsequent steps. This proposed independent synthesis of sparteine and lupanine is in accord with the experiments of Martin and Cho[107,108], who carried out short-term feedings of $^{14}CO_2$ to *Lupinus* species. Their results indicated that sparteine is not a precursor of lupanine. Some years ago Spenser[109] re-examined the biosynthesis of the lupin alkaloids and proposed a new scheme for the formation of sparteine and lupanine, starting with a trimer of Δ^1-piperideine[110]. The results of Spenser's feeding experiments with Δ^1-piperideine are not inconsistent with Hartmann's scheme, since this compound could plausibly react reversibly with the enzyme to afford the second enzyme-bound intermediate (41). The multi-enzyme complex which affords the alkaloids has been isolated from the chloroplasts of *L. polyphyllus* leaves and *L. albus* seedlings[98]. A diurnal fluctuation in the accumulation of the quinolizidine alkaloids was observed in *L. polyphyllus* and related species[101], and could be related to the level of thioredoxin in the chloroplasts[100].

<u>2.4 3-Amino-5-hydroxybenzoic Acid - A Key Intermediate in the Biosynthesis of the Ansamycin and Mitomycin Antibiotics</u>. - It has been previously established that the ansa chain of rifamycin S (52) and part of the naphthaquinone moiety is polyketide in origin, being derived from acetate and propionate units, as illustrated schematically in Scheme 5[111]. Other ansamycin antibiotics, such as geldanamycin[112,113] and actamycin[114], have similar biosynthetic origin. However, the source of the C_7N unit remained an enigma. It was suggested by Hornemann in 1974[115] that the C_7N unit in the

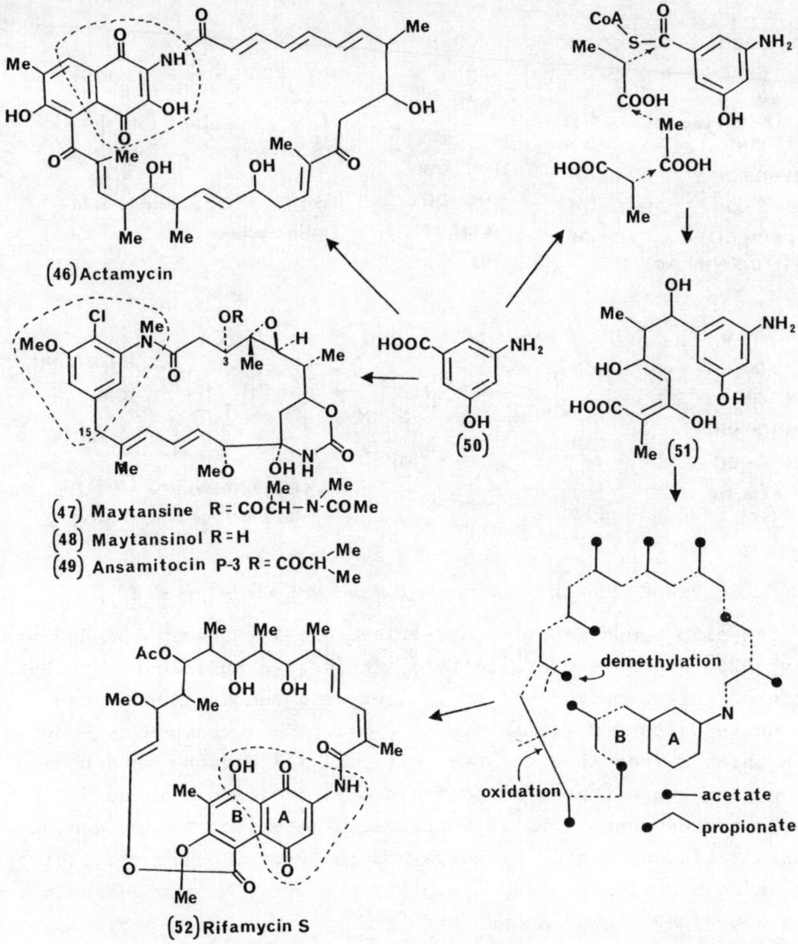

Scheme 5 *Biosynthesis of the Ansamycins*

mitomycins[44,116], circled with a dotted line in the structure of mitomycin C (56) in Scheme 6, and this unit in the ansamycins have a common biogenetic origin. This proposal has now been shown to be correct, and the C_7N unit is derived from 3-amino-5-hydroxybenzoic acid (50)[117,118]. Ghisalba and coworkers,[119] investigating mutants of *Nocardia mediterranei*, were able to show that the C_7N unit of rifamycin is derived from an intermediate of the shikimic acid pathway between 3-deoxy-D-arabinoheptulosonic acid-7-phosphate (53) and shikimic acid. They were able to isolate[120] an early precursor

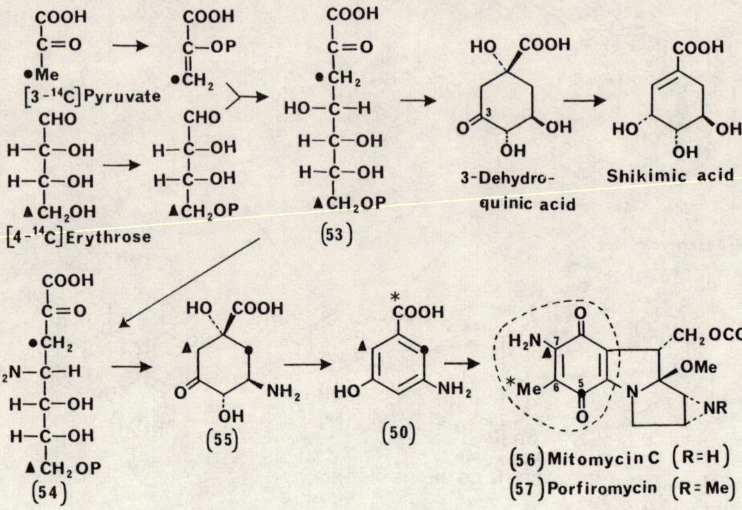

Scheme 6 *Biosynthesis of the C$_7$N unit of the Mitomycins*

of rifamycin: 2,6-dimethyl-3,5,7-trihydroxy-7-(3'-amino-5'-hydroxyphenyl)-2,4-heptadienoic acid (51). The most plausible biosynthetic origin of this compound was recognised as a combination of two propionate units, one acetate, and 3-amino-5-hydroxybenzoyl-coenzyme A, which serves as a starter unit , as illustrated in Scheme 5. It was then shown that 3-amino-5-hydroxybenzoic acid induced rifamycin biosynthesis in an appropriate mutant of *Nocardia mediterranei*[117]. Independently Rickards and coworkers considered that this benzoic acid derivative was a logical precursor of the C$_7$N unit in the ansamycins. It was shown that [*carboxy*-^{14}C]-3-amino-5-hydroxybenzoic acid was an excellent precursor of actamycin (46) in a *Streptomyces* species (strain E/784), all the radioactivity residing in the naphthaquinone residue of the antibiotic.[118] Shikimic acid failed completely to serve as a precursor of actamycin. The direct incorporation of [*carboxy*-^{13}C]-3-amino-5-hydroxybenzoic acid into porfiromycin (57) was established by ^{13}C n.m.r., the methyl group at C-6 being formed by reduction of the carboxyl group[121]. Hornemann had previously[122] shown that D-[4-^{14}C]erythrose and [3-^{14}C]pyruvate (which was converted to phosphoenol pyruvate in the species which produced the mitomycins) were incorporated into the C$_7$N unit of the mitomycins, the pattern of labelling being consistent with the formation of this unit from an aminoheptulosonic acid such as (54). It was also shown that shikimic and 3-dehydroquinic acids were not significant precursors of the mitomycins. The 3-amino-5-hydroxybenzoic acid is thus not formed from these cyclic compounds, and the suggestion of Hornemann that the amino function is introduced prior to the formation of cyclic precursors of the C_7N unit is reasonable. These biosynthetic results are summarized in Scheme 6. Intermediate (55) is currently hypothetical, the structure proposed being analogous to that of 3-dehydroquinic acid. The pattern of labelling found in the C_7N unit of geldamycin derived from D-[U-$^{13}C_6$]glucose was consistent with the intermediate formation of 3-amino-5-hydroxybenzoic acid from phosphoenol pyruvate and erythrose-4-phosphate by this pathway (see Table under *Streptomyces hygroscopicus*)[123]. This experiment is notable since the coupling of the contiguous carbons derived from the glucose was only observable in the ^{13}C n.m.r. when the micro-organism was grown on a medium containing D-glucose which was depleted in ^{13}C (0.12% ^{13}C/carbon).

It seemed probable that maytansine (47), first isolated from higher plants of the *Maytenus* genus[124], would be formed by a biosynthetic route similar to that of the ansamycins. A remarkable discovery was that a micro-organism of the *Nocardia* type produces C-3 esters of maytansinol (48)[125]. Using this organism, it has now been established[126] that [*carboxy*-^{13}C]-3-amino-5-hydroxybenzoic acid is indeed a direct precursor of ansamitocin P-3 (49), labelling C-15. One wonders if the *Maytenus* plants are really producing maytansine; it is possible that the plant is serving as a host for parasitic *Nocardia* species. It is perhaps significant that no maytansine was produced in a cell-suspension culture of *Maytenus buchananii*[127].

<u>2.5 Formation of Alkaloids in Cell-Suspension Cultures</u>.- This subject has been an active area of research in the last few years, and it is potentially a way to obtain economically useful natural products in a controlled environment, rather than by extracting whole plants, which are often difficult to cultivate in hospitable regions of our planet. Several books on this topic have appeared[128-131]. The species *Catharanthus roseus* has attracted much attention since it is a source of the clinically useful antitumor agents vinblastine and vincristine. Extensive work with this species has been carried out by Zenk[56], Kutney[133-138], Scott[139], and their

collaborators. Radioimmune assay proved to be a rapid method for selecting cell lines which were high-alkaloid-producing[56]. However, *Catharanthus roseus* cell-suspension cultures are considered to be 'unstable'. This means that after several generations, the culture loses its ability to produce alkaloids, or the % of alkaloid falls. The alkaloid distribution can be quite different from the intact plant, and new alkaloids not found in the differentiated plant have been discovered. Stöckigt and Soll[132] reported the isolation of Corynanthé, Strychnos, Aspidosperma, and Iboga type alkaloids from cell-cultures of *C. roseus* and *C. ovalis*. A new alkaloid isolated was vinervine (12-hydroxyakuammicine), formerly found in *Vinca erecta*. Kutney and his group examined a large number (458) of cell lines of *C. roseus*. Of these, 312 lines were found to produce alkaloids. These different cell lines could be grouped according to the types of alkaloid which they produced: 6.7% of the lines produced only Corynanthé or Strychnos types, 13.8% produced Strychnos-Aspidosperma, 23.1% Corynanthé-Aspidosperma, and 13.1% contained all three types of alkaloids. Catharanthine, an alkaloid of the Iboga type, was also produced in some cell lines[134]. One particular cell line of *C. roseus*.(No. 943) was selected for detailed chemical investigation[135]. Yields as high as 0.44% of alkaloids (based on the dry weight of cells) were obtained from this particular cell line. In addition to the known alkaloids of the Corynanthé, Strychnos, and Aspidosperma type, four new alkaloids (58-61), illustrated below, were isolated. None of the alkaloids

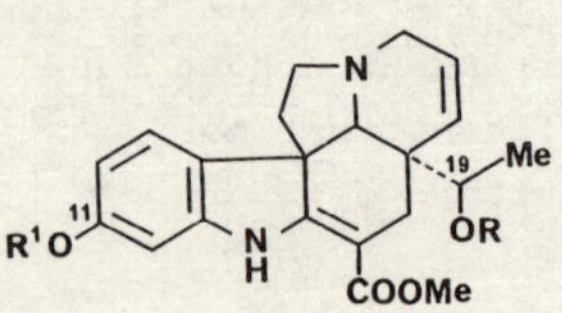

(58) 19-Acetoxy-11-methoxy-tabersonine (R = Ac, R^1 = Me)

(59) 19-Hydroxy-11-Methoxy-tabersonine (R = H, R^1 = Me)

(60) 19-Acetoxy-11-hydroxy-tabersonine (R = Ac, R^1 = H)

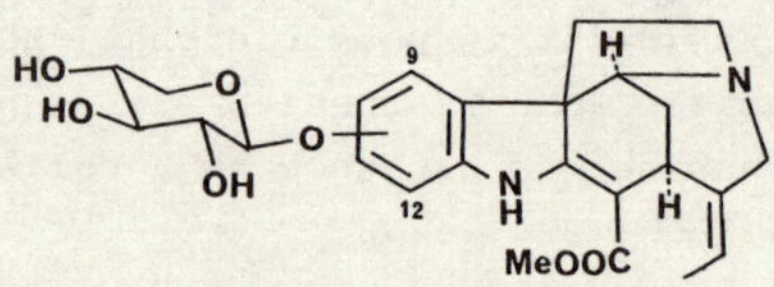

(61) Xyloside of (9 or 12)-hydroxy-akuammicine

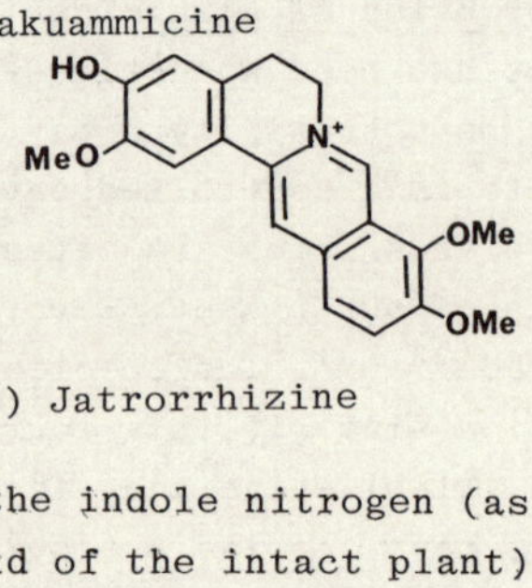

(62) Jatrorrhizine

from this cell line were methylated on the indole nitrogen (as is found in vindoline, the major alkaloid of the intact plant).

Another notable example of the production of alkaloids in a cell-suspension culture was with *Berberis* species (*Berberis stolonifera*) where the yield of berberine alkaloids (eg. jatrorrhizine (62)) was as high as 10% of the dry weight of the cells[140]. Cell cultures of *Peganum harmala* produce harman alkaloids in minute amounts[141].

3 Enzyme Studies

It is now becoming much more routine to investigate the biosynthesis of alkaloids and related compounds in cell-free systems. In several cases the enzymes involved have been obtained in a high state of purity. It has sometimes been found easier to obtain active enzymes from cell-suspension cultures or callus tissues of higher plants (eg. *Catharanthus roseus*). The species from which enzymes have been isolated are listed in alphabetical order, with a brief account of the reactions which they catalyse. Often the easiest way to assay an enzyme is by the use of a radioactive substrate. This information has been incorporated in the Table, which contains more details of some of these enzyme-controlled reactions.

3.1 *Anabasis aphylla*.- A cell-free extract from this species converted [2-^{14}C]lysine to labelled anabasine, aphylline, and lupinine, the incorporation being almost the same as with the intact plant[148].

3.2 *Berberis aggregata* .- A callus tissue of this species yielded a cell-free extract which contained an *N*-methyltransferase catalyzing the transfer of a methyl group from *S*-adenosyl-L-methionine to tetrahydropapaverine to yield laudanosine[142]. Reticuline was converted to palmatine and columbamine in this same extract.

3.3 *Berberis regeliana*.- This is one of several species (others: *Corydalis sempervirens, Dicentra spectabilis, Eschscholtzia pulchella, Fumaria officinalis, Glaucium rubrum, Meconopsis cambrica, Papaver somniferum, Ranunculus flammula,* and *Thalictrum tuberosum*) which have yielded an enzyme which has been named[143] (*S*)-Norlaudanosoline synthase. It catalyzes the reaction between dopamine (64) and 3,4-dihydroxyphenylacetaldehyde (65) yielding norlaudanosoline (69). When the incubation was carried out on a large scale with the enzyme isolated from *Eschscholtzia tenuifolia* , the isolated norlaudanosoline was shown to be chiral, having the *S*-configuration. However, it was only

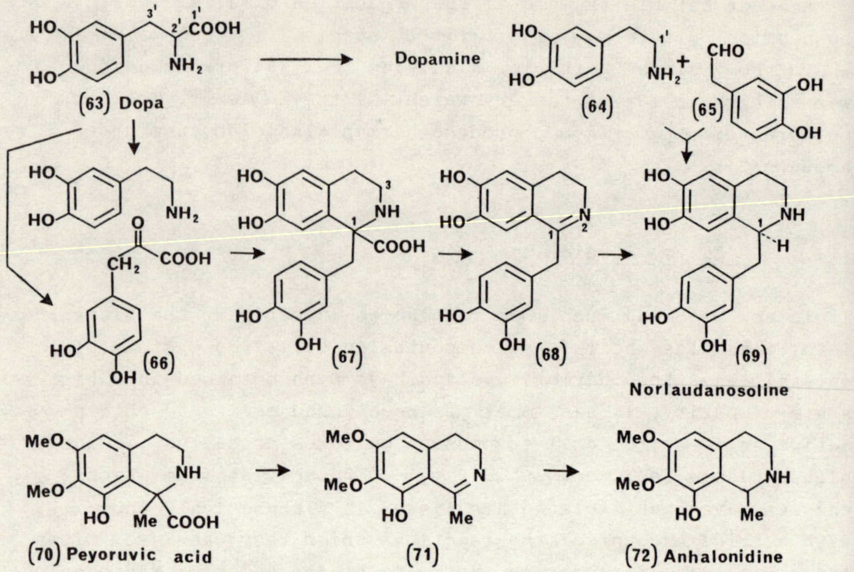

Scheme 7 *Biosynthesis of Norlaudanosoline and Anhalonidine*

about 25% optically pure. There was no evidence for the occurrence of a norlaudanosoline racemase in the enzyme preparation. It was thus proposed that racemic norlaudanosoline is formed by a chemical (non-enzymic) reaction between (64) and (65). This route to norlaudanosoline is different from that previously proposed for the origin of this key compound, from which a large number of alkaloids are derived. It is thus worthwhile to review the experiments which led to the acceptance of this other route which proceeds via norlaudanosoline-1-carboxylic acid (67). Prior to work on the origin of norlaudanosoline it was established that anhalonidine (72), an alkaloid of peyote (*Lophophora williamsii*) is formed from peyoruvic acid (70), probably via the dihydroisoquinoline (71)[144]. The amino-acid (70) resulted from a condensation between a hydroxylated phenylethylamine and pyruvic acid. It was thus natural that a similar pathway should be considered for the 1-benzyltetrahydroisoquinolines. Hence Wilson and Coscia[145] chemically prepared labelled norlaudanosoline-1-carboxylic acid (67) from dopamine and 3,4-dihydroxyphenylpyruvic acid (66). They found that (67) was converted to norlaudanosoline in *Papaver orientale* seedlings and also in the latex expressed from the seed capsules of this species.

Dopa (63) is a precursor of both dopamine (by decarboxylation) and 3,4-dihydroxyphenylpyruvic acid (by transamination). The administration of [2'-^{14}C]dopa to *P. orientale* seedlings yielded labelled amino-acid (67) and 1,2-dehydronorlaudanosoline (68), these compounds being added as carriers . No degradations were carried out to determine the location of the ^{14}C. Battersby et al.[146] had the same idea as Coscia and injected [3-^{14}C]-norlaudanosoline-1-carboxylic acid into the capsules of intact *Papaver somniferum* plants, which yielded labelled morphine, essentially all the activity being present (as expected) in the ethano bridge. It should be noted that the amino-acid (67) was a much poorer precursor of morphine (0.07% inc.) than [2'-^{14}C]dopa (0.78% inc.) which was fed under the same conditions. Scott et al.[147] prepared a cell-free extract from the capsules and stems of *P. somniferum* plants harvested a few days after the fall of the flower petals.[1'-^{14}C]-Dopamine and 3,4-dihydroxyphenylpyruvic acid were added to this cell-free extract, and the formation of labelled (67), (68), and (69) was observed (carrier compounds were added at the end of the incubation). It was stated that no norlaudanosoline-1-carboxylic acid was formed under the incubation conditions when 'boiled enzyme' was used. In none of these experiments was the chirality of the norlaudanosoline-1-carboxylic acid determined. If the reaction yielding this compound is controlled by an enzyme, one would expect this amino-acid to be optically active. The enzyme isolated by Zenk and his co-workers[143] failed to catalyze a reaction between dopamine and 3,4-dihydroxyphenylpyruvic acid. 4-Hydroxyphenylacetaldehyde did serve as a substrate for the enzyme. To summarize, I believe that the evidence favoring the intermediacy of norlaudanosoline-1-carboxylic acid is somewhat flimsy. It is of course possible that there are two independent routes to norlaudanosoline, and this is the most charitable explanation of these conflicting results.

3.4 *Capsicum annuum*.- Cell-free extracts were obtained from the placenta of this species[149]. The partially purified enzyme catalysed the formation of capsaicin and its analogues from vanillylamine and the coenzyme A esters of various iso-type carboxylic acids. 7-Methyloctanoyl-coenzyme A was the most efficient substrate for the enzyme, yielding nordihydrocapsaicin, which is only a minor alkaloid of the intact plant[150]. Decanoyl-, 5-methylhexanoyl-, 6-methylheptanoyl-, and 10-methylundecanoyl-coenzyme A were not utilized by the enzyme to yield amides with vanillylamine.

3.5 *Catharanthus roseus*.- The enzyme strictosidine synthase, which catalyses the stereospecific Mannich reaction between secologanin and tryptamine, affording strictosidine (73) [Scheme 8], has been isolated from several plant cell cultures: *Amsonia salicifolia, Catharanthus roseus, Ochrosia elliptica, Rauwolfia vomitoria, Rhazia orientalis, Stemmadenia tomentosa, Vinca minor,* and *Voacanga africana*[151-153]. Its purification from cultured cells of *C. roseus* has been described in detail[154,155]. The enzyme shows a pH optimum between 5.0 and 7.5.

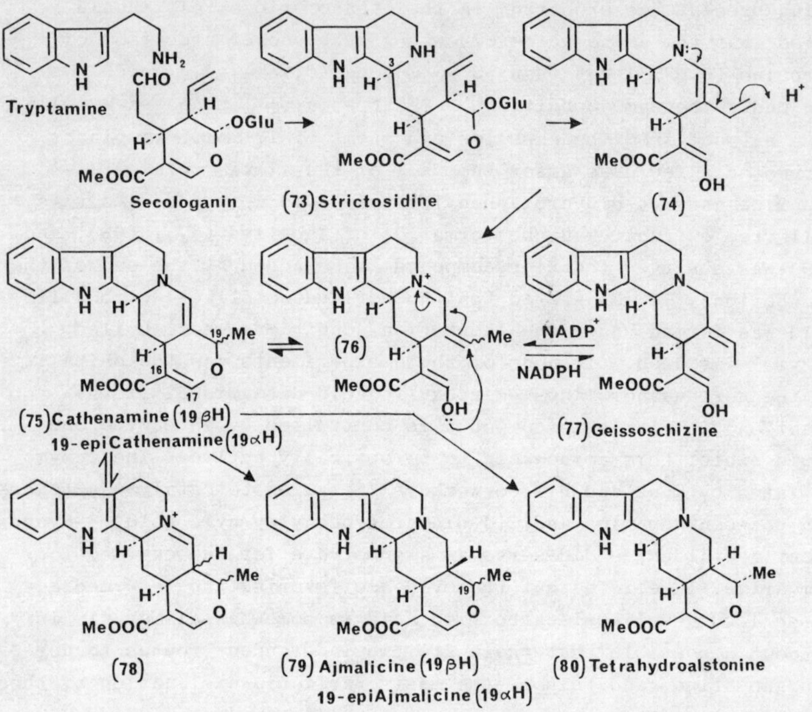

Scheme 8 *The Formation of Heteroyohimbine Alkaloids in Cell-Free Media*

No trace of its 3-epimer, vincoside, is produced by this enzyme. The enzyme is not inhibited by end-products of the biosynthetic pathway, such as ajmalicine, vindoline, or catharanthine. The enzyme was immobilized on a cyanogen-bromide-activated column of Sepharose. With this system it was possible to obtain 6 grams of strictosidine from secologanin and tryptamine[156]. The next step in the biosynthetic sequence is the removal of the glucose from (73).

It had been previously reported that this reaction was non-specific[157]. It has now been established[158] that two highly specific β-glucosidases are present in *C. roseus* cell cultures. These enzymes were specific for the removal of glucose from strictosidine and were also found in species of the Apocynaceae family which produce indole alkaloids (*Amsonia salicifolia*, *Catharanthus ovalis*, *C. pusilus*, *C. trichophyllus*, *Rauwolfia vomitoria*, *Rhazia orientalis*, and *Vinca major*), but not in species which do not contain indole alkaloids. All species examined also contained non-specific glucosidases (tested with *p*-nitrophenyl-β-D-glucoside as a substrate). Subsequent steps to ajmalicine (79) and tetrahydroalstonine (80) are considered to proceed via 20,21-didehydrocorynantheine aldehyde (74), 4,21-dehydrogeissoschizine (76), and the cathenamine isomers (75) as illustrated in Scheme 8 [159-161]. It is proposed that cathenamine is in equilibrium with the iminium structure (78)[162]. Geissoschizine (77),which was thought to be on the main pathway to ajmalicine and tetrahydroalstonine[163,164], is apparently only connected to the sequence by an oxidation-reduction system involving the pyridine nucleotides. The enzyme involved, geissoschizine dehydrogenase,has been purified[165]. Geissoschizine was converted to another indole alkaloid, isositsirikine, in a cell-free extract obtained from a *C. roseus* callus tissue[166]. The 17-hydroxy-compound formed by the addition of water to 19-*epi*-cathenamine, which can be converted to 19-*epi*-ajmalicine under biomimetic conditions[167], is not a precursor of the heteroyohimbine alkaloids in a cell-free extract of *C. roseus*[168]. Cell-free systems have also been obtained which catalyze the formation of vindoline (from tryptamine)[169], the dimeric indole alkaloids anhydrovinblastine and leurosine (from a mixture of catharanthine and vindoline)[169], and interconversions of these dimeric alkaloids[169-173] (see Table).

3.6 *Cephalosporium acremonium*.- After much effort, the cell-free conversion of δ-(L-α-aminoadipyl)-L-cysteinyl-D-valine (LLD-ACV-tripeptide) to isopenicillin N has been achieved[174-178]. This transformation was observed in the probe of an n.m.r. spectrometer, using the tripeptide labelled at specific positions with ^{13}C[176]. It is significant that the only enhanced ^{13}C signals were those of the tripeptide and isopenicillin N. If enzyme-free intermediates are involved in this cell-free conversion, their concentrations were not high enough to reach detectable levels.by this technique. A cell-free extract was also obtained from *Penicillium chrysogenum* which achieved the same conversion of the tripeptide to isopenicillin

N[179]. Structural analogs of the tripeptide were converted to analogs of isopenicillin N in *C. acremonium*[180]. No authentic intermediates between the LLD-ACV-tripeptide and isopenicillin N have been obtained[181,182]. It has been proposed that free radicals are involved in this conversion, and biomimetic experiments support this suggestion[183]. The subsequent conversion of isopenicillin N to penicillin N, and thence to desacetoxycephalosporin C, desacetylcephalosporin C, and cephalosporin C , has been acieved in this species[184-186] and *Streptomyces clavuligerus*[187].

3.7 *Claviceps*.- An *N*-methyltransferase which catalyzes the transfer of a methyl group from *S*-adenosyl-L-methionine to the primary amino-group of 4-(γγ-dimethylallyl)tryptophan to yield the corresponding *N*-methyl derivative has been obtained from a strain of *Claviceps*[188]. The enzyme also catalyzed the *N*-methylation of other substrates with the indicated efficiencies: tryptophan (8%), 4-methyltryptophan (10%), 4-(3'-methylbutyl)tryptophan (184%). Four enzymes which are involved in the biosynthesis of the ergot alkaloids have now been isolated from *Claviceps*. The other three are: 4-dimethylallyltryptophan synthase[189], chanoclavine-I cyclase[190], and agroclavine hydroxylase[191].

3.8 *Coffea arabica*.- An *N*-methyltransferase was isolated from the green fruits of this species (coffee) which yields caffeine as the ultimate alkaloid[192]. The methyl group of *S*-adenosyl-L-methionine was transferred to several xanthines; however, the main route to caffeine is considered to proceed from 7-methylxanthosine via 7-methylxanthine and theobromine.

3.9 *Conium maculatum*.- Roberts has continued her work on the enzyme L-alanine:5-oxooctanal transaminase, which catalyzes the formation of 5-oxo-1-aminooctane, the acyclic form of γ-coniceine. Two different enzymes have been found in the chloroplast and mitochondria. It is proposed that the one in the former organ is responsible for alkaloid formation in this species[193].

3.10 *Datura stramonium*.- A dehydrogenase was obtained from sterile root cultures of this species. Tropine, nortropine, and scopine were oxidized to their corresponding ketones in the presence of the pyridine nucleotides. ψ-Tropine (tropan-3β-ol) or ψ-nortropine were not substrates for this enzyme[194].

3.11 *Euglena gracilis*.- A cell-free extract was obtained from this green flagellate which catalyzed the incorporation of labelled ornithine, arginine, putrescine, and *S*-adenosyl-L-methionine into the polyamines spermidine and spermine[195].

3.12 *Hordeum vulgare*.- Two distinct *N*-methyltransferases were obtained from the roots of young plants, one catalyzing the methylation of tyramine to *N*-methyltyramine and the other the methylation of *N*-methyltyramine to hordenine[196]. An enzyme was also obtained from this species which catalyzed the formation of coumarylagmatine from agmatine and *p*-coumaryl-coenzyme A[197].

3.13 *Lathyrus sativa*.- Spermidine (20) is normally formed in bacteria and animals by reaction of putrescine (obtained by the decarboxylation of ornithine) with decarboxylated *S*-adenosyl-L-methionine. Recent work on this subject , carried out in *Escherichia coli*[198], is summarized in the Table. An alternate route apparently operates in *Lathyrus sativa*[199] . A cell-free system was obtained from seedlings which catalyzed the formation of the Schiff base (82) from putrescine and aspartic β-semialdehyde (81); this was then reduced to (83) with NADPH. A second enzyme, in the presence of pyridoxal, catalyzed the decarboxylation to yield spermidine. The same

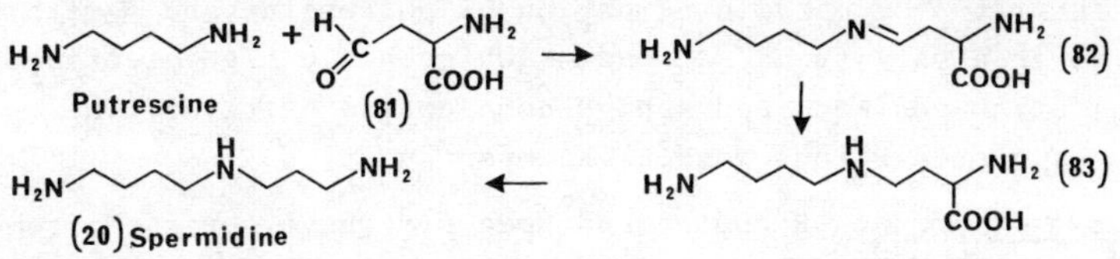

route was also found in the bacteria *Micrococcus denitrificans* and *Rhodopseudomonas spheroides*[200].

3.14 *Lupinus polyphyllus*.- The extensive work of Hartmann and his co-workers, involving cell-free extracts of this species and related ones, is described in Section 2.3 of this chapter.

3.15 *Nicotiana tabacum*.- Saunders and Bush[201] have attempted to relate the relative amounts of nicotine (84) and nornicotine (85) found in different genotypes of *Nicotiana tabacum* with the levels of the enzymes putrescine *N*-methyltransferase (PMT) and *N*-methylputrescine oxidase (MPO) in these different varieties . It is well established that nicotine is formed in tobacco by the route illustrated in Scheme 9[58]. Putrescine is methylated to yield *N*-methylputrescine, which is oxidized to 4-methylaminobutanal. Spontaneous cyclization affords the *N*-methyl-Δ^1-pyrrolinium salt, which reacts with an activated form of nicotinic acid (probably 3,6-dihydronicotinic acid) to ultimately yield nicotine. Nornicotine

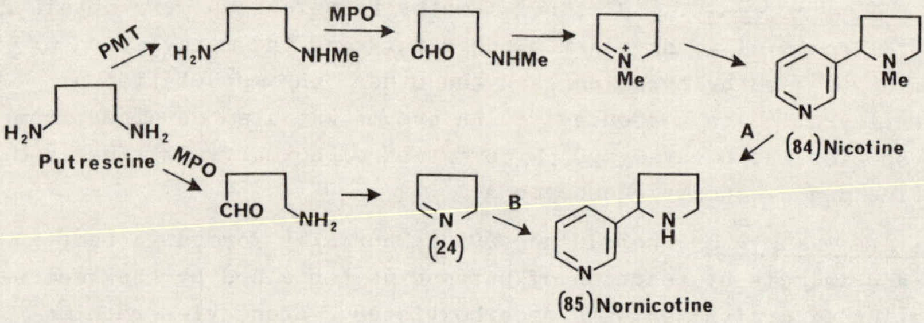

Scheme 9 *Two Potential Routes to Nornicotine*

is then formed by the demethylation of nicotine (Route A). Since the enzyme *N*-methylputrescine oxidase also catalyzes the oxidation of putrescine to Δ^1-pyrroline (24), via 4-aminobutanal, it was considered plausible that nornicotine could be formed directly from putrescine (Route B). However, no difference was observed between the rates of enzymic oxidation of putrescine and *N*-methylputrescine in a variety of *N. tabacum* which had a high nicotine-low nornicotine content, and a high-nornicotine cultivar. The validity of route B thus remains an open question.

3.16 *Papaver somniferum*.- Rapoport and Hodges[202] have obtained from this species an enzyme which catalyzes the reduction of codeinone to codeine, in the presence of NADH as a coenzyme. Another enzyme, partially purified by centrifugal fractionation, was obtained which catalyzed the oxidative coupling of the phenolic rings of reticuline to yield the dienone salutaridine[203]. Hydrogen peroxide was utilized by this enzyme for the oxidation. A crude enzyme from this species converted morphine to the dimeric pseudomorphine and morphine *N*-oxide[204].

3.17 *Phalaris tuberosus*.- This species, which contains *N,N*-dimethyltryptamine and 5-methoxy-*N,N*-dimethyltryptamine as its main alkaloids, served as a source of two *N*-methyltransferases which catalyzed the transfer of a methyl group from *S*-adenosyl-L-methionine to tryptamine and 5-methoxytryptamine[205]. These enzymes also catalyzed the methylation of other indole-ethylamines (5-methyltryptamine, 6-methoxytryptamine) and phenylethylamines, but not tetrahydro-β-carbolines or tetrahydroisoquinolines.

3.18 *Rhodopseudomonas viridis*. - A cell-free extract was obtained from this species which apparently contained enzymes and coenzymes capable of converting putrescine to homospermidine[84].

4 Table of Tracer Work Relating to Alkaloid Biosynthesis

Species are listed in alphabetical order. In order to present the material more concisely than in the previous chapters on this subject in the series *Biosynthesis* , the column which previously listed the incorporation of the administered compound into the alkaloids has been deleted. This information, when relevant, has been included under the heading '*Alkaloids labelled*' . In general the reported incorporation is the absolute incorporation (the total radioactivity or excess isotope found in the alkaloid divided by the total radioactivity or excess isotope administered). Specific incorporation is sometimes recorded, especially in experiments involving stable isotopes. The specific incorporation is the specific activity of the isolated alkaloid divided by the specific activity of the administered precursor. The positions of the isotopes in the administered precursors are defined in accord with the most recent IUPAC rules. That is, the isotopes are listed in alphabetical order, and if more than one isotope of the same element is present they are listed in order of increasing atomic weight. There is considerable confusion in the literature regarding the nomenclature of the aromatic amino-acids phenylalanine, tyrosine, and tryptophan. Often it is neccessary to number both the aromatic positions and the side chain. In this chapter the side-chain positions have been given prime numbers.

There is an occasional entry in the Table to work with unlabelled precursors. This type of experiment is usually only significant when a cell-free system is being used.

Index of Alkaloids and Related Compounds, whose Biosynthesis has been Studied

Alkaloids	*Species*
19-*O*-Acetylchaetoglobosin	*Chaetomium globosum*
Actamycin	*Streptomyces* sp. (strain E/784)
Ajmalicine	*Catharanthus roseus*
19-*epi*-Ajmalicine	*Catharanthus roseus*
Ajmaline	*Rauwolfia vomitoria*
Akuammicine	*Catharanthus roseus*
Akuammidine	*Vallesia glabra, Voacanga africana*
2-Aminophenoxazinone	*Brevibacterium iodinum*
Anabasine	*Anabasis aphylla, Nicotiana glauca*
Anhydrovinblastine	*Catharanthus roseus*
Ansamitocin P-3	*Nocardia* sp. (strain C-15003)
Anthramycin	*Streptomyces refuineus*
Antibiotic A23187	*Streptomyces chartreusis*
Aphyllidine	*Anabasis aphylla*
Aphylline	*Anabasis aphylla*
Aricine	*Rauwolfia canescens*
Aristolochic acid	*Aristolochia bracteata*
Berberine	*Hydrastis canadensis*
Berninamycin	*Streptomyces bernensis*
Betanin	*Opuntia bergeriana*
Bicyclomycin	*Streptomyces sapporonensis*
Bisdethiobis(methylthio)-acetylapoaranotin	*Aspergillus terreus*
Bisdethiobis(methylthio)-acetylaranotin	*Aspergillus terreus*
Bisdethiobis(methylthio)-gliotoxin	*Gliocladium deliquescens*
Bleomycin-morpholino-propylamine deriv.	*Streptomyces verticillus*
Caerulomycin A	*Streptomyces caerulens*
Caffeine	*Coffea arabica*
Caffeoylputrescine	*Nicotiana tabacum*
Camptothecin	*Camptotheca acuminata*
Capsaicin	*Capsicum annuum*
3-Carboxy-6,7-dihydroxy-1,2,3,4-tetrahydroisoquinoline	*Stizolobium hassjoo*
Catharanthine	*Catharanthus roseus*
Catharine	*Catharanthus roseus*
Cathenamine	*Catharanthus roseus, Rauwolfia verticillata*
Cephalotaxine	*Cephalotaxus harringtonia*
Cephalotaxinone	*Cephalotaxus harringtonia*
Cephamycin C	*Cephalosporium acremonium*
Cezomycin	*Streptomyces chartreusis*
α-Chaconine	*Solanum tuberosum*
Chaetoglobosin A	*Chaetomium globosum*
Chanoclavine-I	*Penicillium roqueforti*
Chelidonine	*Chelidonium majus*
Clavicipitic acid	*Claviceps* sp. (strain SD-58)
Clavulanic acid	*Streptomyces clavuligerus*
Cocaine	*Erythroxylon coca*
Cocculinine	*Cocculus laurifolius*
Coccuvine	*Cocculus laurifolius*
Coccuvinine	*Cocculus laurifolius*
Coclaurine	*Annona reticulata*
Codeine	*Papaver somniferum*
Columbamine	*Berberis aggregata*

Alkaloids	*Species*
Corydaline	*Corydalis solida*
Coumarylagmatine	*Hordeum vulgare*
Cuscohygrine	*Erythroxylon coca*
Cycloheximide	*Streptomyces griseus*, *Streptomyces naraensis*
Cyclo-(L-phenylalanyl-D-phenylalanyl)	*Aspergillus terreus*
α-Cyclopiazonic acid	*Penicillium griseofulvum*
β-Cyclopiazonic acid	*Penicillium griseofulvum*
Cytisine	*Sophora griffithii*
Cytochalasin B	*Phoma exigua*
Cytochalasin D	*Zygosporium masonii*
Deacetoxycephalosporin C	*Cephalosporium acremonium*
Decinine	*Decodon verticillatus*
Decodine	*Decodon verticillatus*
4,21-Dehydrogeissoschizine	*Catharanthus roseus*
Demethylcephalotaxine	*Cephalotaxus harringtonia*
Demethylcephalotaxinone	*Cephalotaxus harringtonia*
Dihydrocapsaicin	*Capsicum annuum*
1-(3',4'-Dimethoxybenzyl)-6-ethoxy-7-methoxyisoquinoline	*Papaver somniferum*
1-(3',4'-Dimethoxybenzyl)-6-methoxy-7-ethoxyisoquinoline	*Papaver somniferum*
3α,6β-Ditigloyloxytropane	*Datura innoxia*, *Datura meteloides*
3α,6β-Ditigloyloxytropan-7β-ol	*Datura meteloides*
Echinuline	*Aspergillus amstelodami*
Eleagnine	*Eleagnus angustifolia*
Elymoclavine	*Claviceps* sp. (strain SD-58)
Ergostine	*Claviceps purpurea*
Ergotamine	*Claviceps purpurea*
1-(3'-Ethoxy-4'-methoxybenzyl)-6,7-dimethoxyisoquinoline	*Papaver somniferum*
N-Ethylnorcodeine	*Papaver somniferum*
N-Ethylnormorphine	*Papaver somniferum*
Feruloylputrescine	*Nicotiana tabacum*
Festuclavine	*Penicillium roqueforti*
5-Fluoroanabasine	*Nicotiana glauca*
Fumitremorgin A	*Aspergillus fumigatus*
Fumitremorgin B	*Aspergillus fumigatus*
Geissoschizine	*Catharanthus roseus*
Geldanamycin	*Streptomyces hygroscopicus* var. *geldanus* var. *nova*
Gelsemine	*Gelsemium sempervirens*
Gliotoxin	*Gliocladium deliquescens*
Harman	*Passiflora edulis*
Harringtonine	*Cephalotaxus harringtonia*
Hasubanonine	*Stephania japonica*
Heliosupine	*Cynoglossum officinale*
Hinnuliquinone	*Nodulisporium hinnuleum*
Hitachimycin	*Actinomycete* sp. (strain KM-4927)
Homobrevicolline	*Carex brevicollis*
Homocapsaicin II	*Capsicum annuum*
Homodihydrocapsaicin	*Capsicum annuum*
Homospermidine	*Rhodopseudomonas viridis*, *Senecio isatideus*
Hordenine	*Hordeum vulgare*

Alkaloids	*Species*
Hydrastine	*Hydrastis canadensis*
9'-Hydroxydihydroergosine	*Sphacelia sorghi*
2-Hydroxyphenazine-1-carboxylic acid	*Pseudomonas aureofaciens*
Hygrine	*Nicandra physaloides*
Hyoscyamine	*Datura innoxia*, *Datura meteloides*, *Duboisia leichhardtii/myoporoides*
Indole-3-aldehyde	*Hordeum vulgare*
Indole-3-carbinol	*Hordeum vulgare*
Indole-3-carboxylic acid	*Hordeum vulgare*
Indolmycenic acid	*Streptomyces griseus*
Indolmycin	*Streptomyces griseus*
Iodinin	*Brevibacterium iodinum*
Isococculidine	*Cocculus laurifolius*
Isococculine	*Cocculus laurifolius*
Isoharringtonine	*Cephalotaxus harringtonia*
Isopenicillin N	*Cephalosporium acremonium*
N'-Isopropylnornicotine	*Nicotiana tabacum*
Isoreserpiline	*Rauwolfia canescens*
Isoreserpinine	*Rauwolfia canescens*
Isositsirikine	*Catharanthus roseus*
Isotetrandine	*Cocculus laurifolius*
Isovallesiachotamine	*Rauwolfia verticillata*
Laudanosine	*Berberis aggregata*
Laurifinine	*Cocculus laurifolius*
Laurifoline	*Cocculus laurifolius*
Leurosine	*Catharanthus roseus*
Lupanine	*Lupinus angustifolius*, *Lupinus polyphyllus*
Lupinine	*Anabasis aphylla*
Macromerine	*Coryphantha macromeris*
Matrine	*Sophora microphylla*, *Sophora tetraptera*
Meteloidine	*Datura innoxia*, *Datura meteloides*
1-(3'-Methoxy-4'-ethoxybenzyl)-6,7-dimethoxyisoquinoline	*Papaver somniferum*
3-Methylaminomethylindole	*Hordeum vulgare*
1-Methyl-3-carboxy-6,7-dihydroxy-1,2,3,4-tetrahydroisoquinoline	*Stizolobium hassjoo*
N-Methyl-crotsparine	*Croton sparsiflorus*
N-Methyl-crotsparinine	*Croton sparsiflorus*
δ-*N*-Methylornithine	*Atropa belladonna*
N-Methylphenethylamine	*Dolichothele sphaerica*
N-Methyl-sparsiflorine	*Croton sparsiflorus*
N-Methyltyramine	*Hordeum vulgare*, *Opuntia clavata*
Mitomycin C	*Streptomyces verticillatus*
Monobactam SQ 26,180	*Chromobacterium violaceum*
Monobactam SQ 26,445	*Acetobacter* sp. (strain ATCC 21780)
Monobactam SQ 26,812	*Acetobacterium ratiobacter*
Monocrotaline	*Crotolaria retusa*
Morphine	*Papaver somniferum*
Morphine 3-ethyl ether	*Papaver somniferum*
Morphine *N*-oxide	*Papaver somniferum*

Alkaloids	*Species*
Naphthyridinomycin	*Streptomyces lusitanus*
Narcotine	*Papaver somniferum*
Neopine	*Papaver somniferum*
Nicotine	*Nicotiana glutinosa, Nicotiana tabacum*
Nordihydrocapsaicin	*Capsicum annuum*
Norgramine	*Hordeum vulgare*
(*S*)-Norlaudanosoline	*Berberis regeliana, Corydalis sempervirens Dicentra spectabilis, Eschscholtzia pulchella, Fumaria officinalis, Meconopsis cambrica, Papaver somniferum, Ranunculus flammula, Thalictrum tuberosum*
Normacromerine	*Coryphantha macromeris*
Normetanephrine	*Coryphantha macromeris*
Nornicotine	*Nicotiana glutinosa, Nicotiana tabacum*
Nornuciferine-I	*Croton sparsiflorus*
Nororientaline	*Aristolochia bracteata*
Noscapine	*Papaver somniferum*
Ochotensimine	*Corydalis ochotensis*
Orientaline	*Aristolochia bracteata*
Oripavine	*Papaver orientale, Papaver somniferum*
Oxaline	*Penicillium oxalicum*
17-Oxosparteine	*Lupinus polyphyllus*
Pactamycin	*Streptomyces pactum*
Palmatine	*Berberis aggregata, Cissampelos pariera, Cocculus laurifolius*
Papaverine	*Papaver somniferum*
Paspalicine	*Claviceps paspali*
Paspalinine	*Claviceps paspali*
Penitrem A	*Penicillium crustosum*
Penicillin N	*Cephalosporium acremonium*
Penicillin V	*Penicillium chrysogenum*
Perloline	*Festuca arundinacea*
Phenazine-1-carboxylic acid	*Pseudomonas aureofaciens*
Phenethylamine	*Dolichothele sphaerica*
Phomamide	*Phoma lingam*
Phomazarin	*Pyrenochaeta terrestris*
Pipecolic acid	*Decodon verticillatus, Lupinus angustifolius, Nicotiana glauca*
Porfiromycin	*Streptomyces verticillatus*
Protopine	*Chelidonium majus, Corydalis ochotensis, Corydalis solida*
Protostephanine	*Stephania japonica*
Pseudomorphine	*Papaver somniferum*
Pseurotin A	*Pseudeurotium ovalis*
Putrescine	*Euglena gracilis, Lathyrus sativus*
Pyrazofurin	*Streptomyces candidus*
Pyrrolnitrin	*Pseudomonas aureofaciens*
Quebrachidine	*Rauwolfia vomitoria*

Alkaloids	*Species*
Reserpiline	*Rauwolfia canescens*
(*S*)-Reticuline	*Cocculus laurifolius*
Retronecine	*Heliotropium spathulatum*, *Senecio isatideus*, *Senecio vulgaris*
Retrorsine	*Senecio isatideus*
Rifamycin B	*Nocardia mediterranei*
Rifamycin L	*Nocardia mediterranei*
Rifamycin O	*Nocardia mediterranei*
Rifamycin S	*Nocardia mediterranei*
Roquefortine A	*Penicillium roqueforti*
Roquefortine B	*Penicillium roqueforti*
Roquefortine C	*Penicillium crustosum*, *Penicillium roqueforti*
Roquefortine D	*Penicillium roqueforti*
Rutacridone	*Ruta graveolens*
Saframycin	*Streptomyces lavendulae*
Salutaridine	*Papaver somniferum*
Sarpagine	*Rauwolfia vomitoria*
Scopolamine(Hyoscine)	*Datura innoxia*, *Datura meteloides*, *Duboisia leichhardtii/myoporoides*
Sedamine	*Sedum acre*
Senecionine	*Senecio magnificus*
Sibiromycin	*Streptosporangium sibiricum*
Sirodesmin PL	*Phoma lingam*
Slaframine	*Rhizoctonia leguminicola*
Sparteine	*Lupinus polyphyllus*
Spermidine	*Escherichia coli*,*Euglena gracilis*, *Lathyrus sativus*
Spermine	*Euglena gracilis*, *Lathyrus sativus*
Stephanine	*Strychnos nux-vomica*
Stizolobic acid	*Stizolobium hassjoo*
Streptonigrin	*Streptomyces flocculus*
Streptothricin F	*Streptomyces* sp. (strain L-1689-23)
Strictosidine lactam	*Catharanthus roseus*
Strychnine	*Strychnos nux-vomica*
Stylopine	*Chelidonium majus*
Supidine	*Heliotropium spathulatum*
Swainsonine	*Rhizoctonia leguminicola*
Tetrahydroalstonine	*Catharanthus roseus*
22-*N*-Tetrahydroalstonine	*Rauwolfia verticillata*
Tetrandrine	*Cocculus laurifolius*
Tetrahydropalmatine	*Cocculus laurifolius*, *Stephania glabra*
Tetraphyllicine	*Rauwolfia vomitoria*
Thalicarpine	*Cocculus laurifolius*, *Thalictrum minus* var. *elatum*
Thebaine	*Papaver orientale*, *Papaver somniferum*
Theobromine	*Coffea arabica*
Theophylline	*Coffea arabica*
9'-Thiaergosine	*Claviceps purpurea*
Tiliacorine	*Tiliacora racemosa*
Tiliacorinine	*Tiliacora racemosa*
Tomaymycin	*Streptomyces achromogenes*
Trachelanthamidine	*Heliotropium spathulatum*
3,4,5-Trihydroxy-octahydro-1-pyrindine	*Rhizoctonia leguminicola*
Tylophorine	*Tylophora asthmatica*

Alkaloids	*Species*
Tylophorinine	*Tylophora asthmatica*
Tyramine	*Hordeum vulgare*
Vallesiachotamine	*Rauwolfia verticillata*
Verrucarin E	*Myrothecium verrucaria*
Vinblastine	*Catharanthus roseus*
Vincadifformine	*Vinca major*
Vindoline	*Catharanthus roseus*
Virginiamycin M_1	*Streptomyces virginiae*
α-Yohimbine	*Rauwolfia canescens*

Species / *Compound Administered*	*Alkaloids labelled*	*Ref.*

Actinomycete sp. (strain KM-4927)
Sodium [1-^{13}C]Acetate — Hitachimycin — 206
Sodium [1-^{13}C]Propionate
DL-[1-^{13}C]Phenylalanine

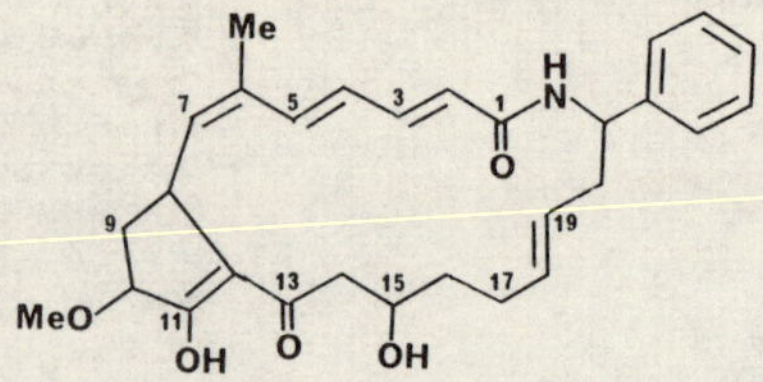

(^{13}C nmr indicated the inc. of [1-^{13}C]acetate into C-1,3,7,9, 11,13,15, and 17. [1-^{13}C]Propionate labeled only C-5. [1-^{13}C]-Phenylalanine labeled C-19).

Acetobacter sp. (strain ATCC 21780)
L-[*methyl*-^{14}C]Methionine — Monobactam SQ 26,445 — 207

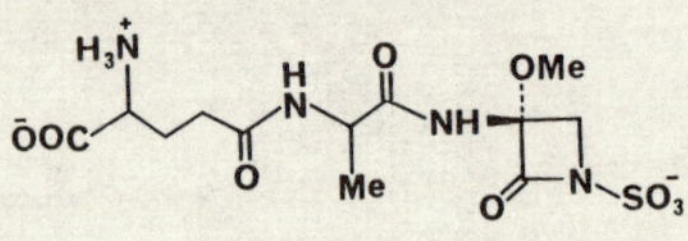

(most of ^{14}C in the methoxyl group)

Acetobacterium ratiobacter
L-[3-^{14}C, 3-^{3}H]Serine — Monobactam SQ 26,812 — 207

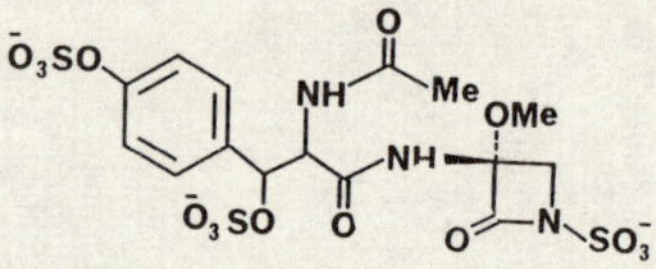

(Incorporation of the serine with little loss of ^{3}H. For related work see under *Chromobacterium violaceum*).

Anabasis aphylla
[^{14}C]Pachycarpine ≡ (+)-Sparteine (obtained by the $LiAlH_4$ reduction of aphylline produced biosynthetically from [1,5-^{14}C]-cadaverine) — Aphylline — 208

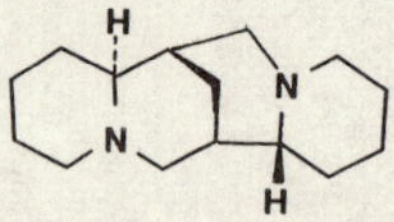

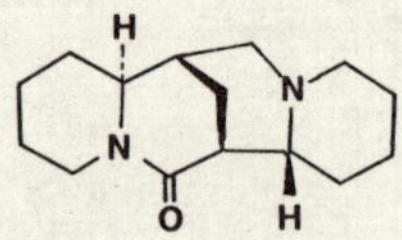

Aphyllidine

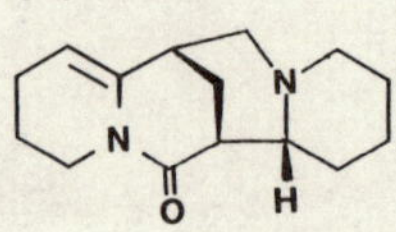

(0.88% inc. into the combined alkaloids when the precursor was fed via the roots of the intact plant)

Species / *Compound Administered*	*Alkaloids labelled*	*Ref.*
Anabasis aphylla (continued) -		
[2-^{14}C]Lysine (fed to cell-free extract)	Anabasine (see under *Nicotiana glauca*)	148
	Lupinine	

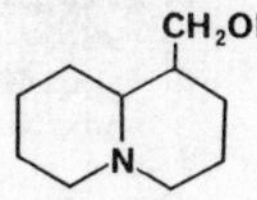

Aphylline

(The rate of alkaloid formation was similar to that previously observed in intact plants, all the above alkaloids being labelled).

Species / *Compound Administered*	*Alkaloids labelled*	*Ref.*
Annona reticulata		
L-[U-^{14}C]Tyrosine	Coclaurine(0.70% inc.)	209

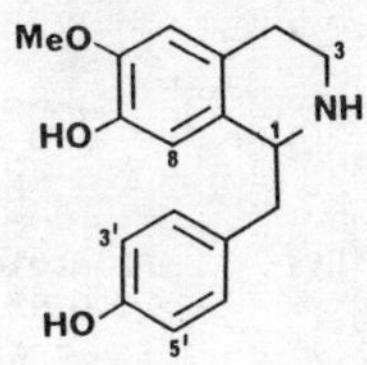

DL-[2'-^{14}C]Tyrosine	Coclaurine(0.65% inc. C-3: 48.9%)	
[3,5-$^{3}H_2$]Tyramine	Coclaurine(0.46% inc. No activity in benzyl residue)	
[3,5,3'-^{3}H]-4-Hydroxyphenylpyruvic acid	Coclaurine(0.025% inc., 25.1% of the activity in the benzylic residue)	
DL-[2'-^{14}C]-3,4-Dihydroxyphenylalanine (DOPA)	Coclaurine(0.42% inc. C-3: 98%)	
[1'-^{14}C]Dopamine	Coclaurine(1.2% inc.)	
(*RS*)-[3-^{14}C]Norcoclaurine-1-carboxylic acid	Coclaurine(0.088% inc. C-3: 98.3%)	
[1-^{14}C]-1,2-Didehydronorcoclaurine	Coclaurine(0.19% inc.)	
(*RS*)-[1-^{3}H]Norcoclaurine	Coclaurine(0.25% inc.)	

(It is claimed that the results support the intermediacy of norcoclaurine-1-carboxylic acid; however, the incorporation of this compound was much poorer than more remote precursors such as tyrosine or dopa).

Species / *Compound Administered*	*Alkaloids labelled*	*Ref.*
Aristolochia bracteata		
L-[U-^{14}C]Tyrosine	Aristolochic acid(0.1% inc.)	210

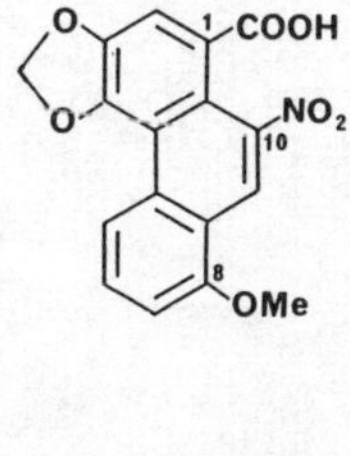

Species *Compound Administered*	*Alkaloids labelled*	*Ref.*
Aristolochia bracteata (continued) -		
DL-[3'-^{14}C]Tyrosine	Nororientaline (0.026% inc.) Orientaline (0.011% inc.) Stephanine (0.135% inc.)	210
(*RS*)-[3'-^{14}C]-3,4-Dihydroxyphenylalanine (DOPA)	Aristolochic acid (0.01% inc.) (96.2% at the COOH. This result is different from that of Spenser and Tiwari[397] who found most of the activity at C-10 after feeding [2'-^{14}C]dopa).	
(*RS*)-[5',8-$^{3}H_2$]Nororientaline	Aristolochic acid (0.046% inc.)	

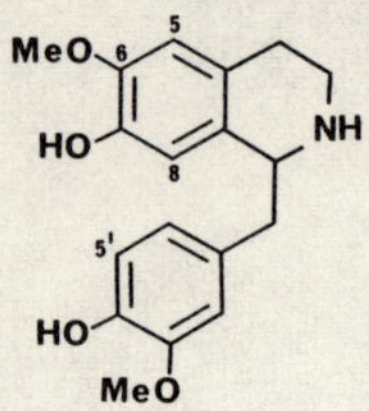

(*RS*)-[6-*methoxy*-^{14}C,5',8-$^{3}H_2$]-Nororientaline	Aristolochic acid (0.029% inc. 82% of the ^{14}C is in the methylenedioxy-group, with 42% ret. of ^{3}H).	
(*RS*)-[5',8-$^{3}H_2$]Orientaline	Aristolochic acid (0.32% inc.)	
(*R*)-[5',8-$^{3}H_2$]Orientaline	Aristolochic acid (0.23% inc.)	
(*S*)-[5',8-$^{3}H_2$]Orientaline	Aristolochic acid (0.06% inc.)	
(*RS*)-[*aryl*-^{3}H]Prestephanine	Aristolochic acid (0.036% inc.)	

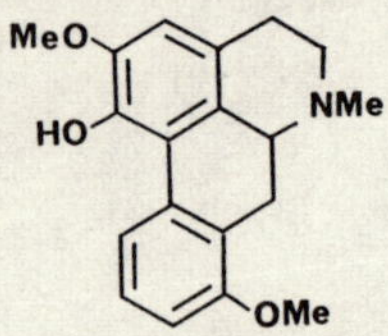

(*RS*)-[*aryl*-^{3}H]Stephanine	Aristolochic acid (0.145% inc.)	

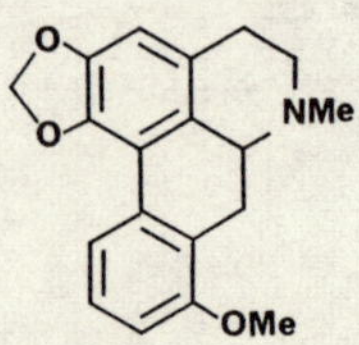

Species / *Compound Administered*	*Alkaloids labelled*	*Ref.*
Aspergillus amstelodami		
[1,2-$^{13}C_2$]Acetate	Echinuline	211

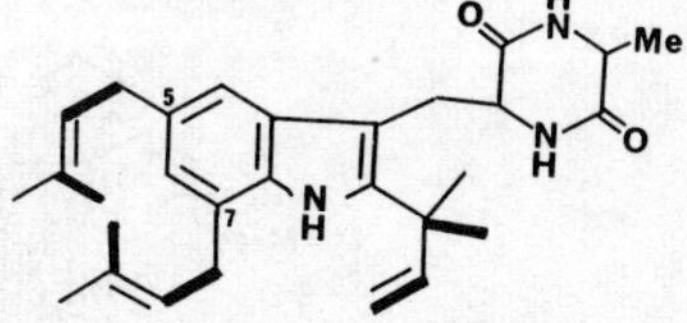

(Thickened bonds in the above formula indicate the pattern of incorporation of the contiguous carbons).

[(5*R*)-^{3}H]Mevalonic acid	Echinuline (0.1—0.5% inc.)	212
[(5*S*)-^{3}H]Mevalonic acid	Echinuline (0.1—0.5% inc.)	

(The stereochemistry of the hydrogens present on the benzylic carbons of the isopentenyl groups at C-5 and C-7 was determined. The results were consistent with inversion at the allylic carbon of dimethylallyl pyrophosphate during the aromatic isoprenylation).

DL-[2-^{14}C]Mevalonic acid	Echinuline (1.3—1.5% inc.)	
L-[3'-^{14}C]Tryptophan	Echinuline (1.95—2.1% inc.)	
1-([1-^{3}H]-3,3-Dimethylallyl)-L-tryptophan	Echinuline (< 0.05% inc.)	

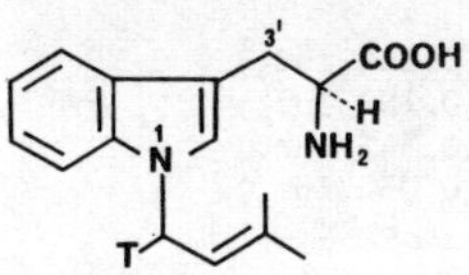

cyclo-L-Alanyl-1-([1-^{3}H]-3,3-dimethylallyl)-L-tryptophan — Echinuline (0.016% inc.) (These results indicated that echinuline biosynthesis does not involve the intermediacy of 1-dimethylallylindole derivatives).

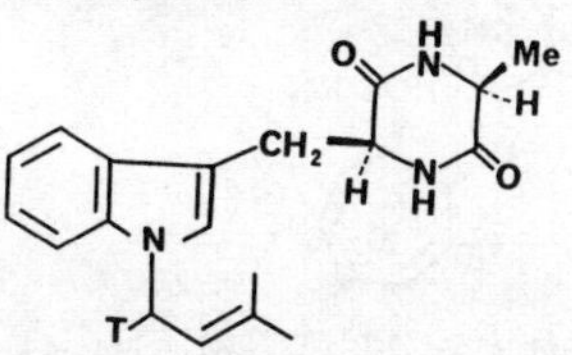

L-(4*S*)-[5-^{14}C, 5-^{3}H]Leucine	Echinuline (1.0% inc. Incorporation occurred with 30% loss of ^{3}H).	213
L-(4*S*)-[2-^{14}C, 5-^{3}H]Leucine	Echinuline (^{3}H: 1.2% inc. 67% loss of ^{14}C) These results are rationalized by postulating formation of β-hydroxy-β-methyl-glutaryl-coenzyme A (1) which undergoes cleavage to aceto-acetate which is then incorporated into mevalonic acid:	

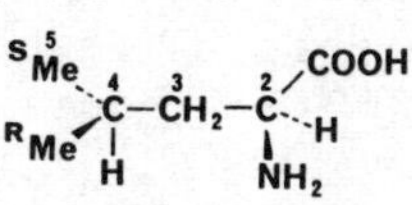

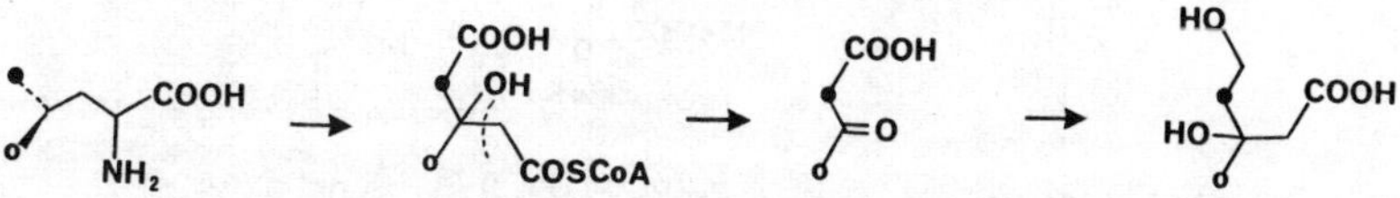

1

Species *Compound Administered*	*Alkaloids labelled*	*Ref.*
Aspergillus fumigatus DL-[3'-^{14}C]Tryptophan	Fumitremorgin A (0.16% inc.) 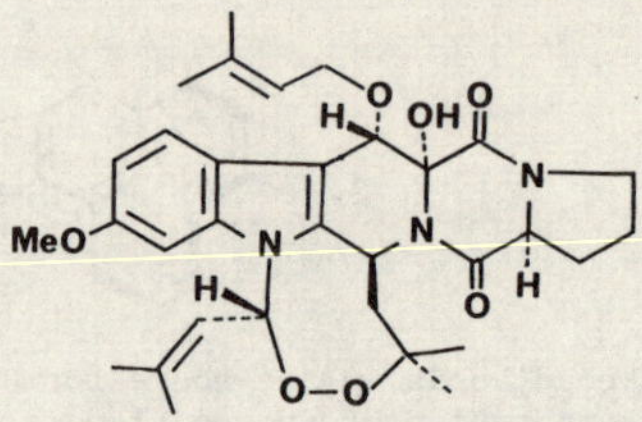 Fumitremorgin B (0.012% inc.) 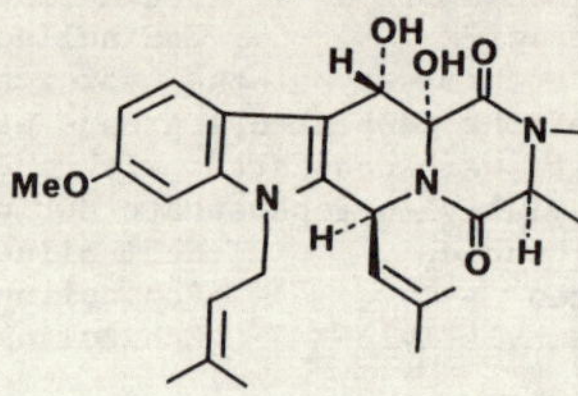	214
(*RS*)-[2-^{14}C]Mevalonolactone	Fumitremorgin A (0.12% inc.) Fumitremorgin B (0.10% inc.)	
L-[U-^{14}C]Proline	Fumitremorgin A (0.34% inc.) Fumitremorgin B (0.05% inc.)	
Aspergillus terreus *cyclo*-(L-[^{15}N]Phenylalanyl-L-[1-^{13}C]phenylalanyl) + *cyclo*(-L-[U-^{14}C]-Phe-L-Phe-) 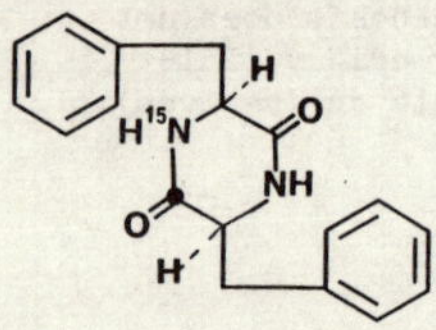	Bisdethiobis(methylthio)-acetylaranotin (BDA) 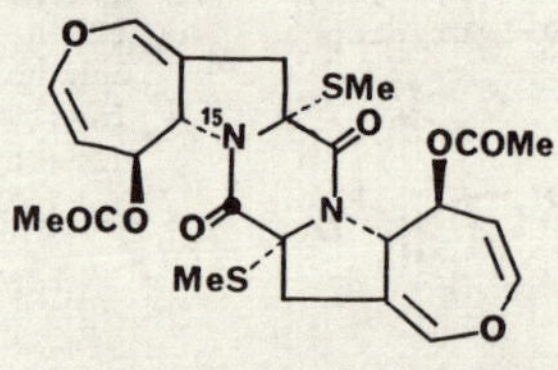 (19.9% inc. and a spec. inc. of 17%. The ^{13}C nmr showed satellites at the signal for the amide C=O due to the contiguous ^{15}N; $^{1}J_{CN}$ = 13 Hz, indicating the direct inc. of the cyclic dipeptide).	215
cyclo-(L-[U-^{14}C]Phenylalanyl-L-phenylalanyl)	BDA(inc. occurred) Bisdethiobis(methylthio)acetyl-apoaranotin (inc. occurred). 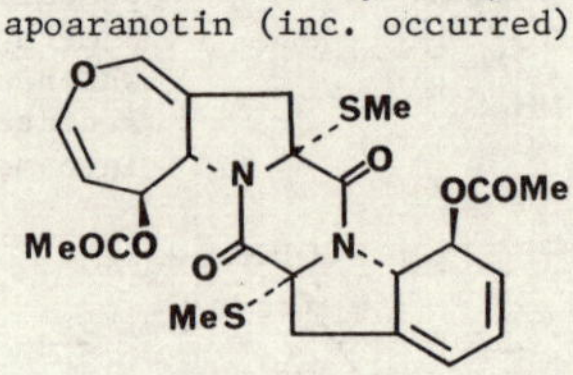 [*cyclo*(-L-Phe-D-Phe-) or *cyclo*(-D-Phe-D-Phe-) were not precursors of BDA]	

Species / *Compound Administered*	*Alkaloids labelled*	*Ref.*
Atropa belladonna		
DL-[5-^{14}C]Ornithine	δ-*N*-Methylornithine (0.37% inc.)	67
DL-[5-^{3}H]Ornithine	δ-*N*-Methylornithine (0.02% inc.)	
(Surprisingly, no activity was detected in the alkaloids hyoscyamine and cuscohygrine isolated from these feeding experiments).		
Berberis aggregata (enzyme system)		
S-Adenosyl-L-[*methyl*-^{14}C]-methionine + tetrahydropapaverine	Laudanosine 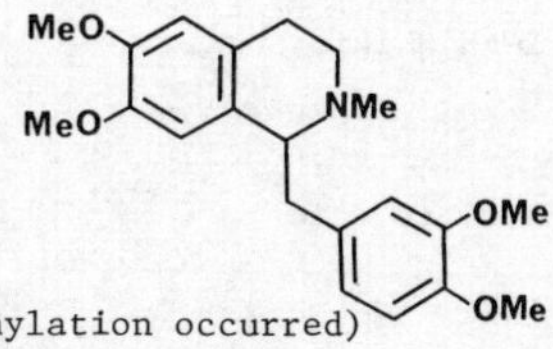 (*N*-methylation occurred)	142
S-Adenosyl-L-[*methyl*-^{14}C]-methionine + (*RS*)-Reticuline	Columbamine 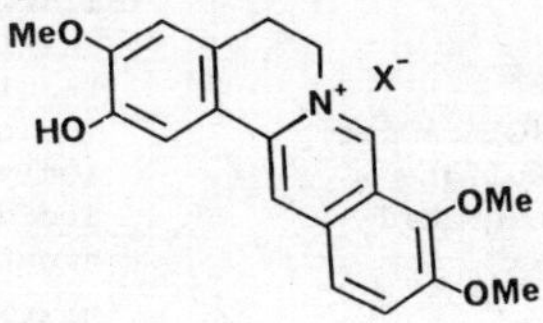 Palmatine (Both these berberine alkaloids were labelled).	
Berberis regeliana		
[2'-^{14}C, 2,6-^{3}H$_2$]Dopamine + 3,4-Dihydroxyphenyl-acetaldehyde (enzyme system) 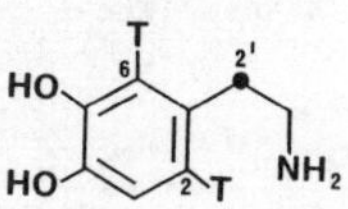	(*S*)-Norlaudanosoline 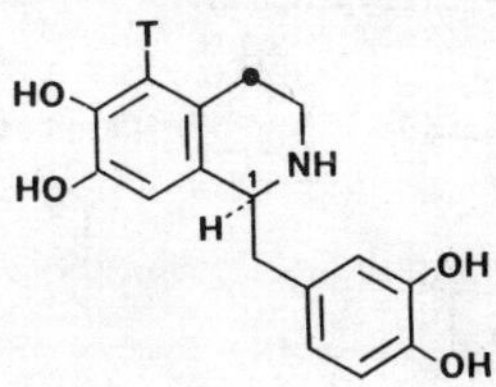	143
Brevibacterium iodinum		
[1-^{14}C]Shikimic acid 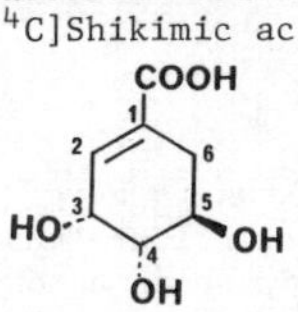	Iodinin 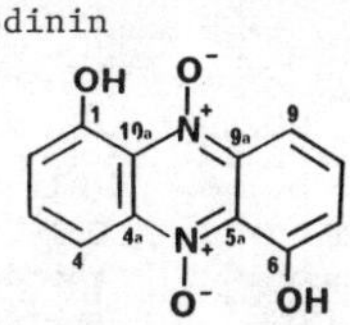	216
(42% inc. All activity at C-1,4,6, and 9)		
[6-^{14}C]Shikimic acid	Iodinin (31% inc. All activity at C-4a, 5a, 9a, and 10a).	
[1,6,7-^{14}C]Shikimic acid	Iodinin (34% inc., allowing for one third loss of activity from the COOH group. Activity equally distributed C-1,4,6,4a,5a,9a, and 10a).	

Species / *Compound Administered*	*Alkaloids labelled*	*Ref.*
Brevibacterium iodinum (continued) -		
[G-^{3}H]Anthranilic acid	Iodinin (0.02% inc., i.e. negligible)	
[^{14}C]-3-Hydroxy-2,3-dihydro-anthranilic acid	Iodinin (0.3-0.7% inc., i.e. not significant)	

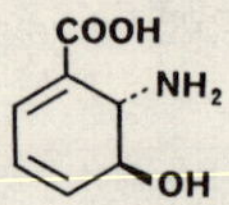

[1,6-^{14}C, 2-^{2}H]Shikimic acid	Iodinin (22% inc. (^{14}C).) (Both isotopes were incorporated to the same extent: d_2 : 7.5%, d_1 : 31%, d_o : 60.5%)	217
[2,4-$^{2}H_2$]Shikimic acid	Iodinin (Incorporation with some loss of the C-4 deuterium, possibly by conversion of the shikimic acid to 3-dehydroshikimic acid and loss of the C-4 hydrogen by enolization. The results obtained with shikimic acid are consistent with the results of Hollstein[218]).	
(*S*)-[*amide*-^{15}N]Glutamine [^{15}N]Ammonium sulfate (*S*)-[^{15}N]Glutamic acid	Iodinin (Only the [*amide*-^{15}N]glutamine yielded iodinin with greater than a statistical amount of the $^{15}N_2$-labelled material).	219,220
	2-Aminophenoxazinone	

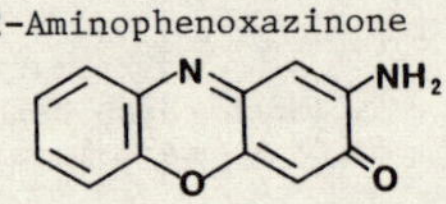

(The [*amide*-^{15}N]glutamine gave rise to the highest inc. of ^{15}N. This metabolite was derived from [^{14}C]shikimic acid, but not from L-[5-^{3}H]tryptophan).

Camptotheca acuminata		
[14-^{3}H, ^{2}H]Strictosamide	Camptothecin (^{3}H: 0.57% spec. inc.)	221

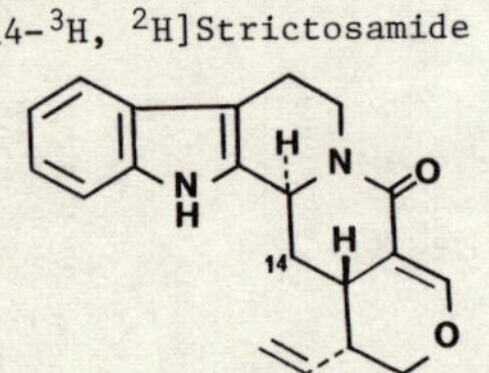

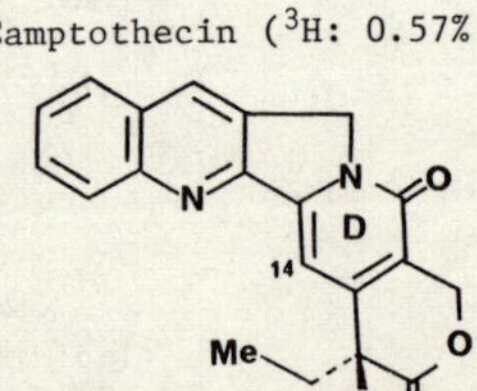

(fed to cut shoots) (Most of the data in this article has been previously recorded: *Biosynthesis* 1976, 4, 114; ibid., 1980, 6, 219).	(All the deuterium was shown to be at C-14 by nmr. The results indicate that the dehydrogenation of ring D is a non-stereospecific process. The biosynthesis of this alkaloid has been reviewed[35,222]).	

Species / *Compound Administered*	*Alkaloids labelled*	*Ref.*
Capsicum annuum		
L-[U-^{14}C]Valine (fed to intact fruits)	Capsaicin	223,224

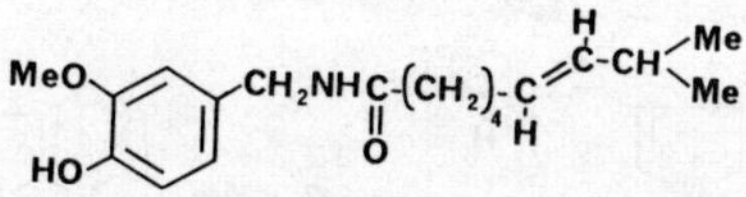

L-[U-^{14}C]Leucine (fed to intact fruits)	Dihydrocapsaicin Nordihydrocapsaicin	

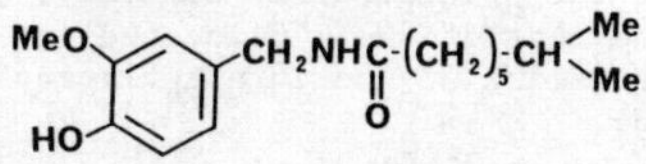

Homodihydrocapsaicin

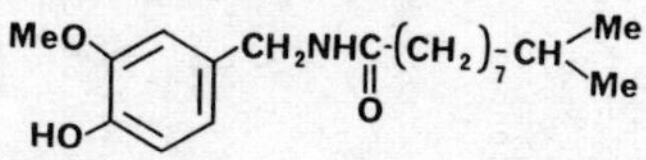

(The even-numbered acyl moieties are derived from valine, while the odd-numbered acyl units are derived from leucine. 2-Methylpropanoic acid and 3-methylbutanoic acids were isolated from incubations with valine and leucine respectively. The results are consistent with earlier work on the biosynthesis of these alkaloids [225,226])

L-[U-^{14}C]Valine (fed to intact plants by wick)	Capsaicin (0.45% inc.) Dihydrocapsaicin (0.56% inc.)	227
L-[U-^{14}C]Leucine (wick feeding)	Nordihydrocapsaicin (0.16% inc.)	
L-[U-^{14}C]Isoleucine (wick feeding)	Homocapsaicin II (0.033% inc.)	

(The isoleucine serves as a precursor of 2-methylbutanoic acid which affords acid moieties containing an odd number of carbons. The acid moiety in homocapsaicin II is 8-methyl-6-decenoic acid).

[*methoxy* -^{14}C]Vanillylamine (incubated in cell-free extract obtained from placenta with coenzyme A esters of iso-fatty acids)	Capsaicin and analogs	149

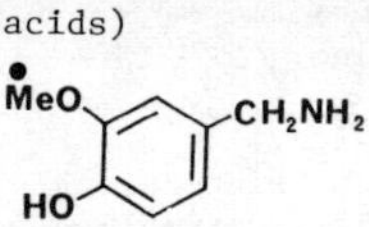

Carex brevicollis		
L-[2-^{14}C]Lysine	Homobrevicolline	228,229

(Material in these articles has been previously reported: *Biosynthesis*, 1980, 6, 219).

Species — *Compound administered* | *Alkaloids labelled* | *Ref.*

Catharanthus roseus

DL-[2-^{14}C, 2',3'-$^{13}C_2$]Tryptophan (fed to intact plants) — Vindoline (0.19% spec. inc.) — 230

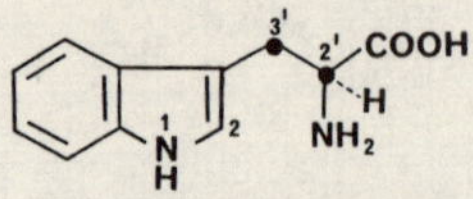

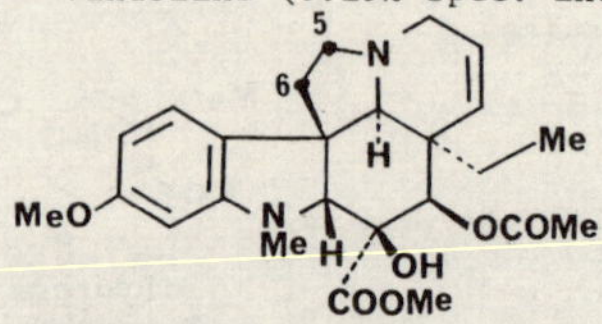

(Since the specific inc. deduced from the ^{13}C nmr was the same as that of the ^{14}C located in the indole nucleus, the incorporation of the intact tryptophan molecule is established).

Catharanthine (0.17% spec. inc.)

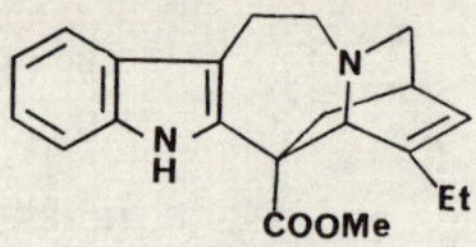

DL-[3'-^{14}C]Tryptophan (fed to a cell-suspension culture) — Akuammicine (0.19% inc.) — 231

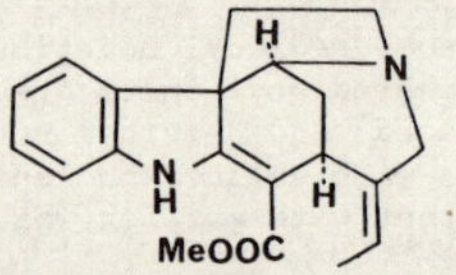

Ajmalicine (0.19% inc.)
Catharanthine (0.15% inc.)
Vindoline (0.0016% inc.)

(The above incorporations were obtained after a 24 hr incubation, somewhat higher inc. were obtained after 48 hrs).

[2'-^{14}C]Tryptamine + secologanin (incubation with a cell-free extract obtained from the leaves of mature plants) — Labelled Ajmalicine, Akuammicine, Tetrahydroalstonine and Vindoline (1.4% inc.) were detected. — 169

Other labelled substrates tested as a precursor of vindoline with this enzyme afforded the following results:

[*methyl*-^{14}C]Methionine — Vindoline (inactive)

S-Adenosyl-L-[*methyl*-^{14}C]-methionine — Vindoline (0.04% inc.)

[*aryl*-^{3}H]Stemmadenine — Vindoline (0.007% inc.)

(This conversion is not considered significant since a similar conversion was obtained with the boiled enzyme).

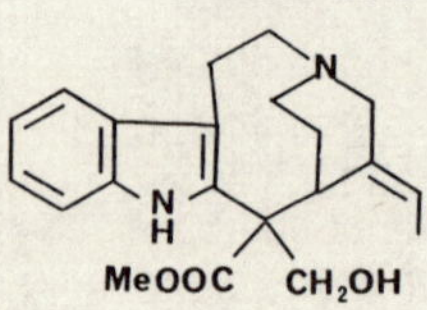

Species — *Compound administered* | *Alkaloids labelled* | *Ref.*

Catharanthus roseus (continued) -

[1,1,7-$^{2}H_3$]-10-Hydroxygeraniol/nerol (fed to cell-suspension culture) — Ajmalicine — 232

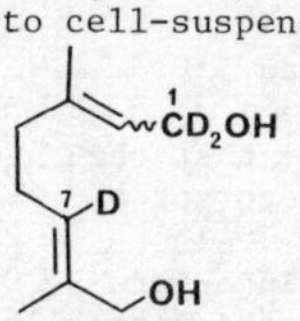

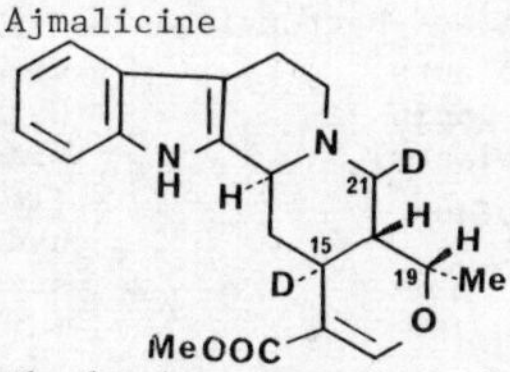

(Mass spectrum consistent with the incorporation of two deuteriums at C-15 and C-21. A 50% inc. of the precursor was obtained).

Strictosidine lactam

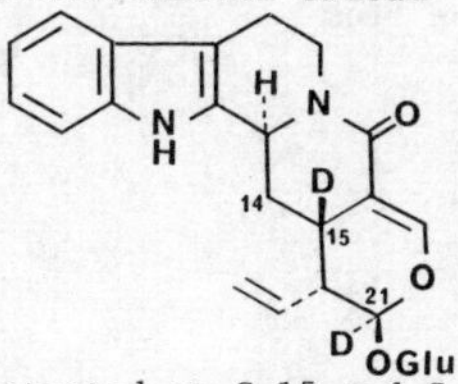

(A 80% inc. with deuterium detected at C-15 and C-21 by mass spectrometry).

[1-^{3}H]-10-Hydroxygeranial/neral (fed to intact plants) — Catharanthine (0.22% inc.), Vindoline (0.14% inc.), Loganin (0.91% inc.) — 233

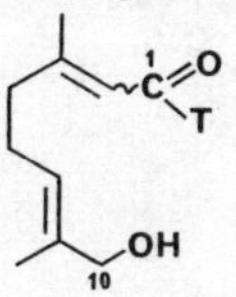

[1-^{3}H]-10-Oxogeraniol/nerol (fed to intact plants) — Catharanthine (0.12% inc.), Vindoline (0.13% inc.), Loganin (0.23% inc.)

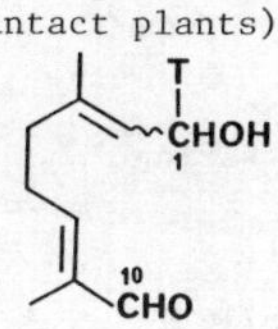

[1-^{3}H]-10-Oxogeranial/neral (fed to intact plants) — Catharanthine (0.09% inc.), Vindoline (0.12% inc.), Loganin (0.28% inc.)

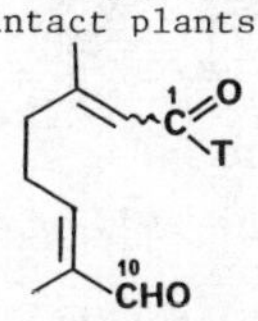

[*O-methyl*-^{14}C, 6-^{3}H]-Hydroxyloganin (fed to intact plants) — Catharanthine (< 0.02% inc.), Vindoline (0.01% inc.) — 234

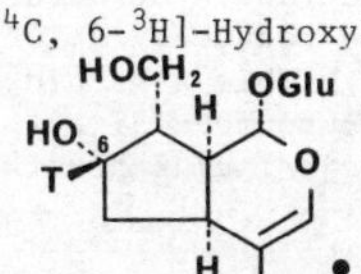

Species *Compound Administered*	*Alkaloids labelled*	*Ref.*
Catharanthus roseus (continued) -		
[*O-methyl*-^{14}C]-6-*epi*-Hydroxyloganin (fed to intact plants)	Catharanthine (< 0.01% inc.) Vindoline (< 0.01% inc.)	234
[*O-methyl*-^{14}C]Loganin (fed to intact plants) 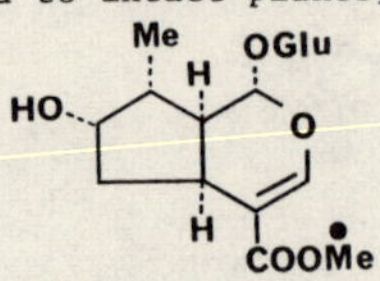	Catharanthine (0.26% inc.) Vindoline (0.13% inc.) (It thus seems unlikely that the hydroxyloganin is a precursor of secologanin. Inouye[235] also failed to observe the incorporation of this hydroxyloganin into secologanin in *Lonicera morrowii*).	
[1'-^{14}C]Tryptamine + inactive secologanin + NADPH (enzyme preparation from cell culture)	Ajmalicine 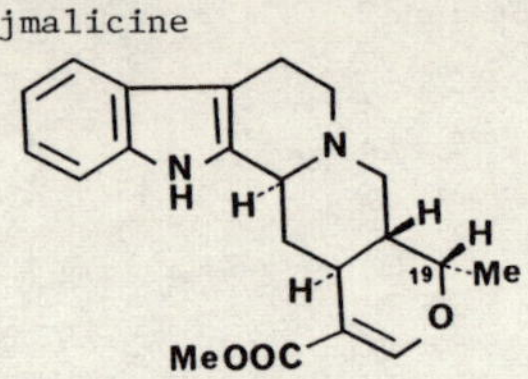 19-*epi*-Ajmalicine 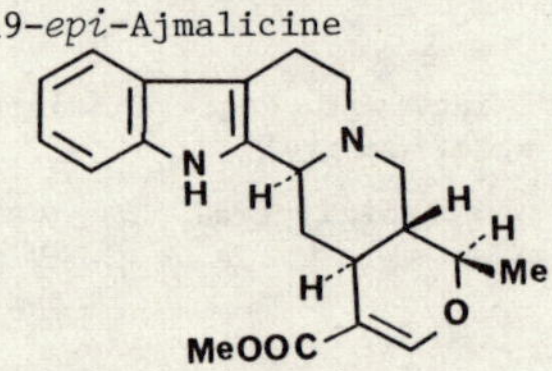 Tetrahydroalstonine 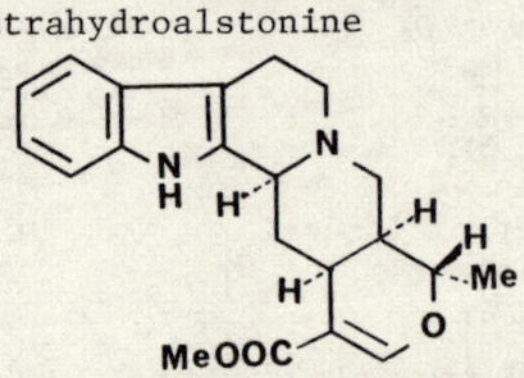 4,21-Dehydrogeissoschizine 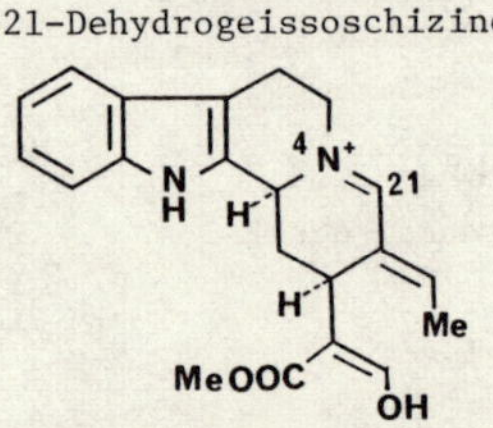	161

(Activity was detected in this alkaloid after a short incubation with the enzyme).

4,21-Dehydrogeissoschizine (with enzyme under NADPH regenerating conditions)	Geissoschizine (above structure with 4-21 bond reduced. A 10% inc. obtained. This alkaloid is not considered to be a direct precursor of ajmalicine or its isomers).	

Species — *Alkaloids labelled* — *Ref.*
Compound Administered

Catharanthus roseus (continued) -

4,21-Dehydrogeissoschizine (enzyme with no NADPH) — Cathenamine (74% inc.) — 161

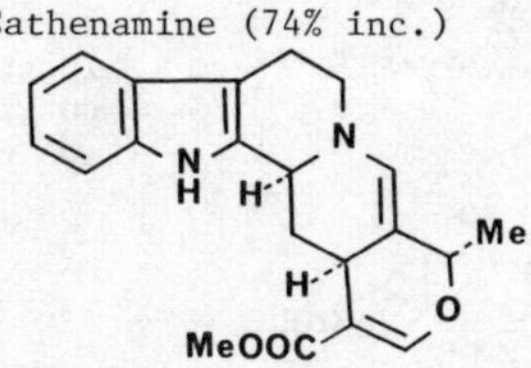

4,21-Dehydrogeissoschizine (enzyme in D_2O + NADPH) — Ajmalicine; 19-*epi*-Ajmalicine; Tetrahydroalstonine

(One deuterium is incorporated into each of the above alkaloids at the C-19 methyl group, indicating that the precursor reverts to 4,21-dehydrocorynantheine prior to incorporation).

[*aryl*-^{3}H]Geissoschizine (incubated with enzyme isolated from cell suspension culture) — In the presence of $NADP^+$, dehydrogenation occurred to yield 4,21-dehydrogeissoschizine which cyclized to cathenamine isomers (epimeric at C-19) on basification and extraction with toluene). — 165

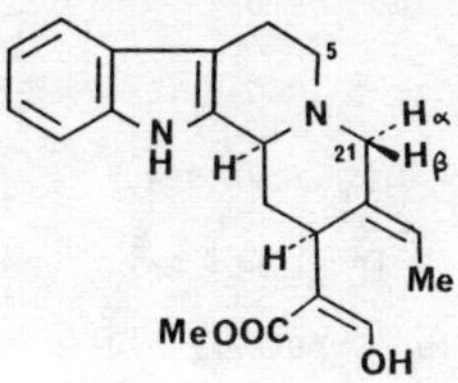

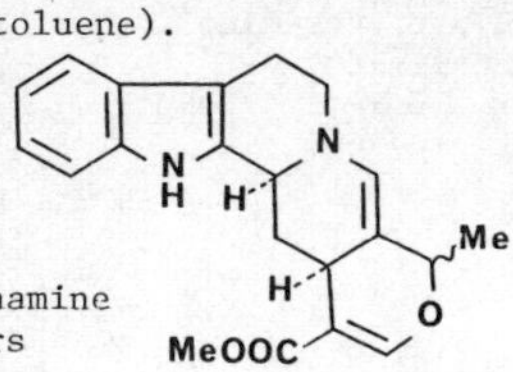

Cathenamine isomers

[5-^{14}C, 21α-^{3}H]Geissoschizine (with enzyme from cell suspension culture + $NADP^+$ + NADPH). — Ajmalicine; Tetrahydroalstonine; 19-*epi*-Ajmalicine — 160

(Incoporation of ^{14}C occurred with > 90% loss of ^{3}H. The same result was obtained in feeding this precursor to *C. roseus* seedlings, indicating that dehydrogenation occurs with loss of the α-hydrogen at C-21 prior to incorporation).

[*aryl*-^{3}H]Geissoschizine (enzyme from callus tissue) — Isositsirikine — 166

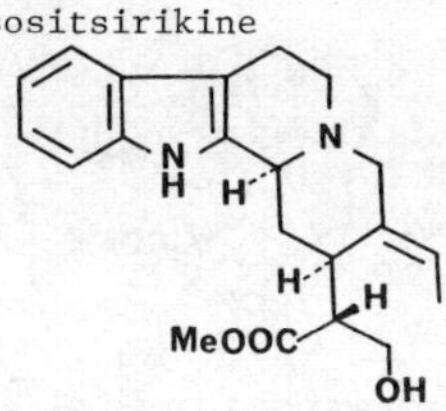

(Incorporations as high as 12.4% were obtained in the presence of NADPH. Removal of low molecular weight compounds with Sephadex G-15 or charcoal reduced the incorporation significantly).

[2'-^{14}C]Tryptamine + Secologanin + NADPH (enzyme from callus tissue) — Isositsirikine (1.8% inc.)

Species / *Compound Administered*	*Alkaloids labelled*	*Ref.*
Catharanthus roseus (continued) -		
[6-^{14}C]Strictosidine (incubate with cell-free extract + NADPH)	Isositsirikine (2.8% inc.) (When this precursor was fed to cut stems of *C. roseus* a 0.1% inc. was obtained).	166

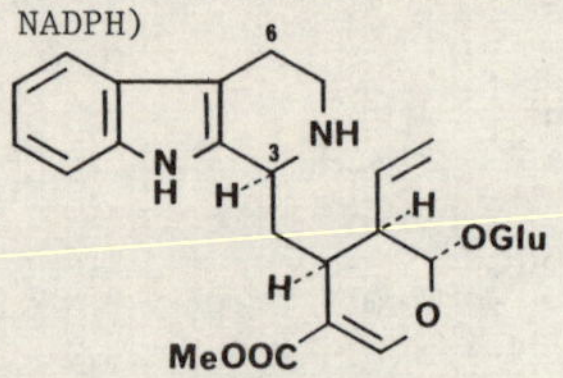

[6-^{14}C, 3-^{3}H]Strictosidine (fed to intact plants)	Ajmalicine (0.46% inc. with 99% ret. of ^{3}H) Catharanthine (0.03% inc. with 103% ret. of ^{3}H) Vindoline (0.19% inc. with 103% ret. of ^{3}H)	236
[6-^{14}C, 3-^{3}H]Vincoside (3β-epimer of strictosidine, fed to intact plants)	Negligible (< 0.001%) inc. into ajmalicine, catharanthine or vindoline.	
[5-^{14}C, 14-^{3}H]Strictosidine (fed to intact plants)	Ajmalicine (0.1% inc., 100% ret. of ^{3}H) Akuammicine (0.93% inc., 94% ret. of ^{3}H) Catharanthine (0.1% inc., 48% ret. of ^{3}H) Vindoline (0.49% inc., 50% ret. of ^{3}H)	173
[*methyl*-^{14}C, 21'α-^{3}H]Anhydro-vinblastine (The carbomethoxy group of the vindoline residue is labelled with ^{14}C. Administered to cell-free prep.).	Vinblastine	170,171

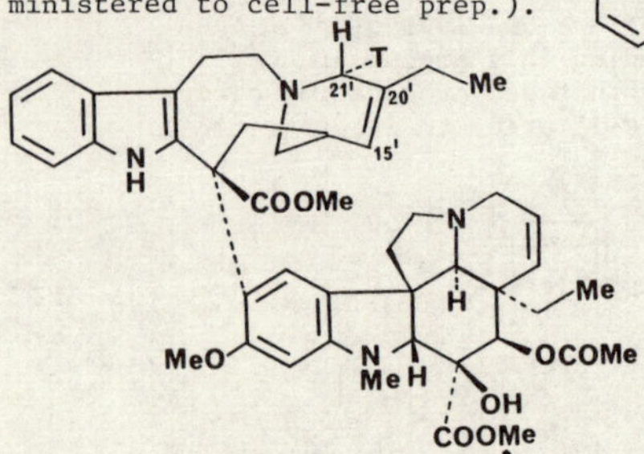

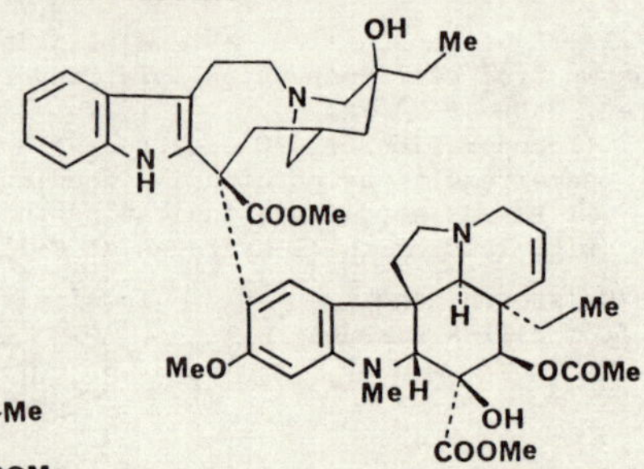

(With incorporations as high as 1.9% all the ^{3}H was retained relative to the ^{14}C. It seems unlikely that the reaction proceeds via an *N*-oxide. A pathway compatible with the results would be direct hydration of the $\Delta^{15'-20'}$ bond, or reduction followed by hydroxylation).

Species / *Compound Administered*	*Alkaloids labelled*	*Ref.*
Catharanthus roseus (continued) -		
	Leurosine (1.4% inc.) 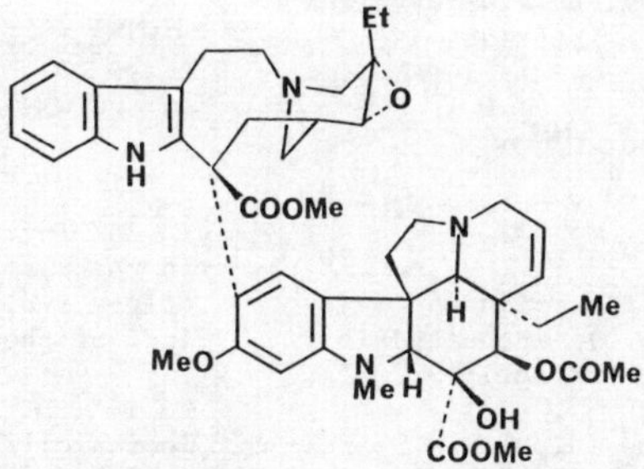	170,171
A similar degree of incorporation (1.2%) was obtained with a boiled enzyme.		
[*aryl*-^{3}H]Anhydrovinblastine (feeding done with cell-free extract from intact plants and a cell-suspension culture [cell-line 916] which normally produces the dimeric indole alkaloids)	Catharine (8-15% spec. inc.) 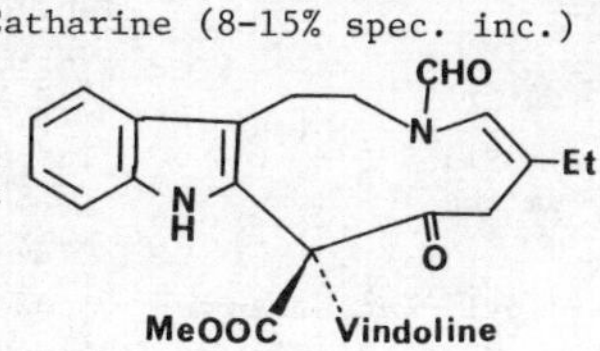 Leurosine (17.0% inc.) Vinblastine (1.84% spec. inc.)	172
[*aryl*-^{3}H]-20'-Deoxyleurosidine (fed in intact plants) 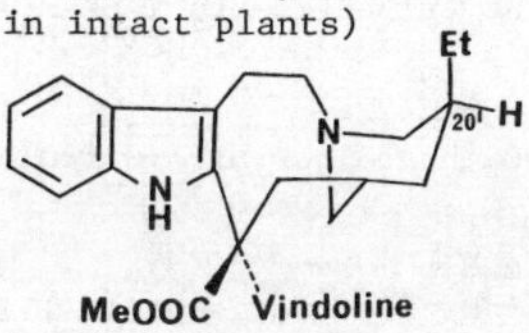	Vinblastine (0.6% inc. The significance of this transformation is not clear.)	237
[*acetyl*-1-^{14}C]-20'-Deoxyleurosidine (label on the acetyl group of the vindoline residue)	Vinblastine (0.4% inc.)	
[*aryl*-^{3}H]Catharanthine + [*acetyl*-1-^{14}C]Vindoline (^{3}H/^{14}C = 23.6) (incubated in cell-free extract from intact plants)	Anhydrovinblastine (^{3}H/^{14}C = 20.8 with an inc. (^{14}C) of 0.32%) Leurosine (^{3}H/^{14}C = 21.1, with an inc. (^{14}C) of 0.30%)	169

Species / *Compound Administered*	*Alkaloids labelled*	*Ref.*
Cephalosporium acremonium		
δ-(L-α-Amino[4,5-^{3}H]adipyl)-L-cysteinyl-D-[4,4-^{3}H]valine (LLD-ACV-tripeptide) 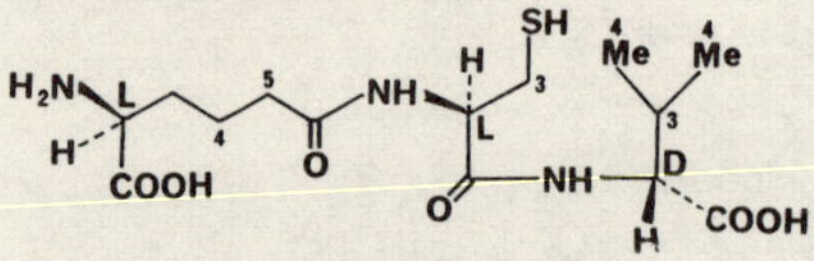 (α-aminoadipic/valine activity = 1.2 ± 0.3) (incubated in cell-free extract).	Isopenicillin N 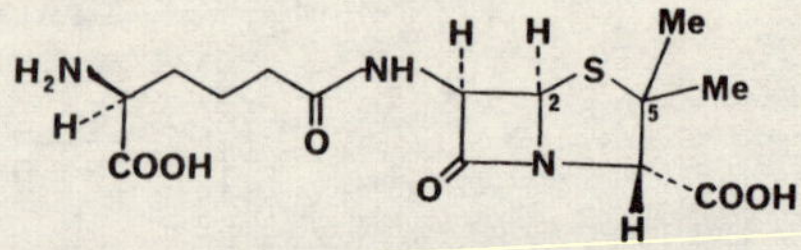 (∿ 20% inc. The ratio of activity in the α-amino-adipic/penicillaminic acid = 1.5, indicative of the direct inc. of the tripeptide).	174,175
δ-(L-α-Amino-adipyl)-L-[3-^{13}C]-cysteinyl-D-[3-^{13}C]valine (fed to cell-free system)	Isopenicillin N (∿ 80% inc. The conversion was carried out in the probe of an n.m.r. spectrometer. The ^{13}C signals due to the enriched carbons in the tripeptide decreased whilst the signals at C-2 and C-5 in the isopenicillin N increased).	176
δ-(L-α-Amino-adipyl)-L-[3-^{13}C]-cysteinyl-D-valine	Isopenicillin N (∿ 80% inc. Enrichment at C-5 observed by n.m.r. spectroscopy).	
δ-(L-α-Amino-adipyl)-L-cysteinyl-D-[2-^{2}H, *methyl*-^{2}H$_6$]valine (fed to cell-free extract)	Isopenicillin N (Incorporation of all seven deuterium atoms established by n.m.r.).	177
LLD-ACV-tripeptide + T_2O	Isopenicillin N (Negligible inc. of ^{3}H. This result and the previous feeding indicate that none of the hydrogens at C-2 or C-4 of the valine moiety are lost in the conversion of the tripeptide to isopenicillin N).	
δ-(L-α-Amino[1,1,6-^{17}O,^{18}O]adipyl)-L-cysteinyl-D-valine (fed to cell-free extract)	Isopenicillin N (conversion with no loss of ^{17}O/^{18}O)	178
LLD-ACV-tripeptide + [O^{17},O^{18}]H_2O	Isopenicillin N (No inc. of ^{17}O/^{18}O into isopenicillin N).	

(These results preclude any mechanism for the formation of isopenicillin N which involves a dehydration-hydration step. Thus the formation of intramolecular thiazolines, ortho-thioamides etc. is inconsistent with the data).

Structural analogs of the LLD-ACV-tripeptide	Isopenicillin N analogs	180

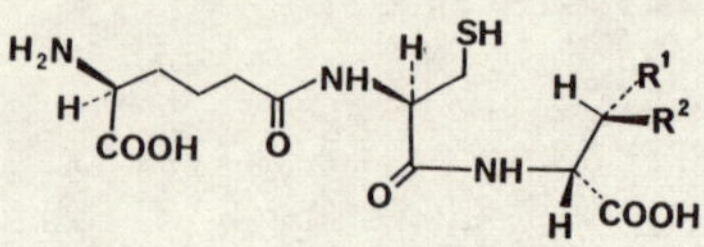

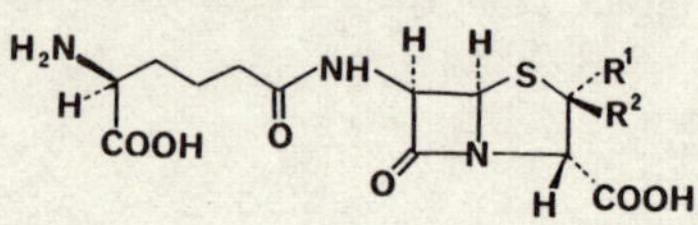

(See the next page for the definition of R^1 and R^2)

Species *Compound Administered*	*Alkaloids labelled*	*Ref.*
Cephalosporium acremonium (continued) -		
R^1 = Me, R^2 = Et; R^1 = Et, R^2 = Me; R^1 = H, R^2 = Me (fed to cell-free extract, non-labelled analogs)	(The three analogs were produced in this cell-free system).	180
L-[1-^{14}C, 3-^{3}H]Valine	LLD-ACV-tripeptide (The L-Valine was incorporated into the D-Valine residue of the tripeptide with complete ret. of ^{3}H).	238
(2*S*,3*S*)-[4-^{13}C]Valine (fed to β-lactam-negative mutant)	LLD-ACV-tripeptide (Hydrolysis of the tripeptide afforded (2*R*,3*S*)-[4-^{13}C]Valine, showing retention of configuration at C-3 of the valine residue).	239
δ-(L-α-Amino-adipyl)-L-cysteinyl-D-3-hydroxyvaline 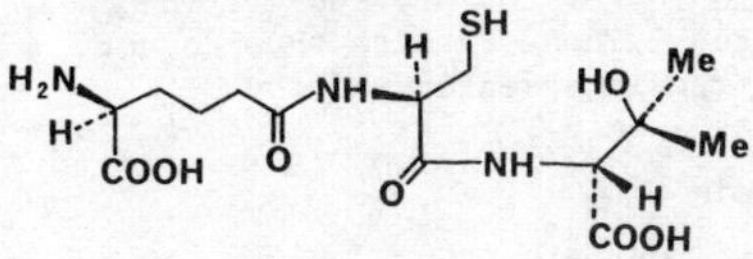	Isopenicillin N (no incorporation)	181
2,2-Dimethyl-6-(L-α-amino-adipyl)-hexahydro-1,4-thiazepin-5-one-3-carboxylic acid 1-oxide 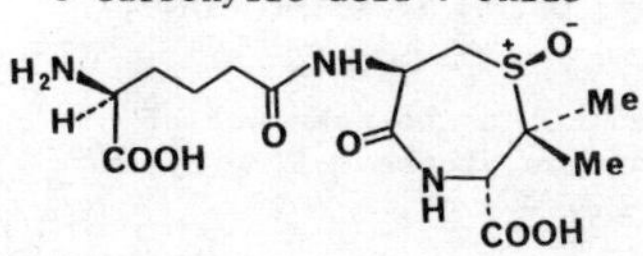	Isopenicillin N (no incorporation)	
2,2-Dimethyl-6-(L-α-amino-adipyl)-hexahydro-1,4-thiazepin-5-one-3-carboxylic acid 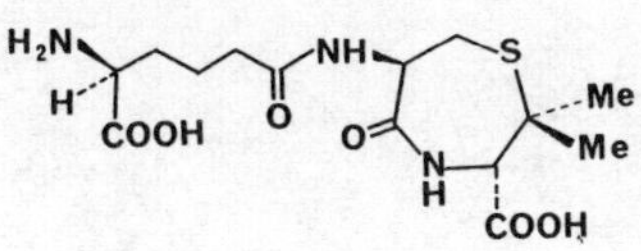	Isopenicillin N (no incorporation)	
Isopenicillin N (fed to cell-free extract)	Penicillin N (No significant ring expansion to deacetoxy-cephalosporin occurred, presumably because Fe^{II}, ascorbic acid, and ATP are needed for this reaction).	185
[6α-^{3}H]Penicillin N 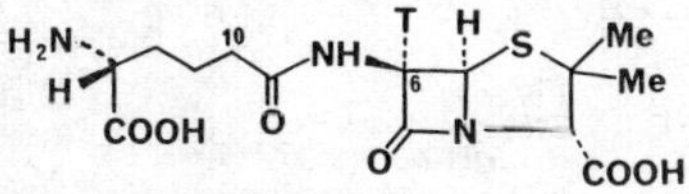	Deacetoxycephalosporin C (~ 10% inc., C-7:97% of the ^{3}H)	186

Species / *Compound Administered*	*Alkaloids labelled*	*Ref.*
Cephalosporium acremonium (continued) -		
[10-^{14}C, 6α-^{3}H]Penicillin N	Deacetoxycephalosporin C (Incorporation with complete ret. of ^{3}H) 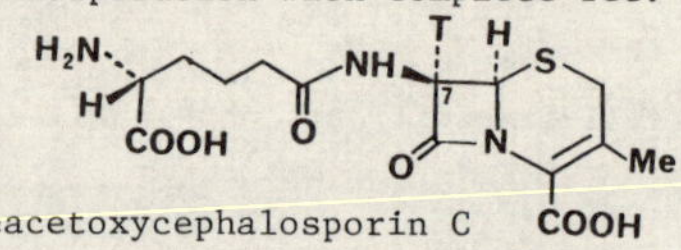	186
[7α-^{3}H]-3-Methylene isomer of deacetoxycephalosporin C 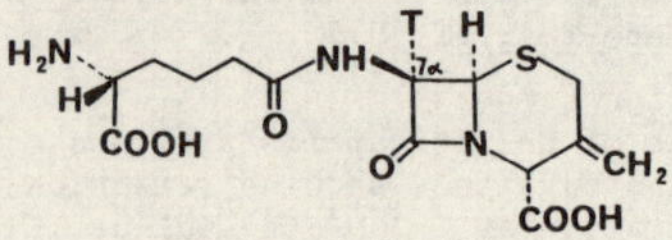	Deacetoxycephalosporin C 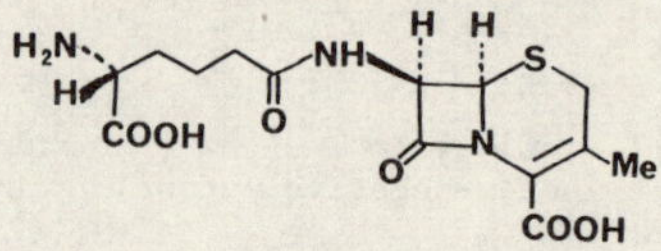 (No incorporation. Also when unlabelled penicillin N was converted to deacetoxy-cephalosporin C in the presence of T_2O, less than 20% of the tritium was incorporated. These results thus do not support the intermediacy of the 3-methylene isomer).	240
[$^{18}O_2$]Oxygen	Cephamycin C 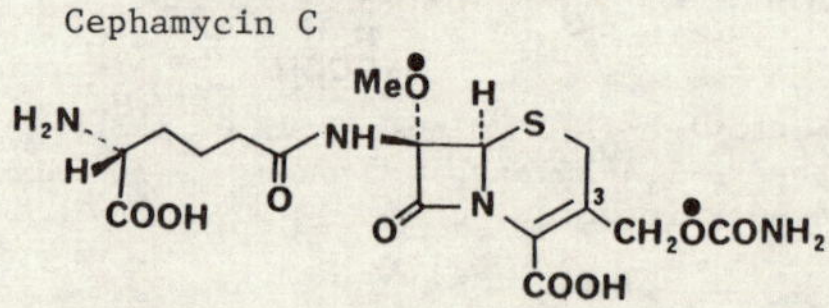	241

(∿ 33% spec. inc. Mass spectrometry indicated the presence of ^{18}O at the 7-methoxy and the oxygen function at the C-3 methylene group)

Cephalotaxus harringtonia		
Some of the data in these publications has been previously reported[242] and abstracted in *Biosynthesis*, 1977, 5, 168.	Cephalotaxine (R = H) 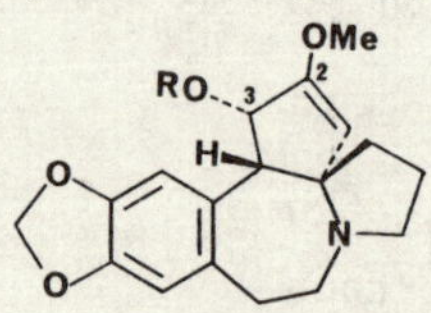	243,244
DL-[9-^{14}C]Deoxyharringtonic acid 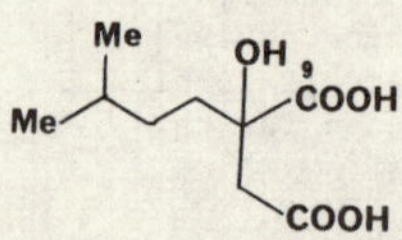	Isoharringtonine 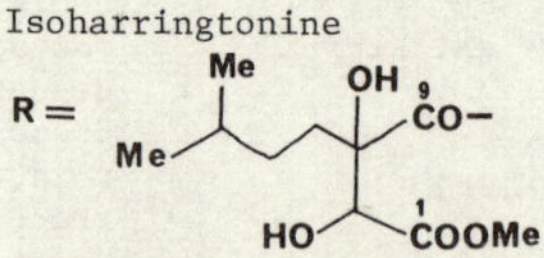 (0.43% inc.,C-9:97%) Harringtonine 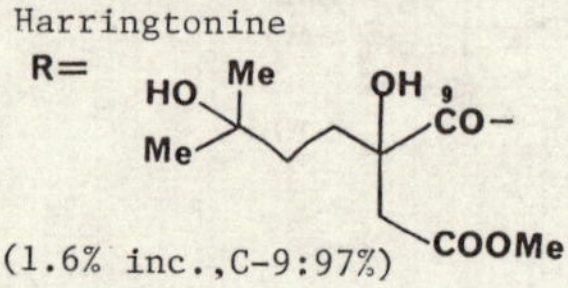 (1.6% inc.,C-9:97%)	

Species *Alkaloids labelled* *Ref.*
Compound Administered

Cephalotaxus harringtonia (continued) -

[9-^{14}C]-2-Hydroxy-3-carboxy-6-methylheptanoic acid — Isoharringtonine (No inc. The lack of incorporation of this compound favors a biosynthetic pathway whereby deoxyharringtonic acid is directly hydroxylated). 243

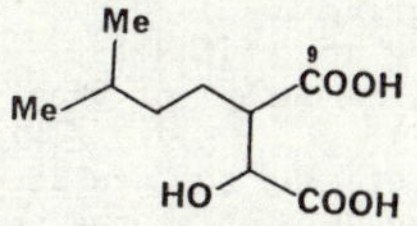

[1'-^{14}C, 3-^{3}H]Deoxyharringtonine — Harringtonine (^{14}C : 0.60% inc.) (^{3}H : ^{14}C = 15.3). One possible explanation for this result is that there is a direct conversion of the acylated precursor to the harringtonine, accompanied by some deacylation of the harringtonine to [3-^{3}H]cephalotaxine followed by reacylation with inactive harringtonic acid. The recovered deoxyharringtonine had almost the same ^{3}H / ^{14}C ratio (6.5) as the administered precursor).

R =

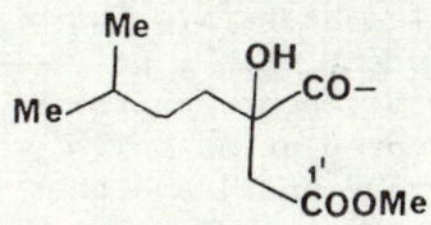

(^{3}H/^{14}C = 6.4)

Most of the data has been previously reported and discussed in *Biosynthesis*, 1977, 5, 140-142; *ibid.*, 1980, 6, 229-230.		
DL-(3'*RS*)-[4-^{14}C, 3'-^{3}H]-Phenylalanine	Cephalotaxine (0.005% inc. 46% ret. of ^{3}H, C-3 : 100%).	244
DL-(3'*RS*)-[3,5-^{14}C, 3'-^{3}H]-Phenylalanine	Cephalotaxine (0.003% inc. 98% ret. of ^{3}H, C-2 : 98.5%).	
(*RS*)-[3-^{14}C]Cephalotaxine	Cephalotaxinone (C-3 is a C=O group and had 95% of activity, 0.44% inc.) Demethylcephalotaxinone (OH at C-2, C-3 is a C=O group, 0.045% inc. with 99% of the activity at C-3).	
(*RS*)-[3-^{14}C]Cephalotaxinone	Cephalotaxine (28% inc. C-3 : 99%). Demethylcephalotaxinone (4.6% inc. C-3 : 102%).	
(*RS*)-[3-^{14}C]Demethylcephalotaxinone	Cephalotaxine (no inc.) Cephalotaxinone (no inc.) Demethylcephalotaxine (13% inc. C-3 : 98%).	
(*RS*)-[3-^{14}C]Demethylcephalotaxine	Cephalotaxine (no inc.).	

Chaetomium globosum

	% inc. into (2)
Sodium [1-^{14}C]Acetate	2.8
Sodium [2-^{14}C]Acetate	4.1
Sodium [1-^{14}C]Propionate	0.13
[*methyl*-^{14}C]Methionine	16.9
DL-[3'-^{14}C]Tryptophan	11.7

Chaetoglobosin A (1) (R = H) 245

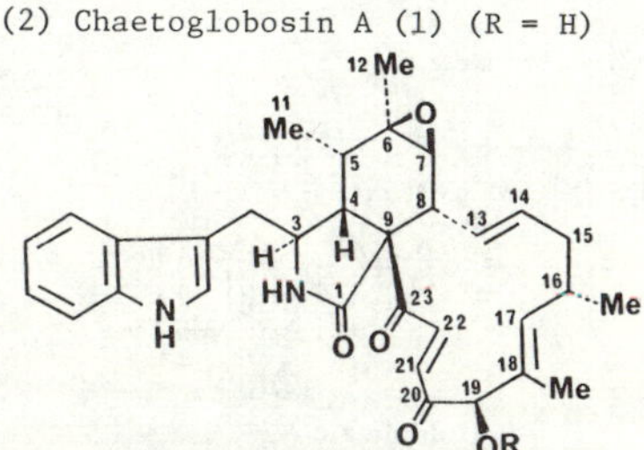

19-*O*-Acetyl-chaetoglobosin A (2) (R = COMe)

Species *Compound Administered*	*Alkaloids labelled*	*Ref.*
Chaetomium globosum (continued) -		
[1-^{13}C]Acetate	(^{13}C n.m.r. indicated enrichment at C-1,5,7,13,15,17,19,21,23 and the C=O of the acetyl group in 2).	245
[2-^{13}C]Acetate	(Enrichment detected at C-6,8,9,11, 14,16,18,20,22 and the methyl of the *O*-acetyl group in 2).	
[1,2-$^{13}C_2$]Acetate	(Well-defined triplets due to "isolated" contiguous ^{13}C atoms were only observed when the labelled material was fed along with unlabelled acetate. The results were consistent with a polyketide formed from 9 acetate units).	
[2-^{13}C,2-$^{2}H_3$]Acetate	(Deuterium is only present at C-11 [the acetate "starter" unit] and the methyl of the acetyl group of 2).	
[*methyl*-^{13}C]Methionine	(enrichments at C-12 and the methyl groups at C-16 and C-18).	
[*methyl*-^{13}C, $^{2}H_3$]Methionine	(The ^{13}C n.m.r. consistent with retention of all three deuterium atoms during the *C*-methylation of the polyketide precursor).	
L-[3'-^{14}C, 2'-^{2}H]Tryptophan	(All the deuterium was retained and located at C-3).	
L-[3'-^{14}C, 2'-*amino*-^{15}N]-Tryptophan	(Both isotopes were incorporated to about the same extent; ^{14}C: 31.2%, ^{15}N: 34.2% spec. inc.).	

These results are consistent with the following biosynthetic scheme:

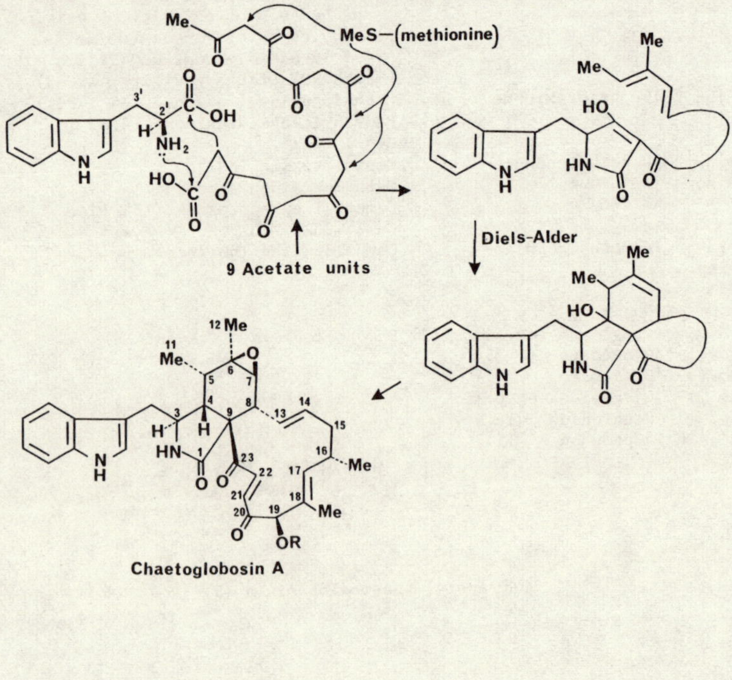

Chaetoglobosin A

Species / *Compound Administered*	*Alkaloids labelled*	*Ref.*
Chelidonium majus		
(−)-(6*RS*, 14*S*)-[6-^{14}C, 6-^{3}H$_1$]-Scoulerine 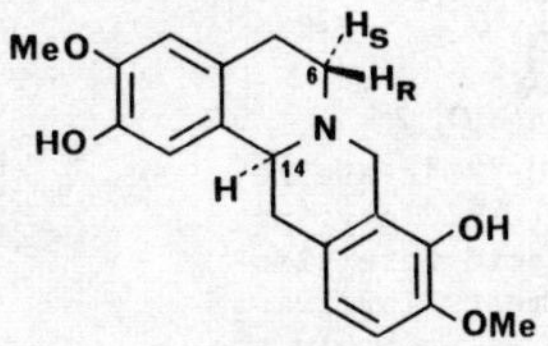	Chelidonine (^{14}C: 0.81% inc. 63% ret. of ^{3}H, somewhat higher than the expected 50%, but consistent with stereospecific loss of one hydrogen from C-6 of scoulerine). 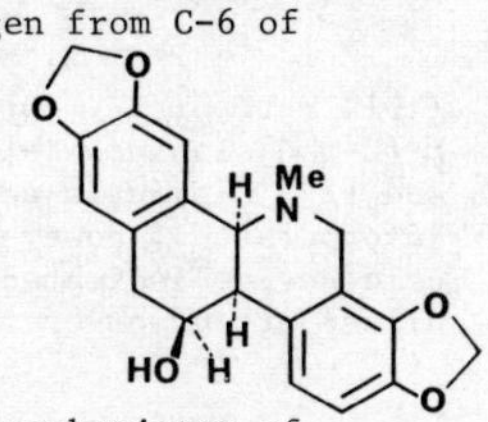 [correct stereochemistry of (+)-chelidonine illustrated [247]]	246
	(*RS*)-Stylopine (^{14}C: 0.18% inc. 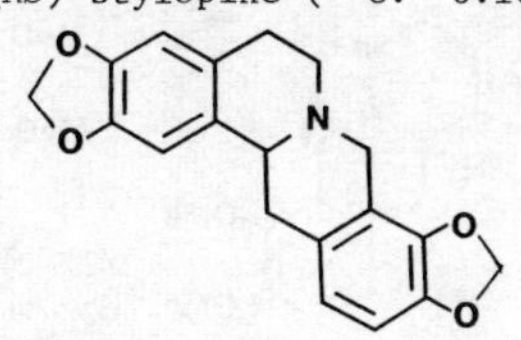 109% ret. of ^{3}H).	
	Protopine 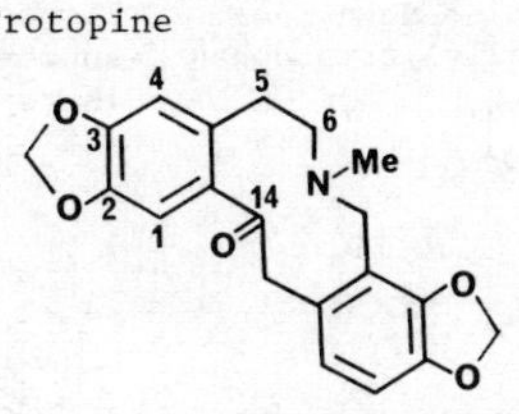 (0.72% inc. 109% ret. of ^{3}H).	
(+)-(6*RS*, 14*R*)-[6-^{3}H]Scoulerine	Chelidonine (<0.01% inc.) (*RS*)-Stylopine (<0.01% inc.) Protopine (<0.01% inc.)	
(6*R*, 14*RS*)-[6-^{14}C, 6-^{3}H]-Scoulerine	Chelidonine (0.22% inc. 100% ret. of ^{3}H) (*RS*)-Stylopine (0.14% inc. 104% ret. of ^{3}H)	
(6*S*, 14*RS*)-[6-^{14}C, 6-^{3}H]-Scoulerine	Chelidonine (5% ret. of ^{3}H) (*RS*)-Stylopine (0.10% inc. 97% ret. of ^{3}H)	
(6*RS*, 14*RS*)-[6-^{14}C, 6-^{3}H]-Scoulerine	Chelidonine (0.29% inc. 64% ret. of ^{3}H) (*RS*)-Stylopine (0.12% inc. 98% ret. of ^{3}H)	

(These results indicate that there is a stereospecific loss of the hydrogen from the *S*-side of C-6 in scoulerine in the conversion to chelidonine).

Species / *Compound Administered*	*Alkaloids labelled*	*Ref.*
Chromobacterium violaceum		
L-[U-^{14}C, 3-^{3}H]Serine	Monobactam SQ 26,180	207

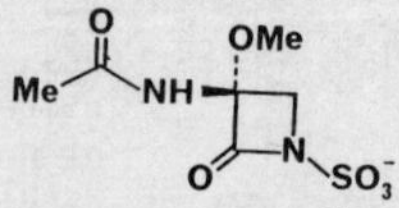

(101% ret. of ^{3}H. Other amino-acids: L-[U-^{14}C]cysteine, L-[U-^{14}C]alanine, L-[U-^{14}C]aspartic acid, [U-^{14}C]glycine, and [2-^{14}C]pyruvic acid, L-[U-^{14}C]-lactic acid were also incorporated. However, competitive experiments involving cold serine indicated that this amino-acid is probably the direct precursor of the β-lactam ring).

Cissampelos pariera
L-[U-^{14}C]Tyrosine — 248

Tetrahydropalmatine(0.013% inc.) Palmatine(0.018% inc.)

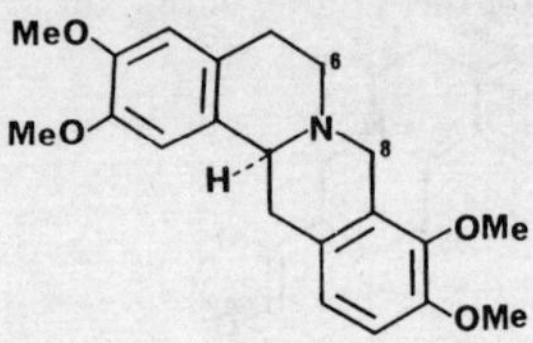

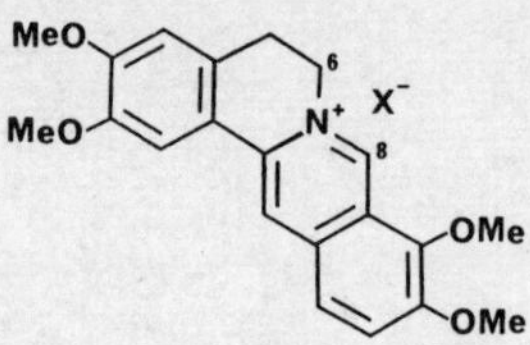

Claviceps species, strain SD-58

[^{2}H$_3$,^{15}N-*N-methyl*]-*N*-Methyl-4-(γ,γ-dimethyl-allyl)tryptophan	Elymoclavine (32.6% inc. in submerged culture, 60.6% inc. in replacement culture)	249,250

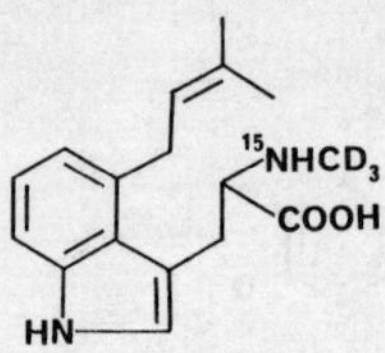

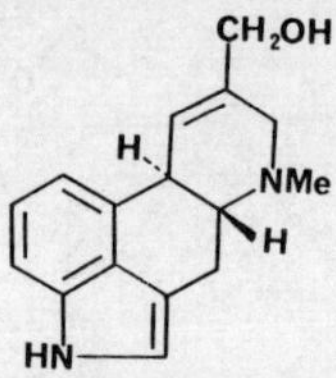

[^{2}H$_3$,^{15}N-*N-methyl*]-*N*-Methyl-4-(γ,γ-dimethylallyl)tryptamine (the above tryptophan derivative lacking the carboxy-group)	Elymoclavine (<2% inc.)	
DL-[1'-^{14}C, 2'-^{3}H]Tryptophan	Clavicipitic acid (a mixture of C-10 epimers, the 10*R* isomer depicted below being the major one)	251

(^{14}C: 4.5% inc. 51% ^{3}H retention, the loss of ^{3}H is attributed to some conversion of the D-tryptophan to L with loss of the α-^{3}H, and some reversible transamination of the L-tryptophan).

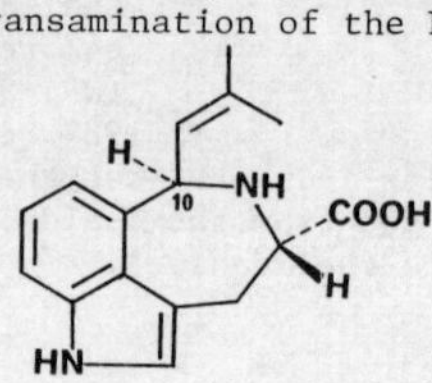

Species / *Compound Administered*	*Alkaloids labelled*	*Ref.*
Claviceps species, strain SD-58 *(continued)* -		
DL-[3'-^{14}C, 2'-^{3}H]Tryptophan	Clavicipitic acid (^{14}C: 4.3% inc., 66% retention of ^{3}H)	251
L-[*methyl*-^{14}C]Methionine	Clavicipitic acid (0.1% inc. The low level of inc. is consistent with the structure of the clavicipitic acid).	
(*RS*)-[2-^{14}C, 5-^{3}H]Mevalonic acid	Clavicipitic acid (^{14}C: 7.1% inc. 75% ret. of ^{3}H)	
(3*RS*)-[2-^{14}C]Mevalonic acid + (3*R*, 5*R*)-[5-^{3}H]Mevalonic acid	Clavicipitic acid (^{14}C: 2.6% inc. 105% ret. of ^{3}H)	
(3*RS*)-[2-^{14}C]Mevalonic acid + (3*R*, 5*S*)-[5-^{3}H]Mevalonic acid	Clavicipitic acid (^{14}C: 43.0% inc. 191% ret. of ^{3}H)	
[^{14}C]Clavicipitic acid (obtained biosynthetically from [2-^{14}C]-mevalonic acid	Elymoclavine (1.4% inc. This incorporation is not considered significant, since genuine precursors of the ergoline ring system usually give incorporations of 20-60%).	
(17*S*)-[17-^{3}H]Chanoclavine I	Elymoclavine (48.8% inc.)	252

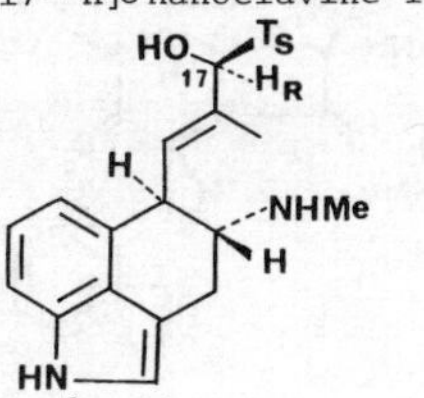

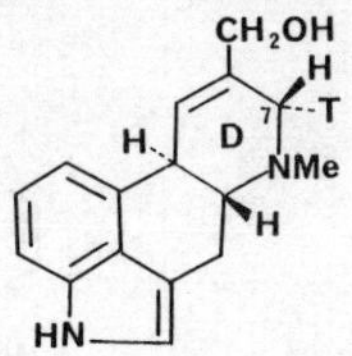

(17*R*)-[17-^{3}H]Chanoclavine-I	Elymoclavine (2.8% inc. These results are consistent with the stereospecific dehydrogenation of chanoclavine-I to chanoclavine-I aldehyde with loss of the *pro-17R* H which then cyclizes with the methylamino-group to generate ring D).	
Claviceps paspali		
[1-^{13}C]Acetate		253,254
[2-^{13}C]Acetate		
[1,2-$^{13}C_2$]Acetate		

Paspaline(1-2% inc.)

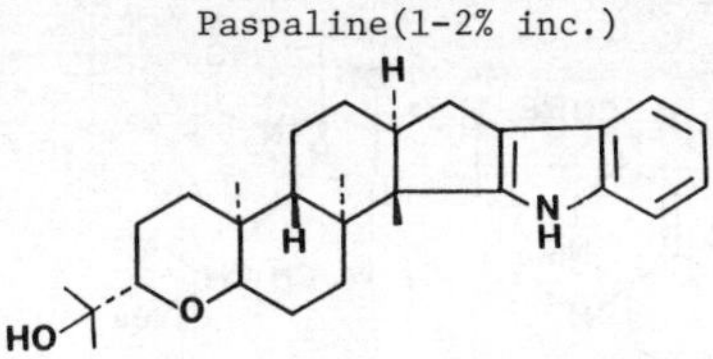

Paspalicine

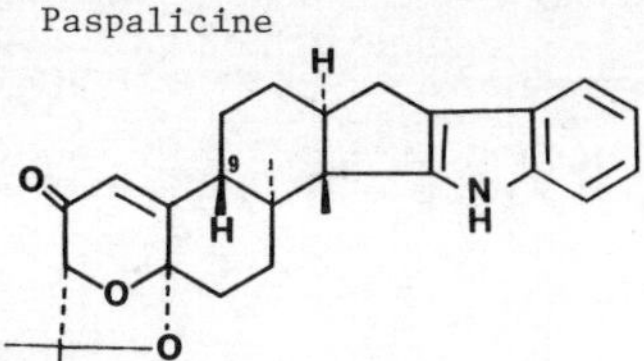

Paspalinine (OH at C-9)

(The pattern of labelling is consistent with the formation of these alkaloids from tryptophan (with loss of the side-chain) and geranyl-geraniol pyrophosphate. Plausible cyclizations and rearrangements of the acyclic diterpene moiety afford the above structures).

Species *Compound Administered*	*Alkaloids labelled*	*Ref.*
Claviceps purpurea L-[U-^{14}C]Phenylalanine (fed to isolated protoplasts)	Ergotamine (0.025% inc. all activity in phenylalanine moiety)	255

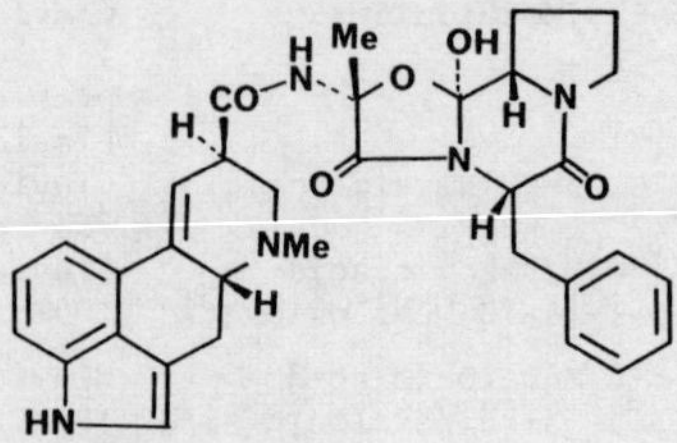

L-[U-^{14}C]Proline	Ergotamine (0.01% inc. These low incorporations were obtained after 4 hr incubations)	
L-[U-^{14}C]Alanine	Ergotamine (inactive)	
(*RS*)-[3-^{14}C]-2-Aminobutanoic acid	Ergostine	256

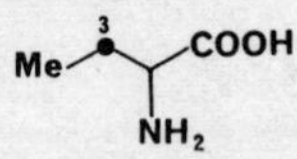

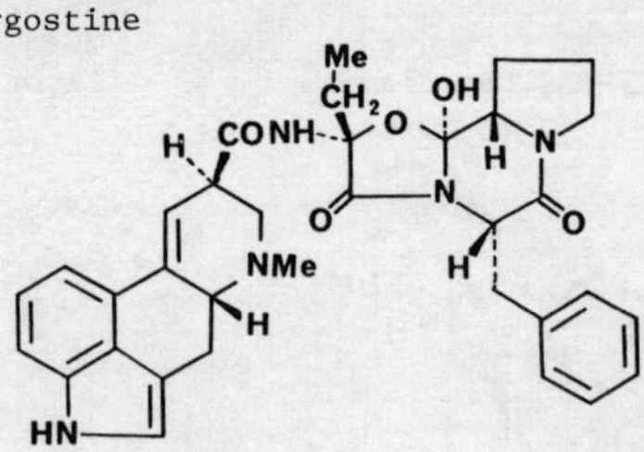

(6.6% of the administered activity was found in this alkaloid which is apparently aberrant, since it had the same specific activity as the precursor).

(*RS*)-[3-^{13}C, 3-^{2}H$_2$]-2-Amino-butanoic acid	Ergostine (Mass spectral analysis indicated the presence of ^{13}C with two deuteriums,	257

indicating that the amino-acid is not incorporated via a 2,3-dehydro derivative).

L-Thiazolidine-4-carboxylic acid (not labelled)	9'-Thiaergosine	258

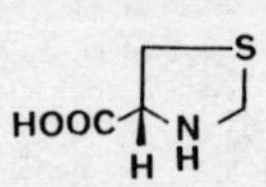

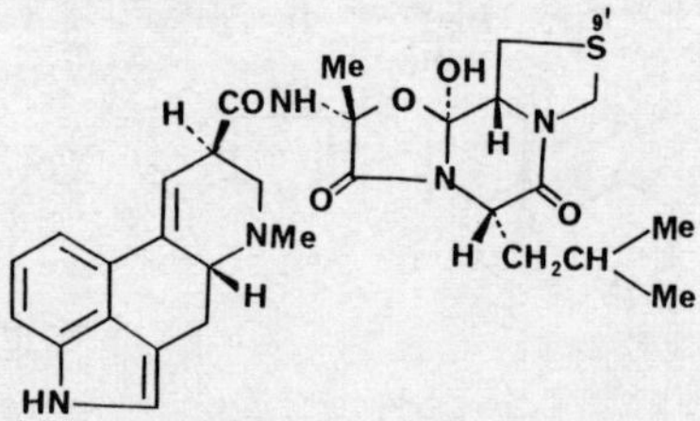

(This sulfur analog of ergosine is produced along with ergosine . A related aberrant biosynthesis is described under *Sphacelia sorghi*).

Species / *Compound Administered*	*Alkaloids labelled*	*Ref.*
Cocculus laurifolius		
L-[U-^{14}C]Tyrosine		
	Magnoflorine(0.079% inc.) 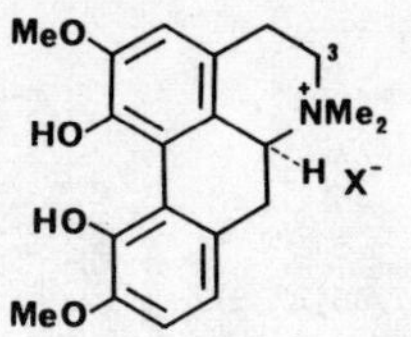 Laurifoline(0.06% inc.) 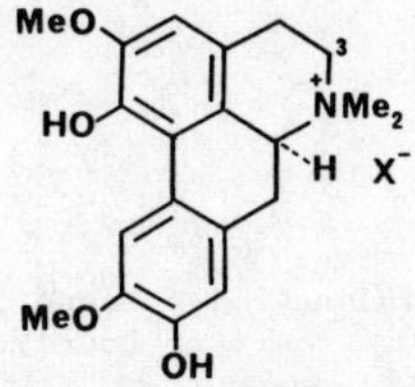 (+)-(*S*)-Reticuline(0.34% inc.) 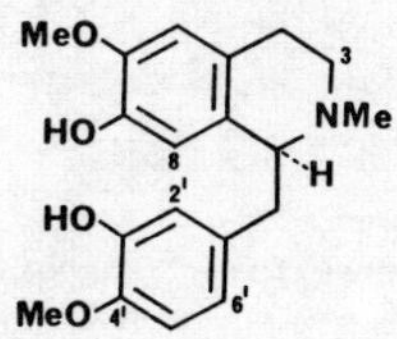	259
[2'-^{14}C]Dopamine	Magnoflorine(0.21% inc.) Laurifoline(0.18% inc.)	
(*RS*)-[2',6',8-$^{3}H_3$]Reticuline	Magnoflorine(1.24% inc.) Laurifoline(0.69% inc.)	
(*RS*)-[*N-methyl*-^{14}C]Reticuline	Magnoflorine(1.17% inc. NMe : 97.6%) Laurifoline(0.52% inc. NMe : 97%)	
(+)-(*S*)-[2',6',8-$^{3}H_3$]Reticuline	Magnoflorine(1.46% inc.) Laurifoline(0.87% inc.)	
(-)-(*R*)-[2',6',8-$^{3}H_3$]Reticuline	Magnoflorine(0.016% inc.) Laurifoline(0.011% inc.)	
(*RS*)-[4'-*methoxy*-^{14}C, 1-^{3}H]-Reticuline	Magnoflorine(^{14}C : 1.7% inc. 93% ret. of ^{3}H) Laurifoline(^{14}C : 0.82% inc. 97% ret. of ^{3}H)	
(*RS*)-[3-^{14}C]Norreticuline	Magnoflorine(0.84% inc. C-3 : 97%) Laurifoline(0.36% inc. C-3 : 98%)	
(*RS*)-[2',6',8-$^{3}H_3$]Reticuline methiodide	Magnoflorine(0.075% inc.) Laurifoline(0.082% inc.)	
(*RS*)-[1-^{3}H]Norlaudanosoline 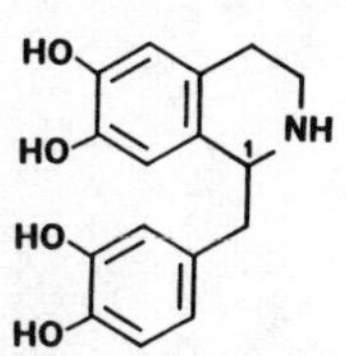	Magnoflorine(0.32% inc.) Laurifoline(0.21% inc.)	

Species *Compound Administered*	*Alkaloids labelled*	*Ref.*
Cocculus laurifolius (continued)		
(*RS*)-[*aryl*-^{3}H]Norprotosinomenine	Magnoflorine(0.005% inc.) Laurifoline(0.098% inc.)	259
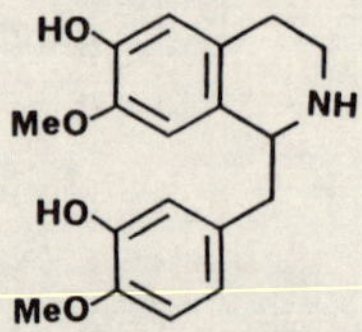		
(*RS*)-[*aryl*-^{3}H]Nororientaline	Magnoflorine(0.0014% inc.) Laurifoline(0.001% inc.)	
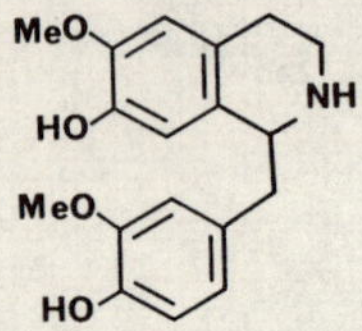		
(*RS*)-[2',6',8-^{3}H$_3$]Laudanosoline	Magnoflorine(0.004% inc.) Laurifoline(0.002% inc.)	
(These results favor the biosynthetic route: tyrosine → norlaudanosoline → (*S*)-reticuline → magnoflorine or laurifoline).		
L-[U-^{14}C]Tyrosine (for alkaloid structures, see under *Cissampelos pariera*)	Tetrahydropalmatine(0.028% inc.) Palmatine(0.032% inc.)	248
[2'-^{14}C]Dopamine	Tetrahydropalmatine(0.18% inc.) Palmatine(0.20% inc.)	
(*RS*)-[2',6',8-^{3}H$_3$]Reticuline	Tetrahydropalmatine(0.34% inc.) Palmatine(0.48% inc.)	
(*RS*)-[*N-methyl*-^{14}C]Reticuline	Tetrahydropalmatine(0.90% inc. C-8: 98%) Palmatine(0.82% inc. C-8: 102%)	
(*RS*)-[4'-*methoxy*-^{14}C, 1-^{3}H]-Reticuline	Tetrahydropalmatine(0.80% inc. 97% ret. of ^{3}H)	
(*RS*)-[3-^{14}C]Reticuline	Tetrahydropalmatine(C-6: 98%) Palmatine(C-6: 97%)	
(+)-(*S*)-[2',6',8-^{3}H$_3$]Reticuline	Tetrahydropalmatine(1.22% inc.) Palmatine(1.26% inc.)	
(-)-(*R*)-[2',6',8-^{3}H$_3$]Reticuline	Tetrahydropalmatine(0.013% inc.) Palmatine(0.02% inc.)	
(*RS*)-[*aryl*-^{3}H]Norprotosinomenine	Tetrahydropalmatine(0.0015% inc.) Palmatine(0.0026% inc.)	
(*RS*)-[*aryl*-^{3}H]Nororientaline	Tetrahydropalmatine(0.0026% inc.) Palmatine(0.0032% inc.)	
(*RS*)-[2',6'-^{3}H$_2$]Norlaudanidine	Tetrahydropalmatine(0.0033% inc.) Palmatine(0.0042% inc.)	
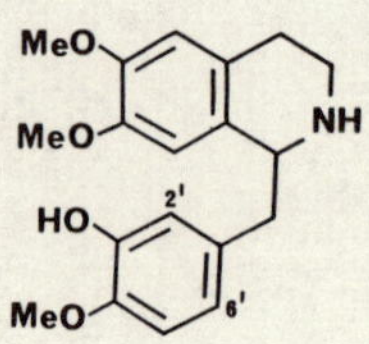		

Species / *Compound Administered*	*Alkaloids labelled*	*Ref.*
Cocculus laurifolius (continued)		
(*RS*)-[1-^{3}H]Norlaudanosoline	Tetrahydropalmatine(0.24% inc. Palmatine(0.32% inc.)	248
(*RS*)-[2',6',8-$^3H_3$]Laudanosine	Tetrahydropalmatine(0.0045% inc.) Palmatine(0.0052% inc.)	
(*RS*)-[3-^{14}C]Norreticuline	Tetrahydropalmatine(0.72% inc.) Palmatine(0.70% inc.)	
[6-^{14}C]Palmatine	Tetrahydropalmatine(0.0006% inc.)	
(-)-(*S*)-[6-^{14}C]Tetrahydro-palmatine	Palmatine(7.52% inc.)	

(These results thus favor the biosynthetic sequence: - Tyrosine → Norlaudanosoline → (*S*)-reticuline (not (*R*), as stated in the original publication) → Tetrahydropalmatine → Palmatine).

L-[U-^{14}C]Tyrosine	Coccuvine(0.11% inc.)	260

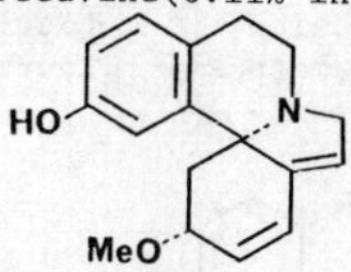

(*RS*)-[1-^{3}H]Norlaudanosoline	Coccuvine(0.20% inc.)
(*RS*)-[1-^{3}H]Norprotosinomenine	Coccuvine(0.38% inc.)
(*RS*)-[7-*methoxy*-^{14}C, 1-^{3}H]-Norprotosinomenine	Coccuvine(0.35% inc. of ^{3}H, no ^{14}C inc.)
(*RS*)-[4'-*methoxy*-^{14}C, 1-^{3}H]-Norprotosinomenine	Coccuvine(0.42% inc. 94% ret. of ^{3}H)
(-)-(*R*)-[*aryl*-^{3}H]Norproto-sinomenine	Coccuvine(0.007% inc.)
(+)-(*S*)-[*aryl*-^{3}H]Norproto-sinomenine	Coccuvine(0.66% inc.)

Late stages in the biosynthesis of these erythrina alkaloids which were established by feeding experiments are depicted below:

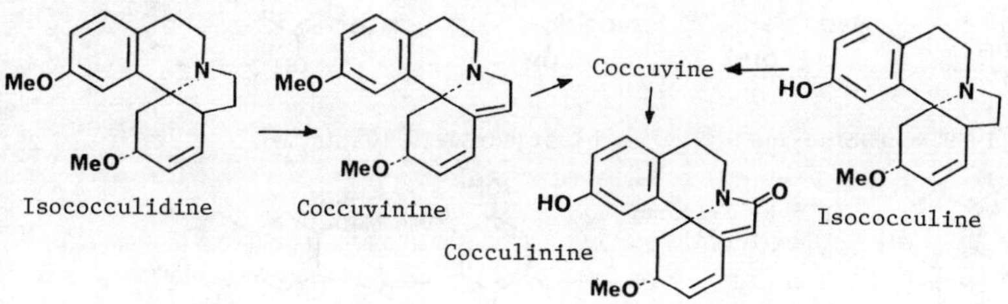

L-[U-^{14}C]Tyrosine	Laurifinine (0.12% inc.)	261

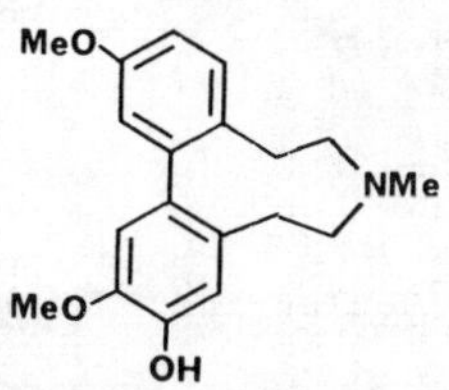

Species / *Compound Administered*	*Alkaloids labelled*	*Ref.*
Cocculus laurifolius (continued) -		
(*RS*)-[1-^{3}H]Norlaudanosoline	Laurifinine (0.32% inc.)	261
(*RS*)-[1-^{3}H]-*N*-Norprotosinomenine	Laurifinine (0.42% inc.)	
(*RS*)-[7-*methoxy*-^{14}C,1-^{3}H]-*N*-Nor-protosinomenine	Laurifinine (0.52% inc., 93% ret. of ^{3}H. However, the validity of this experiment and the following one is questionable because of the high ^{14}C/^{3}H ratio used (27:1)).	
(*RS*)-[4'-*methoxy*-^{14}C,1-^{3}H]-Nor-protosinomenine	Laurifinine (0.59% inc., 94% ret. of ^{3}H)	
(*R*)-[*aryl*-^{3}H]-*N*-Norprotosinomenine	Laurifinine (0.01% inc.)	
(*S*)-[*aryl*-^{3}H]-*N*-Norprotosinomenine	Laurifinine (0.84% inc.)	
(*RS*)-[2',6',8-^{3}H]-N-Norreticuline	Laurifinine (0.006% inc.)	
(*RS*)-[3',8-^{3}H]-N-Nororientaline	Laurifinine (0.003% inc.)	
(*RS*)-[*aryl*-^{3}H]Protosinomenine	Laurifinine (0.002% inc.)	
[*aryl*-^{3}H]-2-(3-Hydroxy-4-methoxy-phenyl)ethyl-2'-(4-hydroxy-phenyl)ethylamine	Laurifinine (0.001% inc.)	

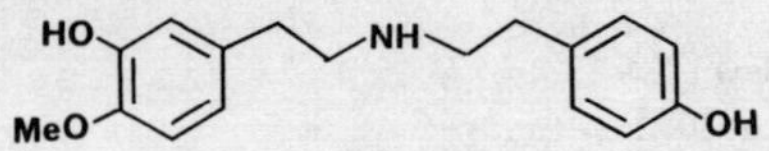

(These experiments are consistent with the biosynthetic pathway: tyrosine → norlaudanosoline → (*S*)-*N*-norprotosinomenine (1) → laurifinine, the terminal steps being illustrated below.)

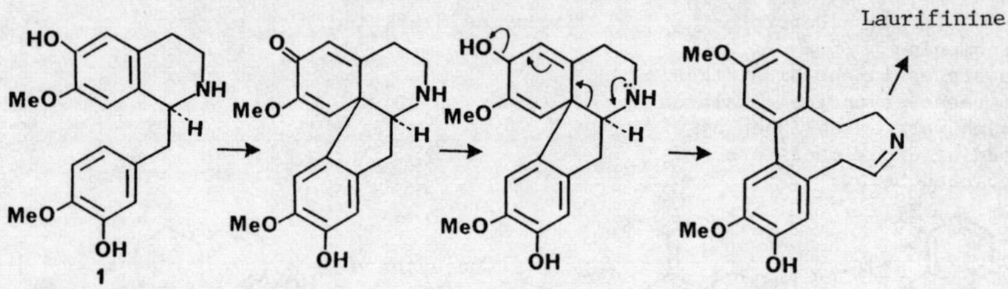

(*RS*)-[1-^{3}H]Norcoclaurine	Tetrandrine(0.10% inc.)	262

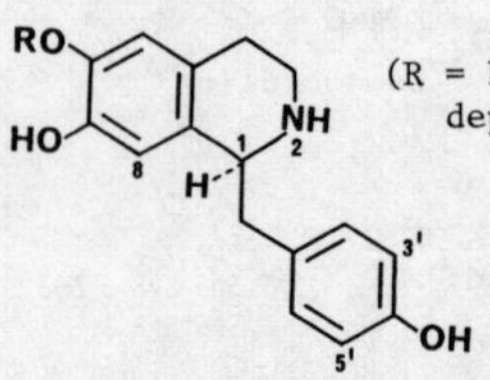

(R = H, *S*-isomer depicted)

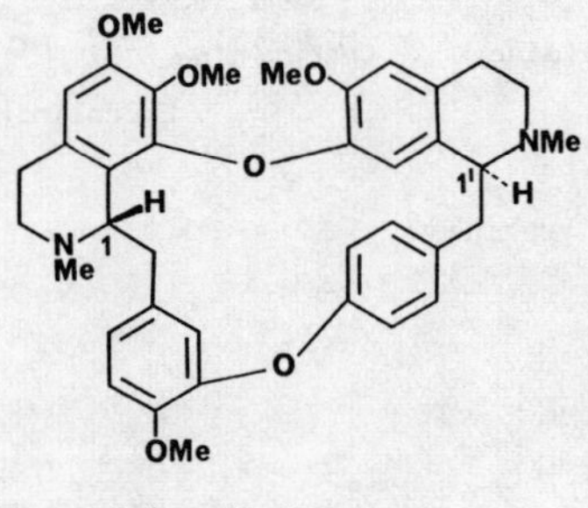

(*RS*)-[3',5',8-^{3}H$_{3}$]Coclaurine (R = Me)	Tetrandrine (0.15% inc.)	

Species / *Compound Administered*	*Alkaloids labelled*	*Ref.*
Cocculus laurifolius (continued) -		
L-[U-^{14}C]Tyrosine	Tetrandrine (0.07% inc.)	262
(*RS*)-[3',5',8-$^{3}H_3$]-*N*-Methylcoclaurine	Tetrandrine (0.21% inc.) (25% of ^{3}H on the left side of molecule and 75% on the right side).	
(*RS*)-[3',5',8-$^{3}H_3$]-*N,O,O*-Trimethylcoclaurine	Tetrandrine (0.002% inc.)	
[*N-methyl*-^{14}C]-1,2-Dehydro-*N*-methylcoclaurinium salt	Tetrandrine (0.17% inc. NMe : 93%)	
(*RS*)-[*N-methyl*-^{14}C, 1-^{3}H]-Coclaurine	Tetrandrine (0.26% inc. 93% ret. of ^{3}H)	
(*S*)-[3',5',8-$^{3}H_3$]-*N*-Methylcoclaurine	Tetrandrine (0.24% inc. 25% of ^{3}H in left side)	
(*R*)-[3',5',8-$^{3}H_3$]-*N*-Methylcoclaurine	Tetrandrine (0.0024% inc.)	
Isotetrandrine is a stereoisomer of tetrandrine in which C-1 and C-1' have the *R* and *S* configurations respectively. Similar tracer work was reported for this alkaloid, also found in *Cocculus laurifolius*, and it was found that the two halves were derived from (*R*) and (*S*)-*N*-methylcoclaurine respectively.		263
(*RS*)-[2',6',8-$^{3}H_3$]Reticuline 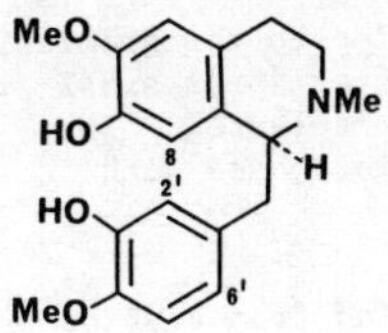 *S*-isomer	Thalicarpine(1.34% inc.) 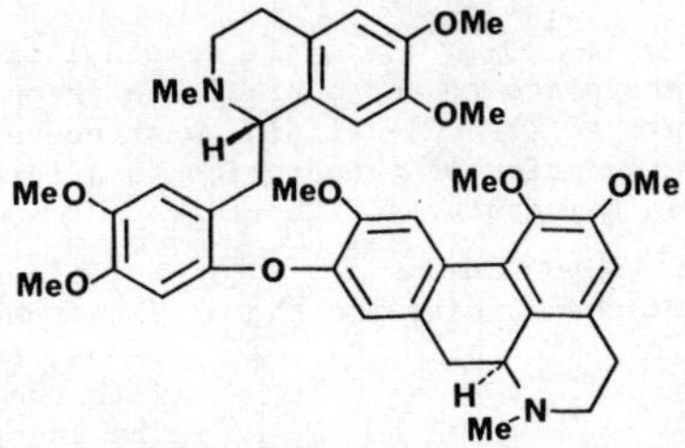 Activity found in both halves, with 2/3 in the benzylisoquinoline part.	264
L-[U-^{14}C]Tyrosine	Thalicarpine(0.10% inc.)	
(*RS*)-[1-^{3}H]Norlaudanosoline	Thalicarpine(0.18% inc.)	
(*S*)-[2',6',8-$^{3}H_3$]Reticuline	Thalicarpine(1.6% inc.)	
(*R*)-[2',6',8-$^{3}H_3$]Reticuline	Thalicarpine(0.016% inc.)	
(*RS*)-[2',6'-$^{3}H_2$]Norlaudanine	Thalicarpine(0.30% inc.)	
(*RS*)-[3',5',8-$^{3}H_3$]-*N*-Methylcoclaurine	Thalicarpine(0.008% inc.)	
(*RS*)-[4'-*O-methyl*-^{14}C, 1-^{3}H]-Norreticuline	Thalicarpine(1.8% inc.) No ^{14}C found in the aporphine half, indicating a selective *O*-demethylation of the 4'-methoxyl group)	
(*RS*)-[3,8-$^{3}H_2$]Boldine 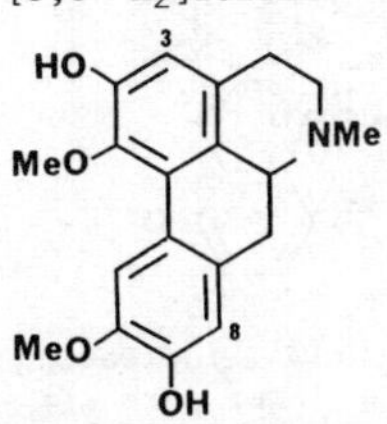	Thalicarpine(0.29% inc.)	

Species / *Compound Administered*	*Alkaloids labelled*	*Ref.*
Cocculus laurifolius (continued) -		
(*S*)-[8-^{3}H]Isoboldine	Thalicarpine(0.51% inc.)	264

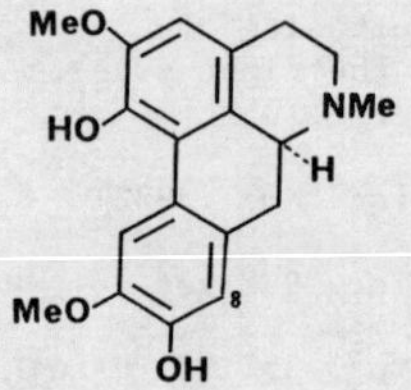

Coffea arabica		
[2-^{14}C]Theobromine (fed to cell-suspension culture)	Caffeine	265

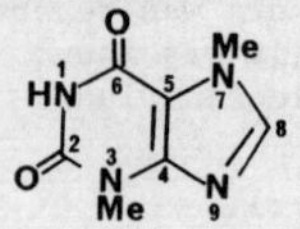

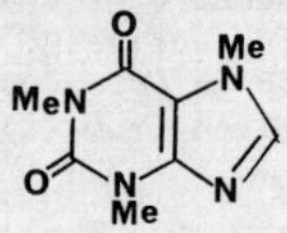

(The reverse reaction, i.e., the demethylation of caffeine to theobromine, does not take place to any appreciable extent. Thus, after 28 days, 84% of administered [2-^{14}C]-caffeine was recovered from the cultures. The rate of methylation of theobromine to afford caffeine increases with increase in temperature of the leaves 266).

L-[*methyl*-^{14}C]Methionine (fed to detached fruits via the petioles for 14-28 hr)	Caffeine (Incorporations varied from 4.9% to 22.5%, the higher level being obtained with the longer feeding time (48 hr). The incorporation was increased by the addition of theobromine and paraxanthine. Theophylline and caffeine reduced the incorporation).	192
Cell-free extract *S*-Adenosyl-L-[*methyl*-^{14}C]-methionine + various xanthines	The following methylations were observed: 7-Methylxanthine → Theobromine 1-Methylxanthine → Theophylline Theobromine → Caffeine Theophylline → Caffeine 7-Methylxanthosine → Theobromine Paraxanthine → Caffeine	
Corydalis ochotensis		
DL-[3'-^{14}C]-3,4-Dihydroxyphenylalanine (DOPA)	Protopine (C-1 to C-6 + C-14 : 103%) Ochotensimine	267

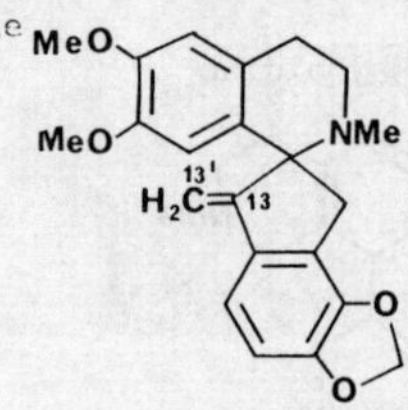

(Only 4% of the activity was found at C-13 + C-13', indicating that DOPA does not serve as a precursor of the lower part of the alkaloid).

Species | *Alkaloids labelled* | *Ref.*
Compound Administered

Corydalis ochotensis (continued) -

L-[*methyl*-^{14}C, ^{3}H]Methionine — Ochotensimine — 267

(The OMe and N-Me groups are formed from the methionine with essentially complete ret. of ^{3}H. The results were also consistent with C-13' being derived from the same source).

Corydalis sempervirens

[2'-^{14}C, 2,6-^{3}H$_2$]Dopamine + 3,4 -Dihydroxyphenyl-acetaldehyde (cell-free system) — (*S*)-Norlaudanosoline (3,4-dihydroxyphenylpyruvate did not serve as a substrate for this enzyme) — 143

Corydalis solida

DL-[3'-^{14}C]-3,4-Dihydroxy-phenylalanine (DOPA) — Corydaline — 267

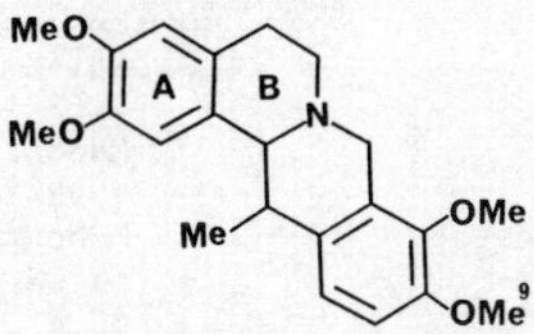

(Rings A and B : 93%)

Protopine (C-1 to C-6 + C-14 : 93%)
(for structure, see under Chelidonium majus)

(*RS*)-[9-*methoxy*-^{14}C]Tetrahydro-palmatine — Corydaline(negligible inc.)

[9-*methoxy*-^{14}C]Palmatine — Corydaline (C-9-*methoxy* : 94%)

Coryphantha macromeris

(*RS*)-[2'-^{3}H]Metanephrine — Normacromerine (4.77% inc.) — 268, 270

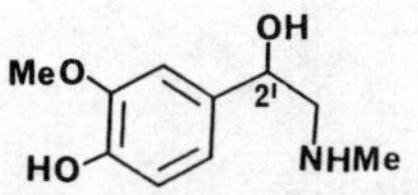

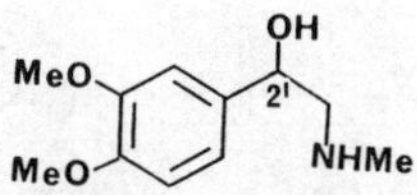

(Fed by direct injection into the cactus with a hypodermic syringe) — (Degradations consistent with all activity at C-2')

(-)-[2'-^{14}C]Normacromerine — Macromerine (NMe$_2$. After a 3-day feeding, most of the normacromerine was recovered and essentially no activity in the macromerine - 0.0005% inc.)

[1'-^{14}C]-*N*-Methyltyramine — Normacromerine (1.1% inc.C-1':97%) — 269

(*RS*)-[2'-^{14}C]Norepinephrine — Normetanephrine (0.11% inc. isolated by dilution) — 270

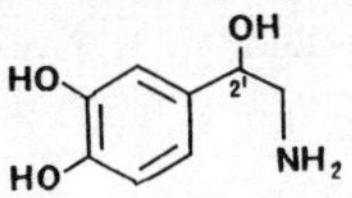

(*RS*)-[2'-^{3}H]Octopamine — Normacromerine (0.01% inc.) (Since the incorporation was so low, this compound is not considered to be a normal intermediate in the biosynthesis of normacromerine from tyrosine.)

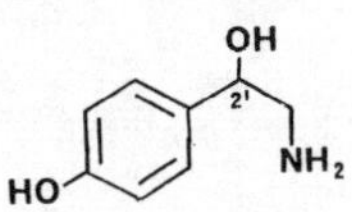

Species / *Compound Administered*	*Alkaloids labelled*	*Ref.*
Crotalaria retusa		
[U-^{14}C]Isoleucine (fed by wick method to intact plants)	Monocrotaline(0.15% inc. 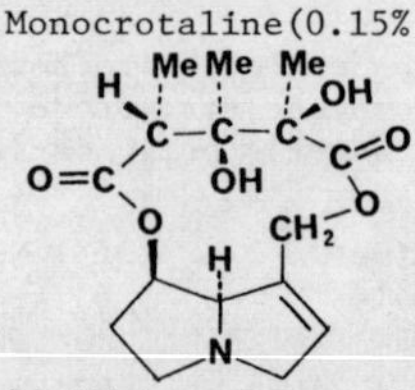 96.8% in monocrotalic acid, 3.9% in retronecine)	271
[U-^{14}C]Valine	Monocrotaline(0.14% inc. 95.4% in monocrotalic acid, 4.6% in retronecine)	
[2-^{14}C]Ornithine	Monocrotaline(0.12% inc. 99.9% in retronecine)	
Sodium [1,2-^{14}C]Acetate	Monocrotaline(0.016% inc. 61.2% in monocrotalic acid, 38.8% in retronecine)	
[U-^{14}C]Threonine	Monocrotaline(0.066% inc. 89.9% in monocrotalic acid, 10.1% in retronecine)	
Croton sparsiflorus		
(*RS*)-[1-^{3}H]Norcoclaurine 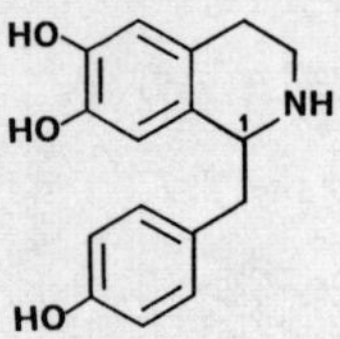	Nornuciferine-I (0.14%) 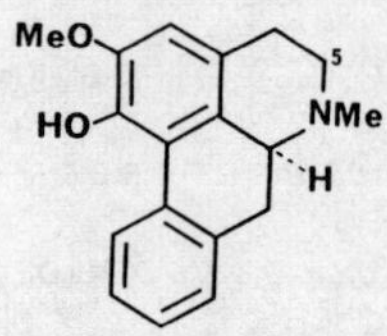	272
DL-[2'-^{14}C]Tyrosine	Nornuciferine-I (0.08% inc.)	
(*RS*)-[3',5'-8-$^{3}H_3$]Coclaurine (6-OMe group)	Nornuciferine-I (0.20% inc.)	
(*RS*)-[1-^{3}H]-*N*-Methylcoclaurine	Nornuciferine-I (0.36% inc.)	
(*RS*)-[*N-methyl*-^{14}C]-*N*-Methylcoclaurine	Nornuciferine-I (0.0004% inc.)	
(*RS*)-*N*,*O*,*O*-Trimethyl[3',5',8-$^{3}H_3$]-Coclaurine	Nornuciferine-I (0.002% inc.)	
(*RS*)-[3-^{14}C]-*N*-Methylcoclaurine	Nornuciferine-I (0.42% inc. C-5 : 98%)	
(+)-(*S*)-[3',5',8-$^{3}H_3$]-*N*-Methylcoclaurine	Nornuciferine-I (0.72% inc.)	
(−)-(*R*)-[3',5',8-$^{3}H_3$]-*N*-Methylcoclaurine	Nornuciferine-I (0.012% inc.)	
[5-^{14}C]-*N*-Methylcrotsparine 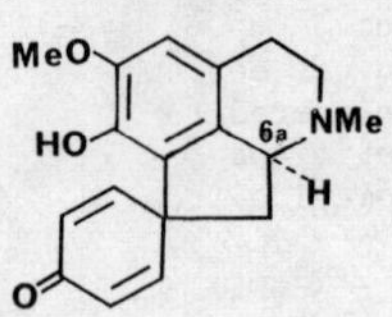	Nornuciferine-I (5.20% inc.)	

Species / *Compound Administered*	*Alkaloids labelled*	*Ref.*
Croton sparsiflorus (continued) -		
[5-^{14}C]-*N*-Methylcrotsparinine (6α-epimer of *N*-methylcrotsparine)	Nornuciferine-I (0.005% inc.)	272
[5-^{14}C]-*N*-methylcrotsparinol-I	Nornuciferine-I (6.12% inc.)	

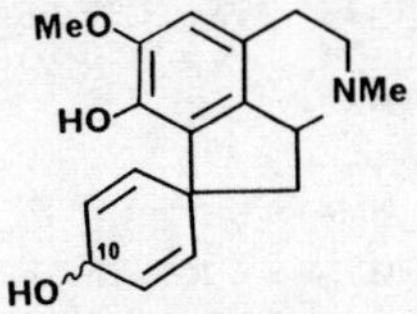

[5-^{14}C]-*N*-Methylcrotsparinol-II (These last two compounds are epimeric at C-10)	Nornuciferine-I (0.32% inc.)	
[5-^{14}C]Crotsparine	Nornuciferine-I (4.6% inc.)	

(This precursor, although not considered to be on the normal route to the nornuciferine-I, is apparently capable of being *N*-methylated to *N*-methylcrotsparine, which is derived from (*S*)-*N*-methylcoclaurine).

N-Methylcrotsparine

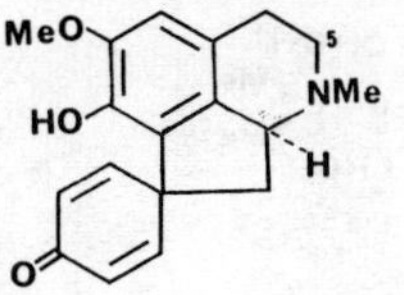

N-Methylcrotsparinine

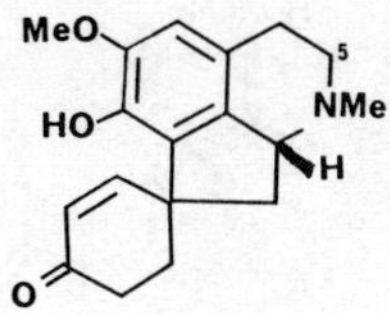

N-Methylsparsiflorine 273

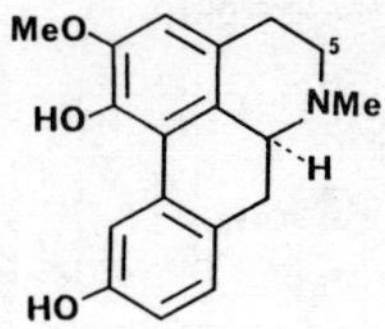

	Incorporation (%)		
	N-Methyl-crotsparine	*N*-Methyl-crotsparinine	*N*-Methyl-sparsiflorine
DL-[2'-^{14}C]Tyrosine	0.30	0.28	0.24
DL-[2'-^{14}C]Dopa	0.36	0.30	0.28
	(C-5: 98%)		(C-5: 97%)
[1'-^{14}C]Dopamine	0.38	0.22	0.26
	(C-5: 97%)		(C-5: 98.5%)
[3,5-^{3}H]-4-Hydroxyphenyl-pyruvic acid	0.20	0.19	0.18
[3,5-^{3}H]Tyramine	0.05	0.03	0.04
(*RS*)-[3-^{14}C]Norcoclaurine-1-carboxylic acid	0.38	0.34	0.34
(*RS*)-[1-^{3}H]Norcoclaurine	0.48	0.38	0.33
(*RS*)-[3',5',8-^{3}H]Coclaurine	0.64	0.54	0.61
(*RS*)-[3',5',8-^{3}H]-*N*-Methyl-coclaurine	0.88	0.83	0.81
(*RS*)-[3',5',8-^{3}H]-*N*,*O*,*O*-Tri-methylcoclaurine	0.0005	0.0007	0.001
(*RS*)-[*N*-*methyl*-^{14}C]Methylnorcoclaurine	0.006	0.004	0.008
(*RS*)-[*N*-*methyl*-^{14}C,1-^{3}H]Methyl-coclaurine	1.08	0.98	0.79
	(Complete ret. of ^{3}H in all alkaloids)		
(*RS*)-[*N*-*methyl*-^{14}C]Methylcoclaurine	0.88	0.67	0.70
[*N*-*methyl*-^{14}C]Didehydromethyl-coclaurinium iodide	0.42	0.36	0.42
(*S*)-[3',5',8-^{3}H]-*N*-Methylcoclaurine	1.97	0.019	1.48
(*R*)-[3',5',8-^{3}H]-*N*-Methylcoclaurine	0.022	1.69	0.0015

Species *Compound Administered*	*Alkaloids labelled*			*Ref.*
Croton sparsiflorus (continued) -	Incorporation (%)			273
	N-Methyl-crotsparine	*N*-Methyl-crotsparinine	*N*-Methyl-sparsiflorine	
[5-^{14}C]-*N*-Methylcrotsparine	-	0.16	7.70	
[5-^{14}C]Crotsparine	6.20	0.12	7.06	
[5-^{14}C]-*N*-Methylcrotsparinine	0.11	-	0.05	

(These experiments favor the following biosynthetic pathway:)

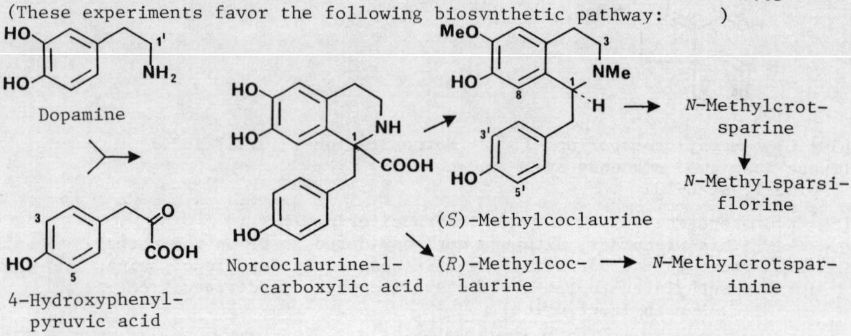

Cynoglossum officinale — Heliosupine (0.86% spec. inc.) — 274

[1-^{14}C]Tiglic Acid

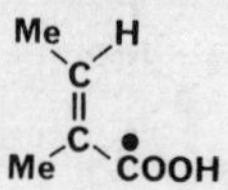

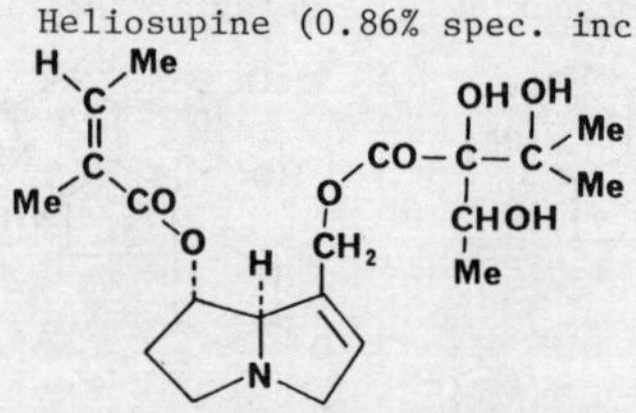

(94.5% of activity in the angelic acid moiety, indicating that the precursor has undergone a *cis* → *trans* isomerism. The reverse reaction (angelic → tiglic acid) did not occur in *Datura*[275]).

Datura innoxia

(2*RS*, 3*S*, 4*S*)-[2-^{14}C, 4-^{3}H]-Isoleucine (fed by wick to intact plants) — Meteloidine (0.004% inc.) — 276

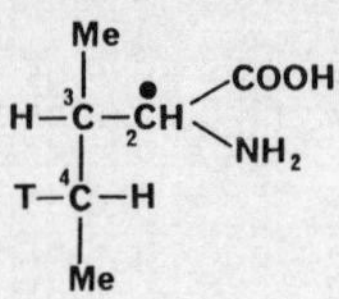

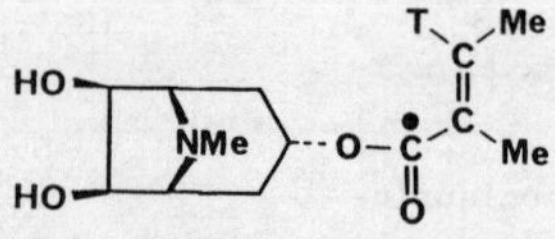

(99% ret. of ^{3}H)

3α,6β-Ditigloyloxytropane (0.13% inc. 98% ret. of ^{3}H at the illustrated positions in the tiglic acid moieties).

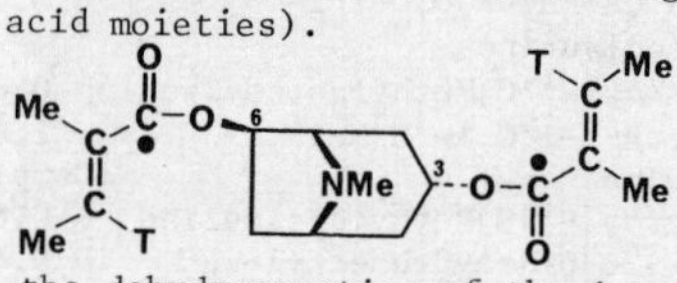

(This result indicates that the dehydrogenation of the intermediate (2*S*)-2-methylbutanoic acid involves an antiperiplanar elimination of the hydrogen at C-2 and the *pro-R* hydrogen at C-3).

Species / *Compound Administered* — *Alkaloids labelled* — *Ref.*

Datura innoxia (continued) -

Hyoscyamine(inactive) 276

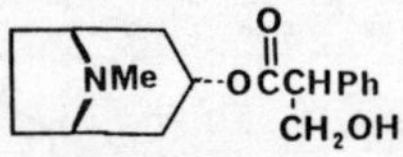

[1-^{13}C, ^{14}C, *methylamino*-^{15}N]-*N*-Methylputrescine — Scopolamine 277

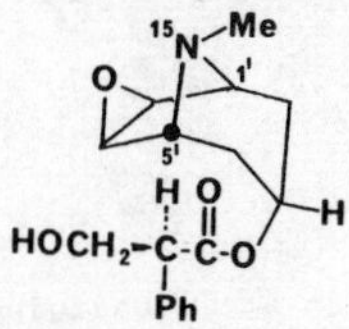

(0.17% spec. inc., C-1' and C-5' have slightly different chemical shifts in the ^{13}C n.m.r ($\Delta\delta$ = 0.09 ppm) due to the chiral tropic acid moiety, and only one of these signals exhibited satellites, J_{CN} = 2.9 Hz, due to the contiguous ^{13}C and ^{15}N atoms).

Datura meteloides

[1",2'-$^{14}C_2$]Tigloylhygroline — Scopolamine(active) 278

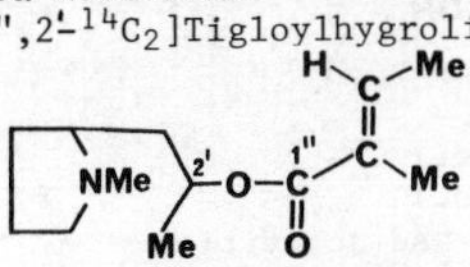

Hyoscyamine(active)
Meteloidine(active)
(tiglic acid/tropane base = 1.90)
3α,6β-Ditigloyloxytropan-7β-ol
(tiglic acid/tropane base = 1.86)

(These results indicate that the administered ester is not incorporated intact. It is proposed (by this reporter) that the resultant hygroline is incorporated via its oxidation product, hygrine).

(*RS*)-[2'-^{14}C]Hygrine — Scopolamine(0.07% spec. inc.)
Hyoscyamine(0.181% spec. in.)
3α,6β-Ditigloyloxytropane
(0.059% spec. inc.)
3α,6β-Ditigloyloxytropan-7β-ol
(0.029% spec. inc.)
Meteloidine(0.088 spec. in.
C-3 : 87%)

(2*S*, 4*R*)-[4-^{14}C, 4-^{3}H]Isoleucine — Meteloidine(0.3% inc. 98% loss of ^{3}H. This result is complementary to related work described under *Datura innoxia*). 276

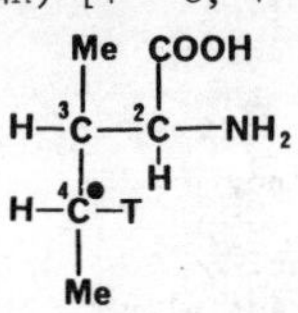

Datura stramonium

[^{14}C]Carbon dioxide — Hyoscyamine
Scopolamine 279

(In general, the alkaloids isolated from the leaves had a higher specific activity than those obtained from the roots. Removal of the flowers increased the level of activity in the root alkaloids. Scopolamine had a lower activity than hyoscyamine, consistent with it being formed from the latter alkaloid).

Species / *Compound Administered*	*Alkaloids labelled*	*Ref.*
Decodon verticillatus		
DL-[6-^{14}C]Lysine	Decodine(0.05% inc.) 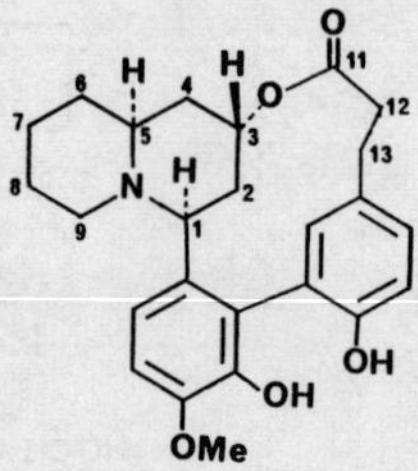 Decinine(0.05% inc.) 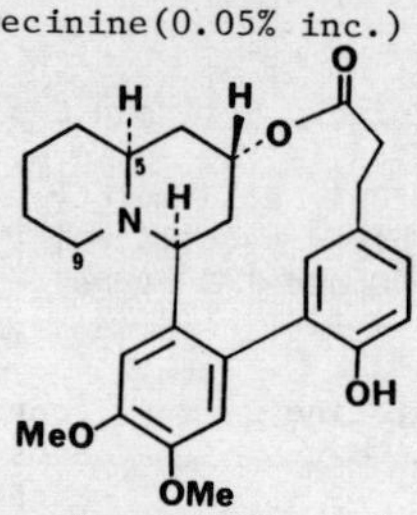	280
DL-[2-^{14}C]Lysine	Decodine(0.12% inc.) Decinine(0.12% inc.)	

(Partial degradations yielded products which had activities consistent with equal labelling of C-5 and C-9 in the alkaloids derived from the [2-^{14}C]- and [6-^{14}C]-lysine).

DL-[6-^{14}C, 4,5-^{3}H]Lysine	Decodine (^{14}C : 0.035%, 104% ret. of ^{3}H) Decinine (^{14}C : 0.025, 105% ret. of ^{3}H) Pipecolic acid (107% ret. of ^{3}H) 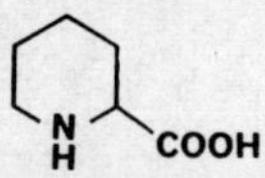	
DL-[6-^{14}C]Lysine + L-[4-^{3}H]Lysine	Decodine (^{14}C : 0.01% inc. 200% ret. of ^{3}H, indicating that the D-lysine does not serve as a precursor of the alkaloid) Decinine (^{14}C : 0.01% inc. 200% ret. of ^{3}H) Pipecolic acid (5.6% ret. of ^{3}H, indicating that essentially all this amino-acid is derived from D-lysine)	
[1,5-^{14}C]Cadaverine	Decodine (0.084% inc. activities of degradation products consistent with activity being at C-5 and C-9)	

Species *Compound Administered*	*Alkaloids labelled*	*Ref.*
Decodon verticillatus (continued) -		
[6-^{14}C]-Δ^1-Piperideine 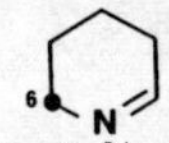	Decodine (0.016% inc. activities of degradation products consistent with all the activity being at C-9).	280
(*RS*)-[6,2'-$^{14}C_2$]Pelletierine 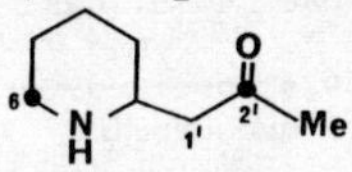	Decodine (no incorporation) Decinine (no incorporation)	
(*RS*)-[2'-^{14}C, 6-^{3}H]Pelletierine	Decodine (no incorporation) Decinine (no incorporation)	
DL-[1'-^{14}C]Phenylalanine	Decodine (0.012% inc. C-3 : 28%, C-11 : 71%) Decinine (0.003% inc. C-3 : 20%, C-11 : 76%)	281
DL-[2'-^{14}C]Phenylalanine	Decodine (0.013% inc. degradations consistent with 62% at C-12 and the rest assumed to be at C-2) Decinine (0.008% inc. degradations consistent with 66% at C-12 and the rest at C-2)	
DL-[3'-^{14}C]Phenylalanine	Decodine (0.017% inc. degradations consistent with 58% at C-13 and 33% at C-1) Decinine (0.013% inc. not degraded)	
DL-[1',3'-$^{14}C_2$]Phenylalanine	Decodine (0.026% inc. degradations consistent with the intact incorporation of the phenylalanine side-chain) Decinine (0.021% inc. degradations consistent with the intact incorporation of the phenylalanine side-chain)	
[2-^{14}C]Malonic Acid	Decodine (0.019% inc. Activity expected at C-4, but results inconclusive, since activity also found in lysine-derived portion of the alkaloid).	
Dicentra spectabilis		
[2'-^{14}C, 2,6-$^3H_2$]Dopamine + 3,4-dihydroxyphenylacetaldehyde (cell-free system)	(*S*)-Norlaudanosoline (3,4-dihydroxyphenylpyruvic acid did not serve as a substrate for this enzyme).	143
Dolichothele sphaerica		
DL-[2'-^{14}C]Phenylalanine	*N*-Methylphenethylamine (1.91% inc. All activity at the α-position). 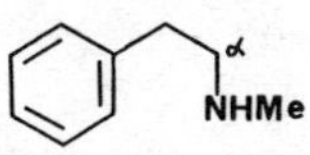 A small amount of activity was also detected in phenethylamine by isotopic dilution.	282

Species / *Compound Administered*	*Alkaloids labelled*	*Ref.*

Duboisia leichhardtii/myoporoides (hybrid)

L-[U-^{14}C]Ornithine (fed by wick to intact plants) — Hyoscyamine (0.17% inc.), Scopolamine (0.13% inc.) — 283

(all activity in the tropane moieties)

[G-^{3}H]Atropine ((*RS*)-Hyoscyamine) — Scopolamine (a slow conversion over 48 hr was observed)

[G-^{3}H]Scopolamine — General metabolism was observed but no conversion to hyoscyamine was detected.

Eleagnus angustifolia

(*RS*)-1-[^{14}C]Methyl-1,2,3,4-tetrahydro-[8-^{3}H]-β-carboline-1-carboxylic acid — Eleagnine — 284

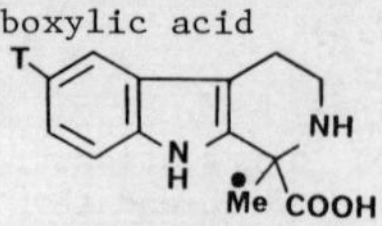

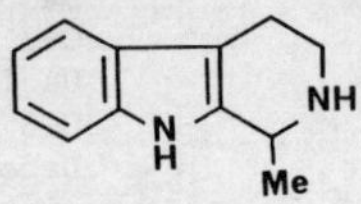

(Amino Acid-1)

(With good inc. of ^{14}C : 0.15, 0.27%, there was 110, 114% ret. of ^{3}H. The excess ^{3}H may be due to catabolism of the precursor to [5-^{3}H]tryptamine which is then used for alkaloid synthesis).

L-[5-^{3}H]Tryptophan — Eleagnine (0.06% inc.), Amino acid-1 (0.19%)

Erythroxylon coca

DL-[5-^{14}C]Ornithine (fed by painting on leaves of intact plant) — Cocaine — 62,63

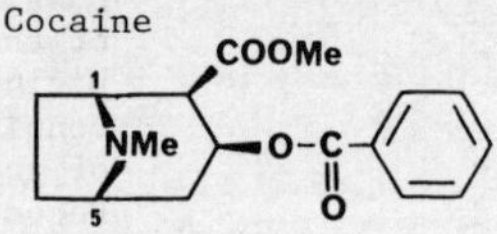

(Activity was located equally at C-1 and C-5, indicating incorporation via a symmetrical intermediate, presumably putrescine).

Cuscohygrine

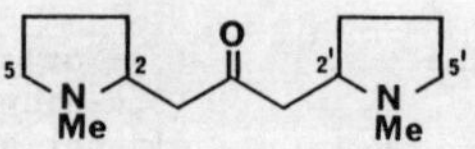

(Activities of degradation products consistent with equal labelling at C-2, 2', 5, and 5').

Sodium [1-^{14}C]Acetate + DL-[4-^{3}H]Phenylalanine (These two precursors were fed together by painting on leaves, ^{3}H/^{14}C = 1.0) — Cocaine (^{3}H/^{14}C = 11.4. The % distribution of ^{14}C and ^{3}H was as follows: — 64

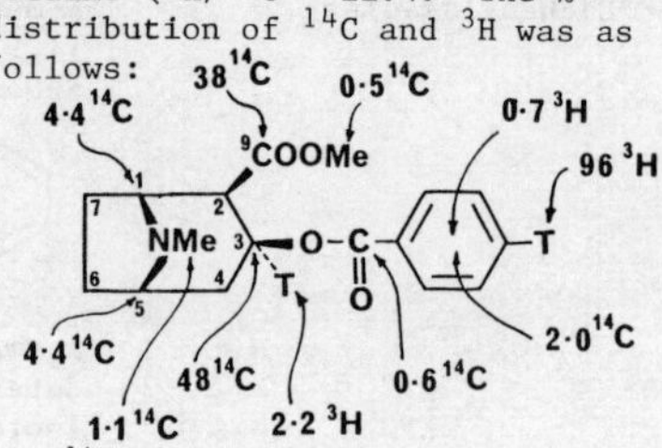

Cuscohygrine (No ^{3}H, 67% of the ^{14}C at the C=O group).

Species / *Compound Administered*	*Alkaloids labelled*	*Ref.*
Escherichia coli		
L-[3,4-$^{13}C_2$]Methionine	Spermidine(∿ 100% spec. inc.)	198

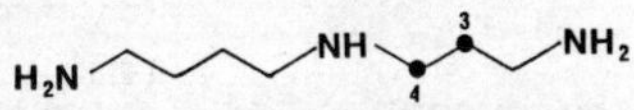

(All excess ^{13}C at C-3, C-4, none at C-2, thus excluding azetidine as an intermediate in the reaction between putrescine and decarboxylated *S*-adenosyl-methionine). It has also been shown that putrescine reacts with the decarboxylated *S*-adenosyl-methionine with inversion, i.e., a typical S_N2 reaction[285]).

Species / *Compound Administered*	*Alkaloids labelled*	*Ref.*
Eschscholtzia pulchella		
[2'-^{14}C, 2,6-$^{3}H_2$]Dopamine + 3,4-Dihydroxyphenyl-acetaldehyde (enzyme preparation)	(*S*)-Norlaudanosoline (material recovered from a large-scale incubation was only 25% optically pure. Racemic material may have arisen by a non-enzymic Mannich reaction).	143
Euglena gracilis		
(Cell-free extracts were used since incorporations in intact cells were poor)		195
DL-[5-^{14}C]Ornithine	Putrescine (38.2% inc.)	
DL-[5-^{14}C]Arginine	Putrescine (2.4% inc.)	
DL-[5-^{14}C]Citrulline	Putrescine (7.9% inc.)	
DL-[5-^{14}C]Ornithine + *S*-adenosyl-L-methionine	Putrescine (7.0% inc.) Spermidine (17.0% inc.) Spermine (2.7% inc.)	
[1,4-^{14}C]Putrescine	Spermidine (74.9% inc.) Spermine (19.4% inc.)	
Festuca arundinacea		
[3'-^{14}C, 5-^{3}H]Tryptophan	Perloline	35

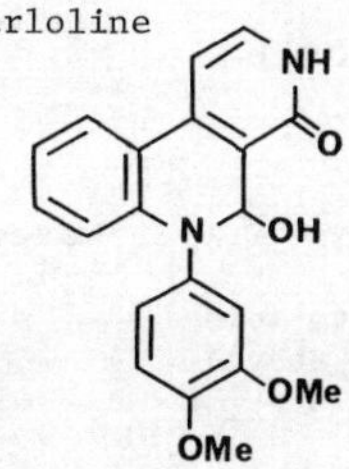

(incorporation with no change in the $^{3}H/^{14}C$ ratio)

[3'-^{14}C, U-^{3}H]Tryptophan	Perloline (incorporation with only 1% ^{3}H loss!)	
[3'-^{14}C, 3,5-$^{3}H_2$]-*p*-Coumaric acid	Perloline (only ^{3}H incorporated into the alkaloid)	

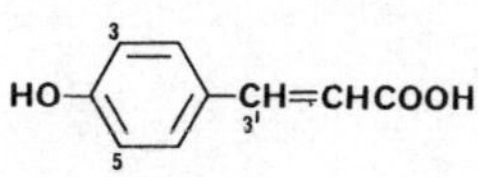

Species *Compound Administered*	*Alkaloids labelled*	*Ref.*
Festuca arundinacea (continued) -		
[3'-^{14}C, 2, 5, 6-$^{3}H_{3}$]Caffeic Acid	Perloline (no incorporation of either isotope) (These results are consistent with the hypothesis[286,287] that this ring system is formed from kynurenine and a C_6 unit derived from coumaric acid, with loss of the C_3 side-chain).	35

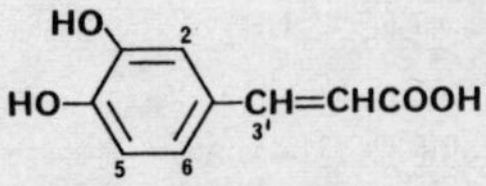

Fumaria officinalis		
[2'-^{14}C, 2,6-^{3}H]Dopamine + 3,4-Dihydroxyphenylacetaldehyde (cell-free system)	(*S*)-Norlaudanosoline (3,4-dihydroxyphenylpyruvic acid was not a substrate for this enzyme)	143
Gelsemium sempervirens		
[6-^{14}C]Strictosidine (see under *Catharanthus roseus*)	Gelsemine (0.47% inc.)	236

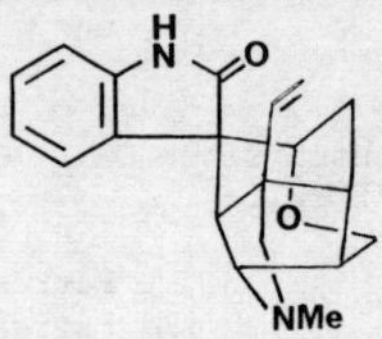

[6-^{14}C, 3-^{3}H]Vincoside	Gelsemine (< 0.001% inc.)	
[6-^{14}C, 10-^{3}H]-5α-Carboxy-strictosidine	Gelsemine (< 0.001% inc.) (Good incorporation of labelled strictosidine into gelsemine was observed under the same feeding conditions).	288

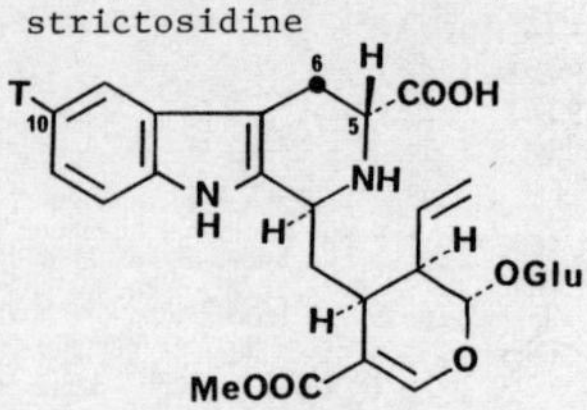

[6-^{14}C, 10-^{3}H]-5α-Carboxy-vincoside (3-epimer of the strictosidine derivative illustrated above)	Gelsemine (< 0.001% inc.)	
Gliocladium deliquescens		
[^{14}C]Gliotoxin	Bisdethiobis(methylthio)gliotoxin (8.6% inc.)	289

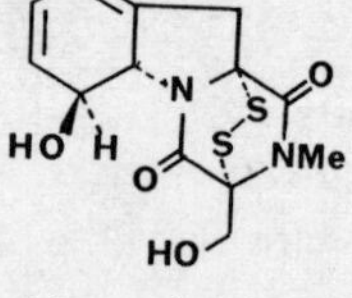

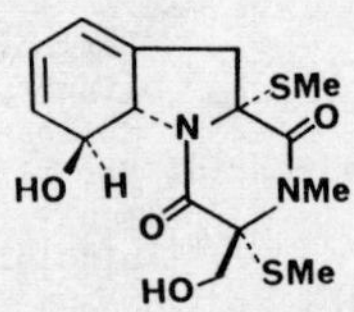

(made biosynthetically from L-[U-^{14}C]phenylalanine)		
[^{14}C]Bisdethiobis(methylthio)-gliotoxin	Gliotoxin (< 0.2% inc. The reduction and methylation of gliotoxin is thus shown to be irreversible.)	

Species / *Compound Administered*	*Alkaloids labelled*	*Ref.*
Heliotropium spathulatum		
[^{14}C]Carbon dioxide	(−)-Trachelanthamidine (−)-Supidine (+)-Retronecine	94

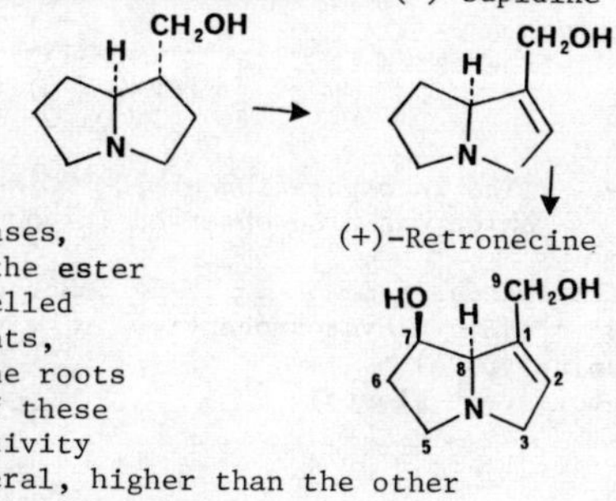

(Activity was found in these necine bases, which were obtained by hydrolysis of the ester alkaloids found in this species. Labelled material was obtained with intact plants, and de-rooted ones, indicating that the roots are not necessary for the formation of these pyrrolizidine bases. The specific activity of the trachelanthamidine was, in general, higher than the other two bases, consistent with the biosynthetic pathway: CO_2 → trachelanthamidine → supidine → retronecine).

Species / *Compound Administered*	*Alkaloids labelled*	*Ref.*
Hordeum vulgare		
[α-^{14}C]Hordenine (fed to intact seedlings)	After 24 hr, essentially all the hordenine absorbed by the 6-day-old plants was metabolized and eliminated as $^{14}CO_2$.	290

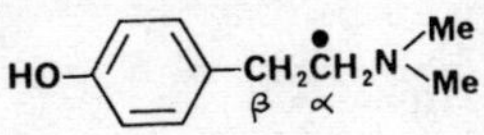

[β-^{14}C]Hordenine (fed to 6-day-old seedlings for 26 hr).	Hordenine (99% spec. inc.) Tyramine (74% spec. inc.) *p*-Hydroxybenzoic acid (53% spec. inc.) Activity was also detected in the lignin .	291
S-Adenosyl-L-[*methyl*-^{14}C] methionine (incubated with enzymes obtained from roots of young plants).	Hordenine, *N*-Methyltyramine (Two distinct enzymes: tyramine-*N*-methyltransferase, and *N*-methyltyramine-*N*-methyltransferase were isolated and separated. They differed in their pH optimum and their stability).	196
[*methylene*-^{14}C]Gramine	When this alkaloid was fed to the roots of 11-and 14-day-old seedlings, most of the radioactivity (90-95%) was recovered in respired CO_2.	292
[*methylene*-^{14}C]Gramine (added to a cell-suspension culture)	Activity found in indole-3-carbinol, indole-3-aldehyde, and indole-3-carboxylic acid. No significant activity found in expired CO_2.	293
[*N-methyl*-^{14}C]Gramine (to cell-suspension culture)	3-Methylaminomethylindole (Norgramine)	

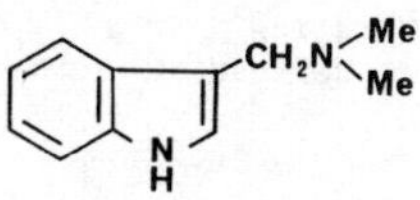

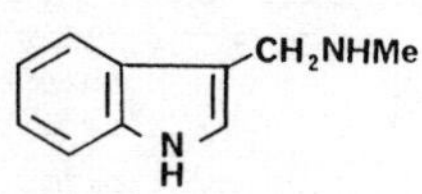

(The formation of this alkaloid represents a reversal of the established biosynthesis of gramine).

Species / *Compound Administered*	*Alkaloids labelled*	*Ref.*
Hordeum vulgare (continued) -		
[U-^{14}C]Agmatine + *p*-Coumaryl-coenzyme A (cell-free extract)	Coumarylagmatine 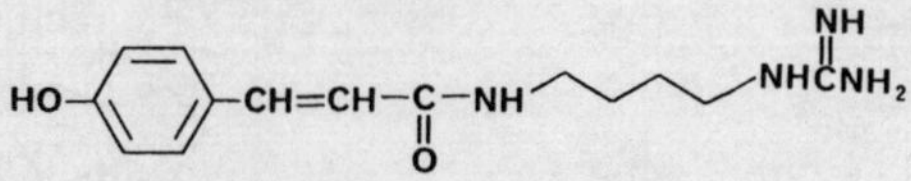	197
	(The incorporation of [U-^{14}C]agmatine into the above amide was also observed in young seedlings).	
Hydrastis canadensis		
DL-[3'-^{14}C]3,4-Dihydroxyphenyl-alanine (DOPA) (fed to intact plants)	Berberine (Rings A + B : 99%) 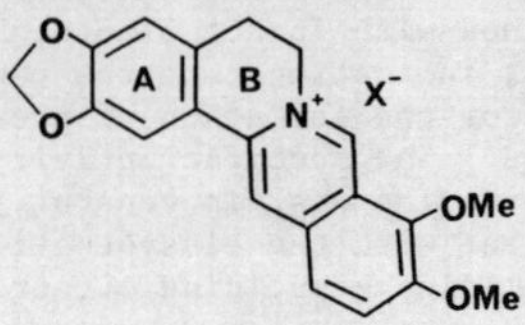 Hydrastine (Rings A + B : 93%) 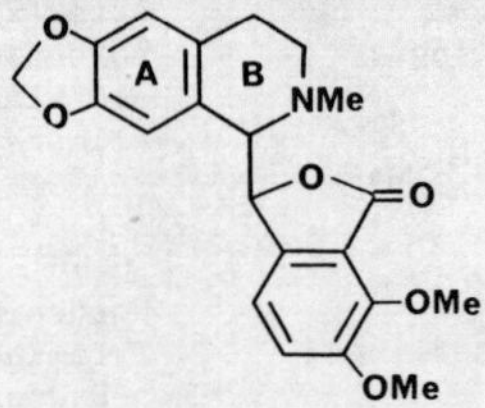	267
Lathyrus sativus		
L-[U-^{14}C]Aspartic Acid	Spermidine (0.075% inc.) Spermine (0.008% inc.) Putrescine (0.02% inc.) (see under *Escherichia coli* for structures)	199
L-[G-^{3}H]Methionine	Spermidine (0.13% inc.) Spermine (0.054% inc.)	
Lupinus angustifolius		
DL-[6-^{14}C]Lysine + L-[4,5-$^{3}H_2$]Lysine	(+)-Lupanine 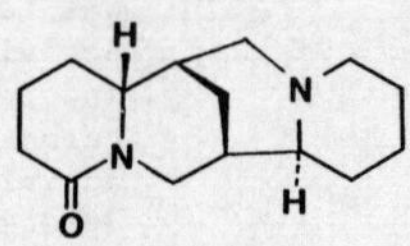 (192% ret. of ^{3}H, indicating that the alkaloids are derived from L-Lysine)	294

Species
Compound Administered — *Alkaloids labelled* — *Ref.*

Lupinus angustifolius (continued) -

L-Pipecolic acid 294

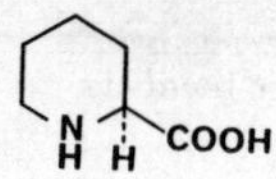

(1.2% ret. of ^3H, indicating that this amino-acid is derived from D-lysine). The publication reports this compound as the L-isomer; however, this is not justified since it was isolated by dilution with DL-pipecolic acid.

D-[6-^{14}C]Lysine + DL-[4,5-$^3H_2$]Lysine — (+)-Lupanine (8.7% ret. of ^{14}C relative to ^3H)

Pipecolic acid (49% ret. of ^3H)

Lupinus polyphyllus

[1,5-^{14}C]Cadaverine (in crude chloroplasts) — Lupanine (1.7% inc.) — 98

Chloroplasts were isolated from the leaves of this species and *Lupinus albus* seedlings.

17-Oxosparteine (0.7% inc.)

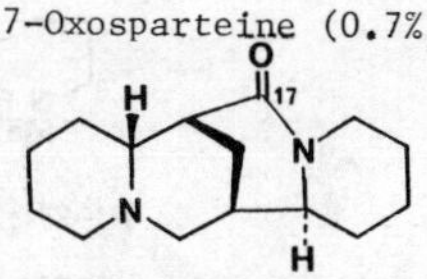

[1,5-^{14}C]Cadaverine (in crude chloroplasts + 2% digitonin) — Lupanine formation was inhibited and sparteine was the main alkaloid formed.

Sparteine:

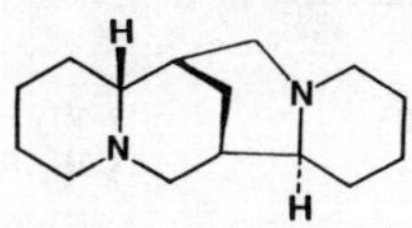

[1,5-^{14}C]Cadaverine (in solubilized chloroplast enzyme) — The major alkaloid formed (95%) was 17-oxosparteine.

17-Oxosparteine (added to crude chloroplasts) — Lupanine
This was the major alkaloid formed and its formation was not affected by NADPH or NADH.

Meconopsis cambrica

[2'-^{14}C, 2,6-$^3H_2$]Dopamine + 3,4-Dihydroxyphenylacetaldehyde (cell-free system) — (*S*)-Norlaudanosoline (3,4-dihydroxyphenylpyruvic acid did not serve as a substrate for this enzyme) — 143

Species — *Alkaloids labelled* — *Ref.*
Compound Administered

Myrothecium verrucaria

Sodium [1-^{13}C]Acetate — Verrucarin E — 295

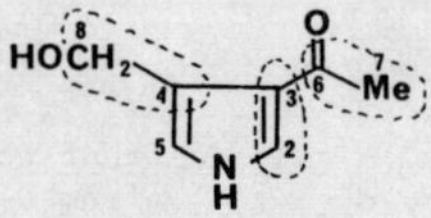

(enrichment at C-2, 4, 6)

Sodium [2-^{13}C]Acetate — Verrucarin E
(enrichment at C-3, 5, 7, 8)

Sodium [1,2-$^{13}C_2$]Acetate — Verrucarin E (∿ 4% spec. inc.
The coupled carbons indicated with dotted lines: 2-3, 6-7, 4-8)

Sodium [2-^{13}C, 2-$^2H_3$]Acetate — Verrucarin E
(Examination of the C-7 signal in the ^{13}C n.m.r. indicated that this is the starter unit for one acetoacetate unit. The authors failed to cite related work on the biosynthesis of 8-desoxy-verrucarin[296]).

Nicandra physaloides

Sodium [3-^{14}C]Acetoacetate — Hygrine — 297

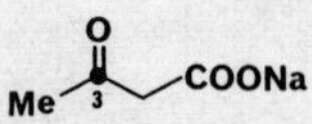

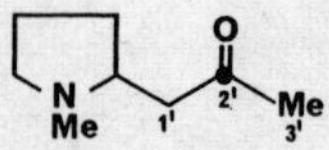

[C-2' : ∿ 100%]

Sodium [4-^{14}C]Acetoacetate — Hygrine
[All activity at C-2' or C-3', indicating that hygrine is formed by reaction of *N*-methyl-Δ^1-pyrrolinium salt with C-2 of acetoacetate].

Nicotiana glauca

5-Fluoro-[5,6-$^{13}C_2$,^{14}C]-nicotinic acid — 5-Fluoroanabasine (16.2% inc., 100% spec. inc.) — 298

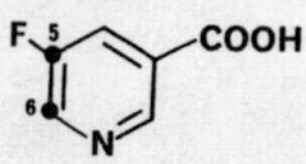

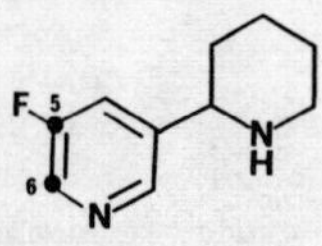

(^{13}C detected at the 5,6-positions by n.m.r. This compound, the product of an aberrant biosynthesis, was racemic. However, it was discovered that the major alkaloid of this species, anabasine, is also racemic).

DL-[4,5-$^{13}C_2$, 6-^{14}C]Lysine — Anabasine — 299

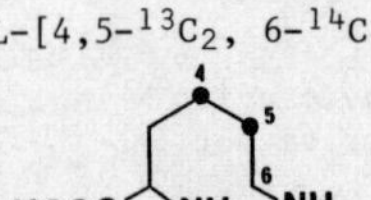

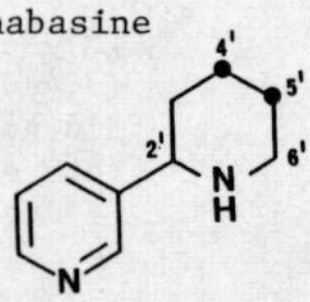

(5.7% abs. inc. 0.25% spec. inc. The excess ^{13}C was located at C-4', 5' by n.m.r. Essentially all the ^{14}C was at C-6' with negligible amounts (0.2%) being at C-2').

Species / *Compound Administered*	*Alkaloids labelled*	*Ref.*
Nicotiana glauca (continued) –		
L-[4,5-^{3}H]Lysine + D-[6-^{14}C]Lysine (^{3}H/^{14}C = 10.2)	L-Pipecolic Acid (^{14}C : 1.8% inc., ^{3}H/^{14}C = 1.3) Anabasine (^{3}H : 0.9% inc. ^{3}H/^{14}C = 200). L-Lysine (^{3}H : 0.6% inc., ^{3}H/^{14}C = 81)	300
D-[1-^{14}C, 6-^{3}H]Lysine	L-Pipecolic Acid (^{14}C : 11.8% inc. 97% ret. of ^{3}H) Anabasine (^{3}H : 0.10% inc., no ^{14}C) L-Lysine (^{14}C : 0.37% inc. 49% ret. of ^{3}H)	

(The primary precursor of anabasine is considered to be L-lysine which is formed from L-pipecolic acid, which originates in this species from D-lysine according to the following route:

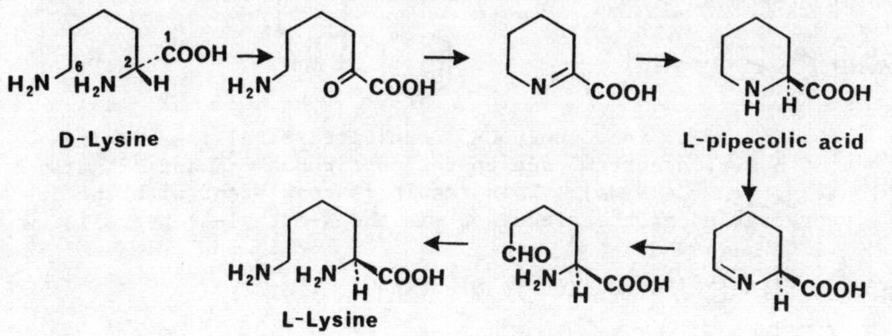

Nicotiana glutinosa		
DL-[2,3-$^{13}C_2$,5-^{14}C]Ornithine	Nicotine (0.14% spec. inc.	301

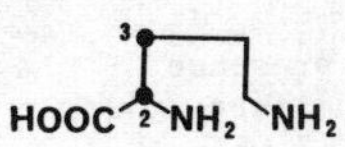

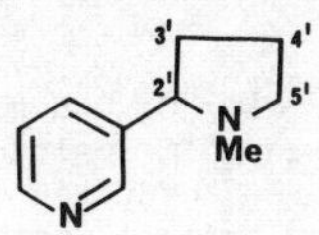

^{13}C n.m.r. revealed satellites at C-2', 3', 4', 5' indicative of symmetrical labelling of the pyrrolidine ring)

Nornicotine (0.051% spec. inc.

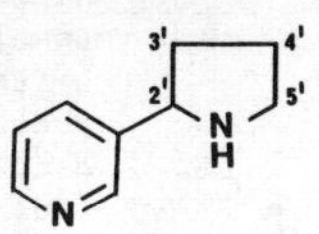

symmetrical labelling of the pyrrolidine ring, established by ^{13}C n.m.r.)

(These results completely validate the previous work on the incorporation of ornithine into the pyrrolidine ring of nicotine and nornicotine, namely that the ornithine is incorporated via a symmetrical intermediate, presumably putrescine. It also demonstrates that precursors labelled with contiguous ^{13}C atoms are especially useful for investigating biosynthetic pathways, since satellites which arise from such precursors in the ^{13}C n.m.r. can be detected at high dilutions ~ 4000)

Species — *Compound Administered* | *Alkaloids labelled* | *Ref.*

Nicotiana tabacum

(*RS*)-[2'-^{14}C]Nicotine (fed to excised leaves which were dried in air for 4 weeks) — Nicotine (3.95% inc.); Nornicotine (3.16% inc.); *N'*-Isopropylnornicotine (0.018% inc.) — 302

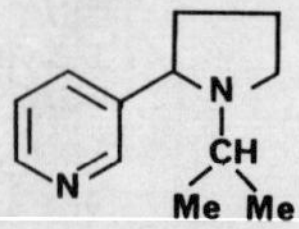

(The degree of inc. of activity into this alkaloid was increased (to 0.035%) if the leaves were also fed 20% acetone. There was negligible formation of this alkaloid in the intact plant).

[1-^{13}C, ^{14}C, *methylamino*-^{15}N]-*N*-Methylputrescine — Nicotine — 277

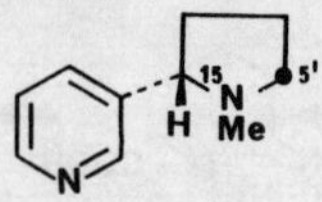

(0.104% spec. inc. Only C-5' exhibited satellites in its ^{13}C n.m.r. spectrum, due to the contiguous ^{13}C and ^{15}N atoms ($J_{^{13}C-^{15}N}$ = 4.2 Hz). This result is consistent with the incorporation of the precursor via the *N*-methyl-Δ^1-pyrrolinium salt).

(*R*)-[1-^{2}H] + [1,4-^{14}C]Putrescine — Nicotine — 303

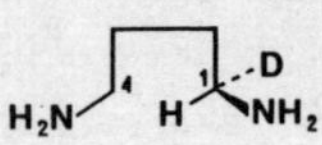

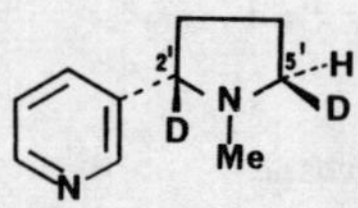

(1.3% spec. inc. of ^{14}C. ^{2}H n.m.r. indicated the presence of ^{2}H at the 2' and 5'*R* positions. This result indicates that it is the *pro-S* hydrogen which is lost from C-1 when 4-methylaminobutylamine (*N*-methylputrescine) is oxidized to 4-methylaminobutanal).

DL-[5-^{14}C]- + DL-[5(*RS*)-^{3}H]-Ornithine — Nicotine (70% ret. of ^{3}H, calc. ret. 75%)

DL-[5-^{14}C]- + L-[5(*RS*)-^{3}H]-Ornithine — Nicotine (140% ret. of ^{3}H, this result indicates that only the L-isomer of ornithine serves as a precursor of nicotine).

DL-[5-^{14}C, 2-^{3}H]Ornithine — Nicotine (50% ret. of ^{3}H. This result indicates that L-ornithine is decarboxylated with retention of configuration, and the tritium at C-1 of the putrescine is in the *pro-S* position).

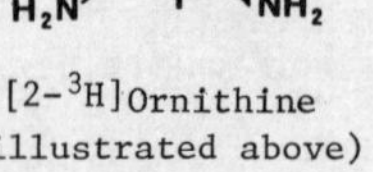

L-[2-^{3}H]Ornithine (illustrated above)

Species / *Compound Administered*	*Alkaloids labelled*	*Ref.*

Nicotiana tabacum (continued)-

L-[U-^{14}C]Phenylalanine
[1,4-^{14}C]Putrescine
L-[5-^{14}C]Ornithine
L-[U-^{14}C]Arginine
(fed to cell-suspension cultures)

Caffeoylputrescine — 304

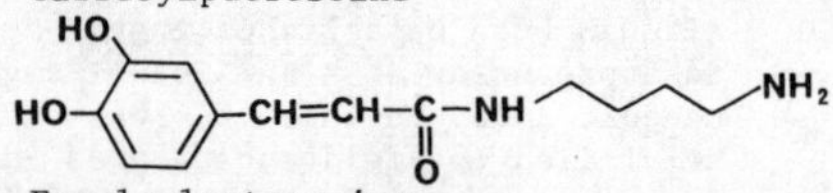

Feruloylputrescine

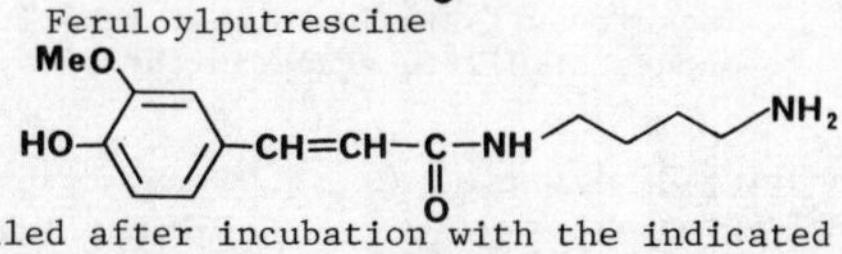

(These amides were labelled after incubation with the indicated precursors. α-Difluoromethylornithine, an enzyme-activated irreversible inhibitor of ornithine decarboxylase, did not reduce the incorporation of ornithine into these cinnamoyl-putrescines, suggesting that the putrescine is formed via arginine. The amounts of these amides in this cell-suspension culture were increased when the concentration of phosphate in the culture medium was reduced[305]).

Nocardia species (strain C-15003)

[*carboxy*-^{13}C]-3-Amino-5-hydroxybenzoic acid

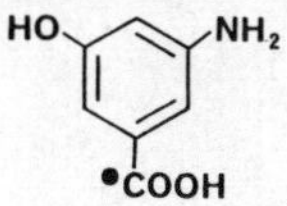

Ansamitocin P-3 — 126

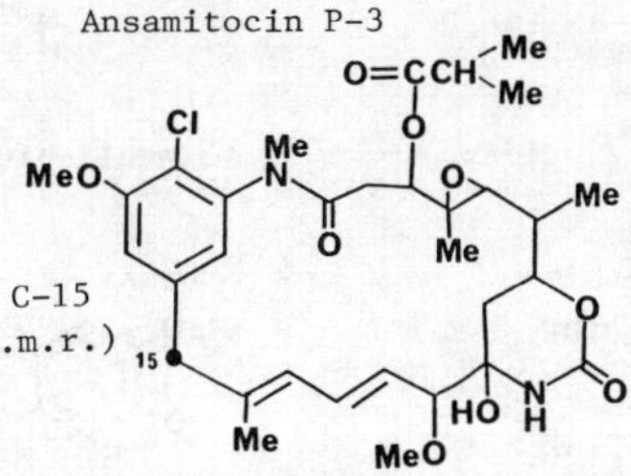

(Enrichment at C-15 was established by ^{13}C n.m.r.)

Nocardia mediterranei — 306

	Rifamycin B	Rifamycin L
	(inc. after 6 hr)	
[U-^{14}C]Glycerol	36%	5%
[U-^{14}C]Pyruvate	4.6%	67%
[1-^{14}C]Pyruvate	0.7%	60%
[2-^{14}C]Pyruvate	3.4%	50%
[3-^{14}C]Pyruvate	4.1%	55%
[U-^{14}C]Serine	low	$<2\%$

(The following modification of the quinone ring of rifamycin S is proposed:

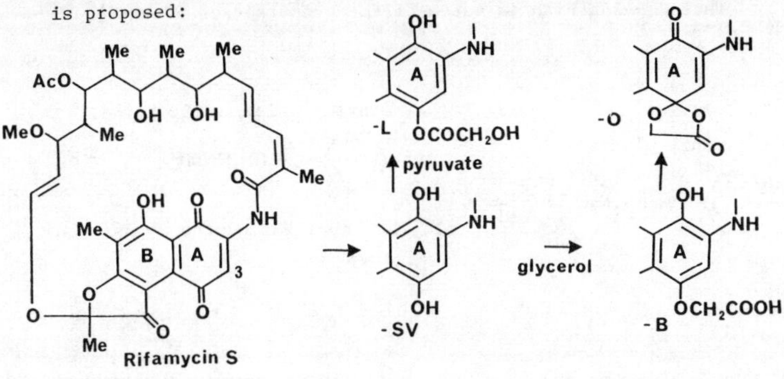

Species *Alkaloids labelled* *Ref.*
Compound Administered

Nocardia mediterranei (continued) -

(It has been established that 3-amino-5-hydroxybenzoic acid is a precursor of ring A of rifamycin S[117]. Analogs of this benzoic acid, in which C-4 was substituted with OH, OMe, or Me, failed to yield substituted derivatives of rifamycin S. These results indicate that substitution at C-3 of rifamycin S occurs at a late stage in the biosynthetic sequence[307]).

Nodulisporium hinnuleum
[U-^{3}H]Tryptophan — Hinnuliquinone (0.8% inc.) — 308

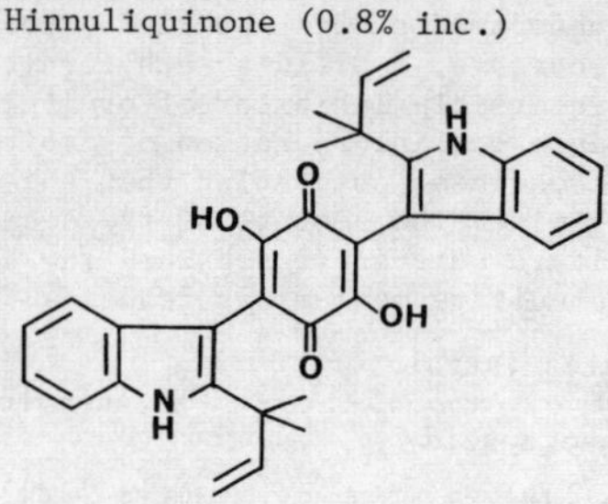

[2-^{14}C]Mevalonic acid — Hinnuliquinone (0.2% inc.)

Opuntia bergeriana
[2'-^{14}C]Dopaxanthine — Betanin — 309

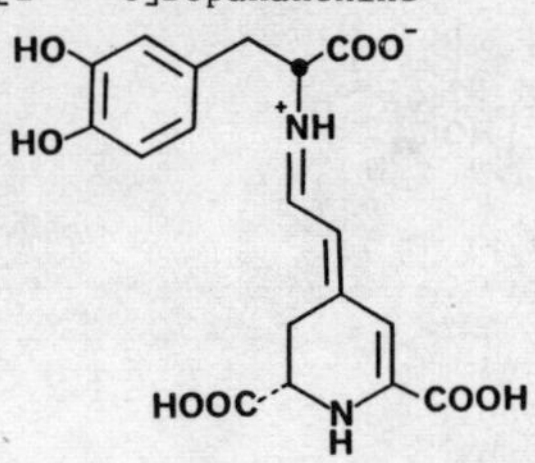

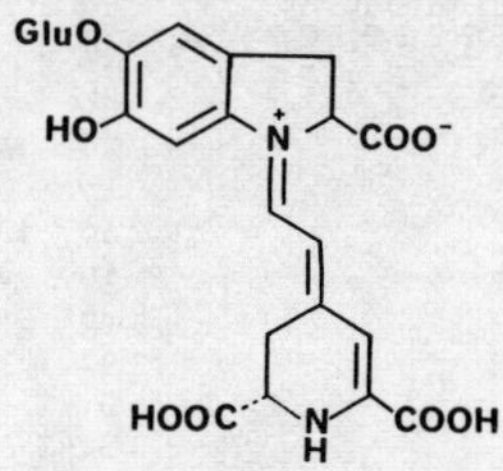

(The incorporation was similar to previous feeding experiments with labelled dopa; however, degradations indicated that most (95%) was in the lower betulamic acid moiety of the betanin. Thus dopaxanthine is not a direct precursor of betanin).

Opuntia clavata
[1'-^{14}C]Tyramine — *N*-Methyltyramine (0.95 % inc. — 269
(fed to intact plants)

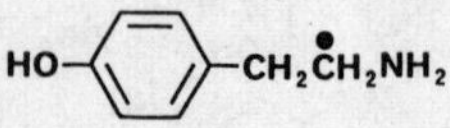

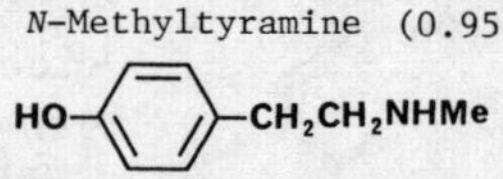

All the label was on the α-position)

Species / *Compound Administered*	*Alkaloids labelled*	*Ref.*

Papaver orientale

(*RS*)-[*N-methyl*-^{14}C, 1-^{3}H]-Reticuline — Thebaine (0.05% inc. 98% of ^{14}C on NMe, 52% loss of ^{3}H) — 310

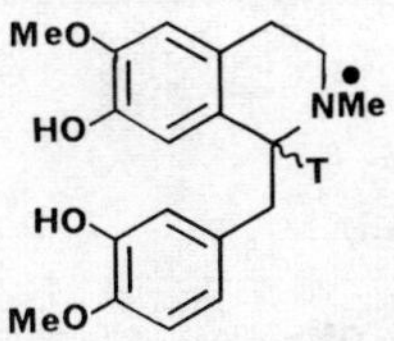

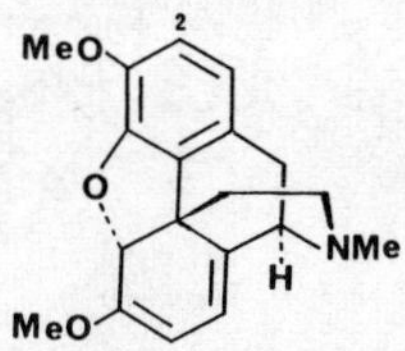

Oripavine (0.31% inc. with 53% loss of ^{3}H, attributed to racemization of the reticuline)

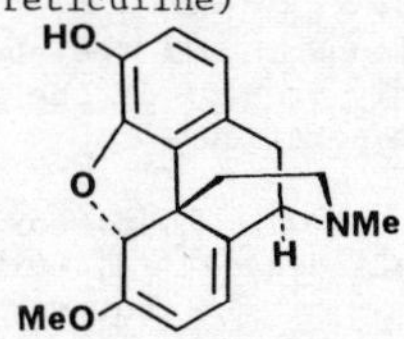

[2-^{3}H] Thebaine — Oripavine (0.17% inc.)

Papaver somniferum

(*2'S*)-[U-^{14}C,2'-^{3}H]Tyrosine — Papaverine (0.22 and 0.26% inc. — 311

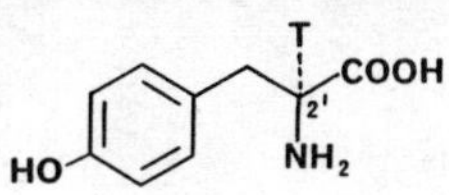

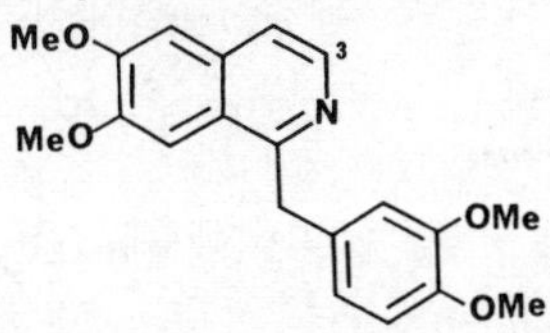

100% of the tritium is lost. Since the formation of papaverine involves the loss of the *pro-S*-hydrogen at C-3 of nor-reticuline,[312] this result indicates that the decarboxylation of tyrosine occurs with retention of configuration).

Morphine

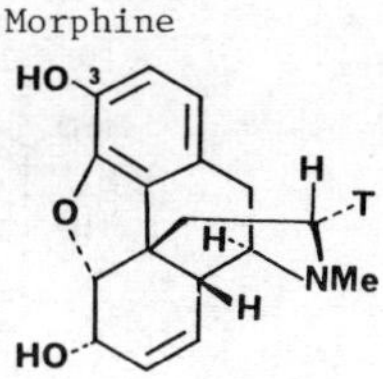

(46% ret. of ^{3}H, corrected for loss of ^{14}C present in the carboxyl group of tyrosine).

[^{3}H]Tyrosine (fed to a cell-suspension culture) — Morphine, Thebaine, Codeine — 313

(Labelled alkaloids were obtained. Protein synthesis inhibitors: puromycin and actinomycin promoted alkaloid formation at low concentrations, but inhibited their production at high concentrations. Cycloheximide also inhibited alkaloid formation).

Species / *Compound Administered*	*Alkaloids labelled*	*Ref.*
Papaver somniferum (continued) -		
L-[3-^{14}C]-3,4-Dihydroxy-phenylalanine (Dopa) (fed to fractions obtained from capsule latex)	Morphine, Codeine, Thebaine, Papaverine Noscapine (Narcotine) 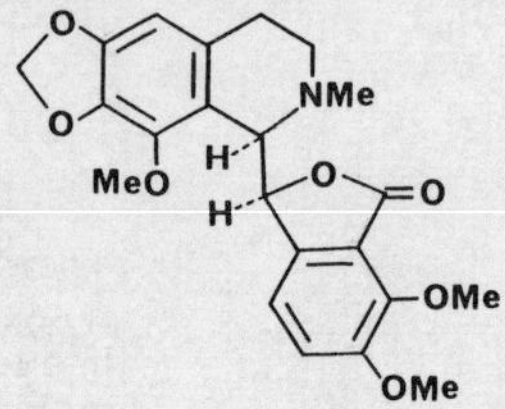	314

(Activity was found in all the above alkaloids. Biosynthetic activity was only present in the 1000 G fraction obtained by a centrifugal separation)

[2'-^{14}C, 2,6-^{3}H$_2$]Dopamine + 3,4-Dihydroxyphenylacetaldehyde (in cell-free system)	(*S*)-Norlaudanosoline (3,4-Dihydroxyphenylpyruvic acid did not serve as a substrate for this enzyme)	143
[16-^{3}H]Codeinone 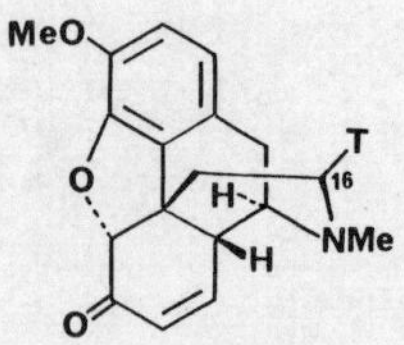 (added to cell-free enzyme with NADH)	Codeine (15-26% inc. 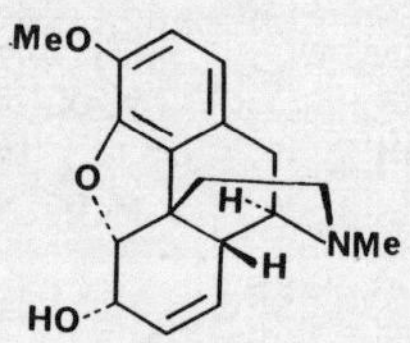 This reaction was also achieved in a cell-free system obtained from *Papaver bracteatum*, even though this species does not contain either alkaloid).	202
(*RS*)-[3-^{3}H]Reticuline 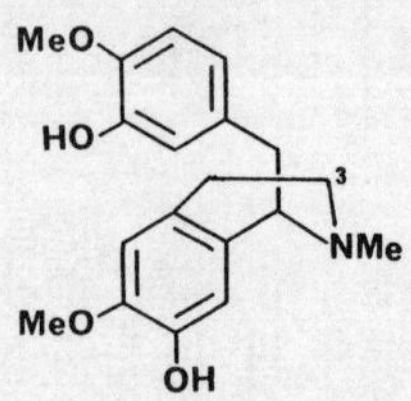 (in a cell-free extract,purified by centrifugal fractionation)	Salutaridine 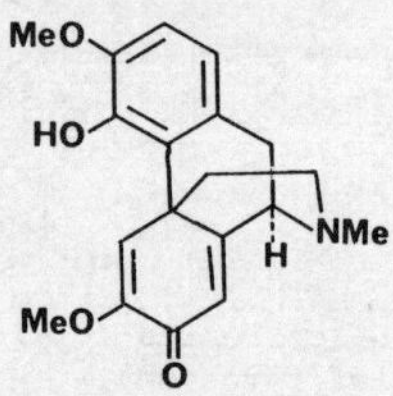 (In optimum experiments at pH 9 an 80-85% conversion was achieved, based on the consumed reticuline. The enzyme utilized H_2O_2 for the reaction. The radiochemical purity of the alkaloid was established by conversion to thebaine).	203

Species / *Compound Administered*	*Alkaloids labelled*	*Ref.*
Papaver somniferum (continued) -		
Morphine (not labelled) (added to crude enzyme fraction)	Pseudomorphine 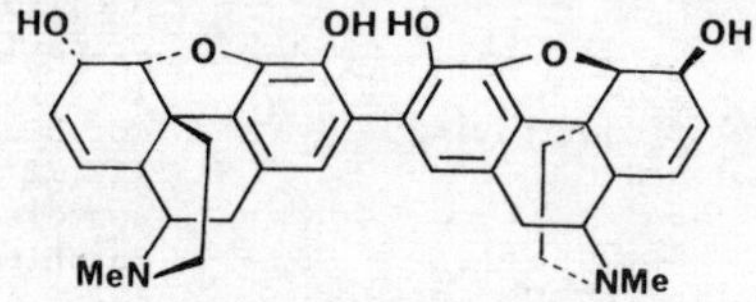 Morphine *N*-oxide	204
Thebaine (not labelled, added to cell-suspension culture)	Neopine 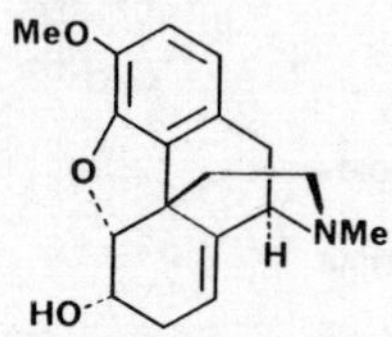	315
Codeinone (not labelled, added to cell-suspension culture)	Codeine (In earlier work[316] the production of codeine by this cell-suspension culture was reported).	
(*RS*)-[3-^{14}C, 1-^{3}H]Reticuline	Morphine (60.7% loss of ^{3}H from the chiral center)	317
(*RS*)-[3-^{14}C, 1-^{3}H]-6-Ethyl-4'-methyllaudanosoline 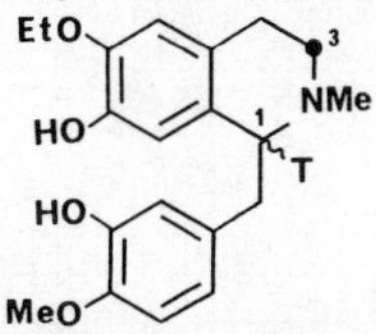	Morphine (3.28% inc. 100% ret. of ^{3}H)	
(*RS*)-[3-^{14}C, 1-^{3}H]-6-Methyl-4'- ethyllaudanosoline	Morphine 3-ethyl ether (0.44% inc. 100% ret. of ^{3}H). Morphine (0.56% inc., 100% ret. of ^{3}H)	
(*RS*)-[3-^{14}C, 1-^{3}H]-*N*-Ethyl-nor-reticuline	*N*-Ethylnormorphine (0.17% inc. 100% ret. of ^{3}H)	
(*RS*)-[3-^{14}C]-*N*-Ethylnor-reticuline	*N*-Ethylnormorphine (0.95% inc.) *N*-Ethylnorcodeine (0.65% inc.)	

(In these aberrant biosyntheses of the morphinan skeleton from reticuline analogs, the retention of tritium at C-1 is remarkable, since reticuline itself loses considerable tritium from this position).

[2-^{3}H]-6-Ethyl-6-desmethyl-thebaine	Morphine (9.4% inc.) Codeine (8.9% inc.)	318

(The facile de-ethylation of this thebaine analog is considered to support the oxidative mechanism proposed for the demethylation of thebaine[319]).

Species / *Compound Administered*	*Alkaloids labelled*	*Ref.*
Papaver somniferum (continued) -		
(*RS*)-[*N-methyl*-^{14}C]Reticuline	Morphine(4.8% inc. NMe: 99.2%) Oripavine (0.003% inc.) (see under *P. orientale*)	320
[*N-methyl*-^{14}C]Oripavine 3-ethyl ether	Morphine 3-ethyl ether (12.7% inc., NMe:98.5%. This conversion is an aberrant reaction) Morphine (3.6% inc.)	
(*RS*)-[3-^{14}C]-1-(3',4'-Dimethoxybenzyl)-6-ethoxy-7-methoxy-1,2,3,4-tetrahydroisoquinoline	1-(3',4'-Dimethoxybenzyl)-6-ethoxy-7-methoxyisoquinoline (11.25% inc.)	321

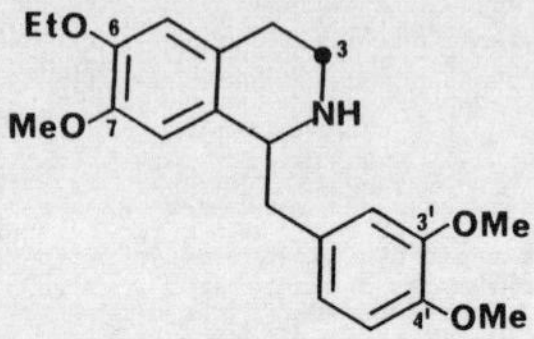

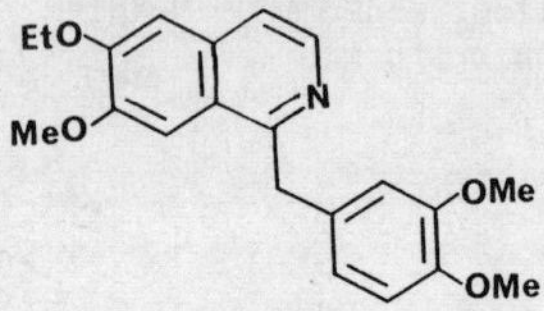

(*RS*)-[3-^{14}C]-1-(3',4'-Dimethoxybenzyl)-6-methoxy-7-ethoxy-1,2,3,4-tetrahydroisoquinoline	1-(3',4'-Dimethoxybenzyl)-6-methoxy-7-ethoxyisoquinoline (10.0% inc)	
(*RS*)-[3-^{14}C]-1-(3'-Ethoxy-4'-methoxybenzyl)-6,7-dimethoxy-1,2,3,4-tetrahydroisoquinoline	1-(3'-Ethoxy-4'-methoxybenzyl)-6,7-dimethoxyisoquinoline (1.91% inc.)	
(*RS*)-[3-^{14}C]-1-(3'-Methoxy-4'-ethoxybenzyl)-6,7-dimethoxy-1,2,3,4-tetrahydroisoquinoline	1-(3'-Methoxy-4'-ethoxybenzyl)-6,7-dimethoxyisoquinoline (1.50% inc.)	

(These transformations are examples of aberrant biosyntheses. Apparently, the dehydrogenation involved is more sensitive to bulky groups in the benzyl moiety)

Passiflora edulis

(*RS*)-1-[^{14}C]Methyl-1,2,3,4-tetrahydro-[8- ^{3}H]-β-carboline-1-carboxylic acid	Harman	284

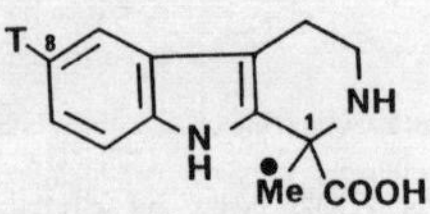

(Amino Acid-1)

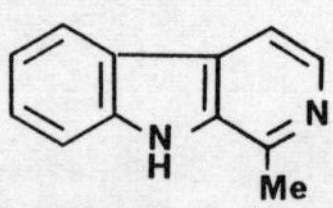

(0.5% inc. in winter, 1.5% inc. in spring, with almost complete ret. of tritium)

L-[5-^{3}H]Tryptophan	Harman (0.002% inc.) Amino acid-1 (0.014% inc.)	
N-Acetyl-[5-^{3}H]Tryptamine	Harman (The inc. of ^{14}C was 8 times that of ^{3}H, indicating that the ^{3}H-labelled precursor was probably incorporated via tryptamine, and not directly)	

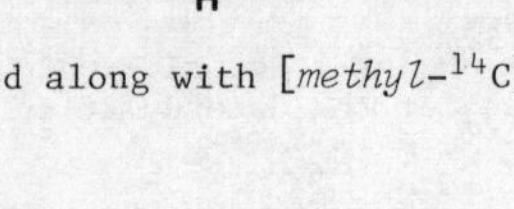

(fed along with [*methyl*-^{14}C]-Amino acid-1)

Species / *Compound Administered*	*Alkaloids labelled*	*Ref.*
Penicillium chrysogenum		
δ-(L-α-Aminoadipyl)-L-[3-^{3}H]-cysteinyl-D-[1-^{14}C]valine (to cell-free preparation)	Isopenicillin N (^{14}C: 6% inc. with 50% loss of ^{3}H)	179
δ-(L-α-Aminoadipyl)-L-cysteinyl-D-[1-^{14}C, 2-^{3}H]valine (to cell-free preparation)	Isopenicillin N (100% ret. of ^{3}H. Evidence is presented for the intermediacy of 1-[(1*R*)-carboxy-2-methylpropyl]-(3*R*)- (5*S*)-amino-5-carboxypentanamido -(4*R*)-mercapto-azetidin-2-ône (1) in the biosynthesis of isopenicillin N. However, this compound	

proved to be very unstable in neutral and basic solution[182], and it seems unlikely that it was isolated. The disulfide corresponding to this mercaptan failed to serve as a precursor of isopenicillin N in either this species or *Cephalosporium acremonium*.

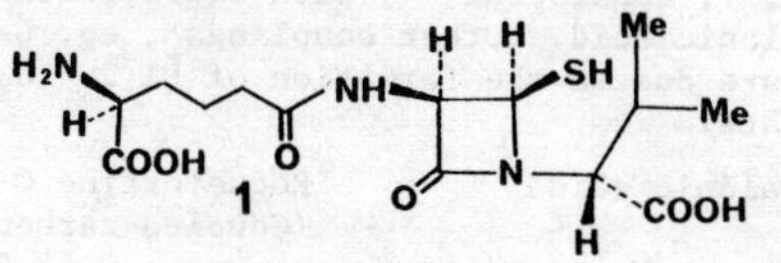

L-[U-^{14}C, 3-^{3}H]Valine	Penicillin V	322

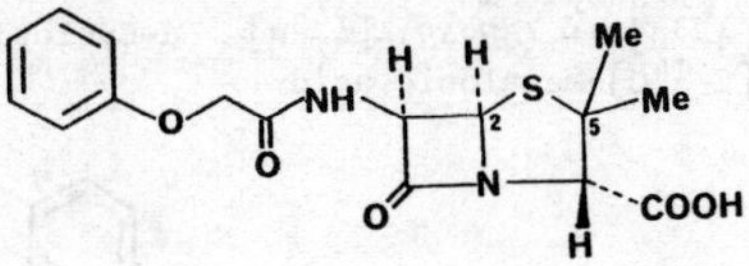

(There was no incorporation of ^{3}H. This result eliminates mechanisms for the formation of the penicillin molecule which would involve the migration of the C-3 hydrogen of valine to the SH group of the cysteine residue (in the LLD-ACV-tripeptide) followed by a Michael addition to an unsaturated valine residue).

Penicillium crustosum		
Sodium [1-^{13}C]Acetate Sodium [2-^{13}C]Acetate Sodium [1,2-^{13}C]Acetate	Penitrem A	323,324

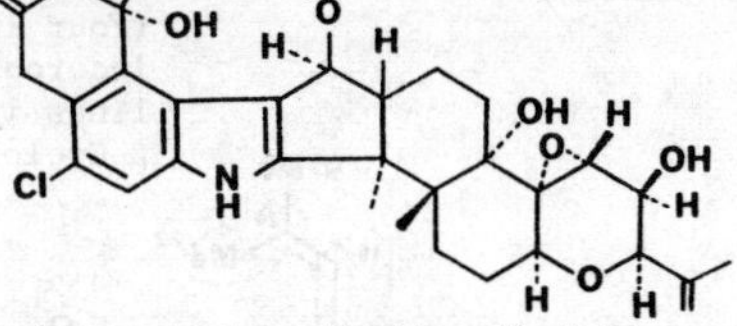

(The terpenoid origin of the non-indole portion of this alkaloid was apparent from the ^{13}C n.m.r. spectra of material derived from the labelled acetate. The penitrem A derived from [2-^{13}C]acetate exhibited several satellites due to carbon-carbon coupings which arose from [1,2-$^{13}C_2$]-acetate. This was formed from the methyl-labelled acetate via the Krebs cycle).

Species / *Compound Administered*	*Alkaloids labelled*	*Ref.*

Penicillium crustosum (continued) -

[1,2-$^{13}C_2$]Acetate — Roquefortine C — 325

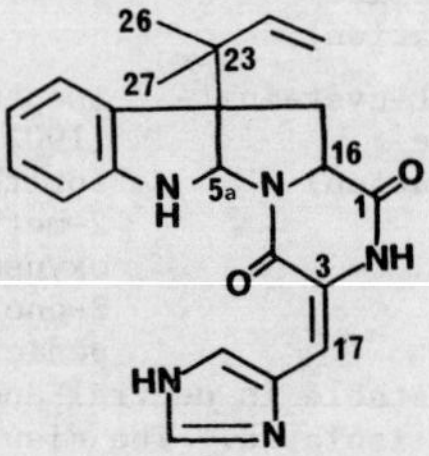

(In the ^{13}C n.m.r., $^{1}J_{CC}$ couplings were observed for C-23,24, and for C-23,27, consistent with equilibration of C-2 and C-3' of mevalonic acid. Other couplings , eg. C-1,16, were observed, and are due to the formation of [1,2-$^{13}C_2$]serine from the acetate)

[2,3-$^{13}C_2$]Mevalonic acid — Roquefortine C (Coupled carbons observed in the n.m.r. : C-23,26 and C-23,27).

Penicillium griseofulvum

(3*S*,4*S*)-[4-^{3}H]- + (3*R*,4*R*)-[4-^{3}H]- + (3*RS*)-[2-^{14}C]-Mevalonic acid lactone — α-Cyclopiazonic acid — 326

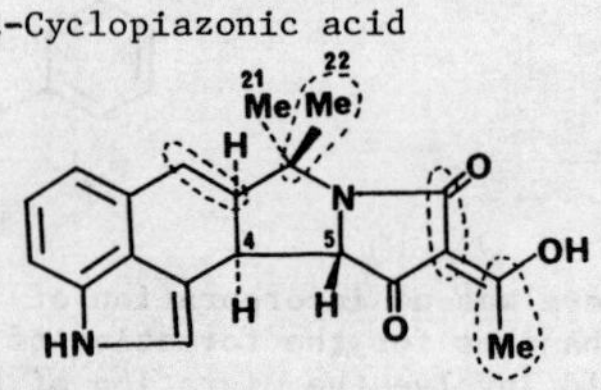

(104% ret. of ^{3}H)

(3*S*,4*S*)-[4-^{3}H]- + (3*R*,4*R*)-[4-^{3}H]- + (3*R*)-[3-^{14}C]-Mevalonic acid lactone — α-Cyclopiazonic acid (50% ret. of ^{3}H, indicating that the 3*R* isomer of mevalonic acid is the precursor, and the 4-*pro-R* hydrogen which is retained.

[1,2-$^{13}C_2$]Acetate — α-Cyclopiazonic acid (Four intact acetate units are incorporated, indicated with dotted lines in the above structure)

β-Cyclopiazonic acid

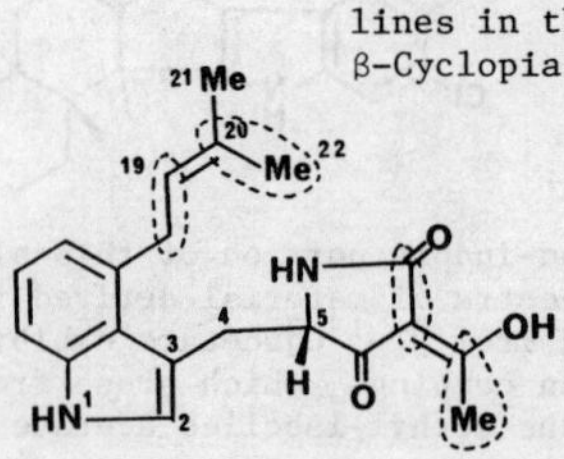

(Four intact acetate units are incorporated,indicated with dotted lines. The results are consistent with a *syn*-addition of the NH group to the 19-20 double bond).

Species / *Compound Administered*	*Alkaloids labelled*	*Ref.*
Penicillium griseofulvum (continued) -		
(2'*S*,3'*R*)-[3'-^{14}C,3'-^{3}H]-Tryptophan 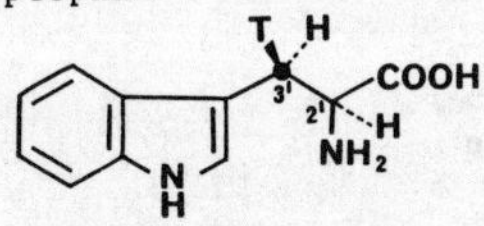	α-Cyclopiazonic acid (98% ret. of ^{3}H, which was located at C-4. Also [3'-$^{2}H_2$]tryptophan yielded material labelled with deuterium at C-4, established by ^{2}H n.m.r.) β-Cyclopiazonic acid (98% ret. of ^{3}H)	327
(2'*RS*,3'*SS*)-[3'-^{14}C,3'-^{3}H]-Tryptophan	α-Cyclopiazonic acid (95% loss of ^{3}H) β-Cyclopiazonic acid (98% ret. of ^{3}H)	
(2'*RS*,3'*RR*)-[3'-^{14}C,3'-^{3}H]-Tryptophan	α-Cyclopiazonic acid (94% ret. of ^{3}H) β-Cyclopiazonic acid (98% ret. of ^{3}H)	
(2'*S*,3'*RS*)- + (2'*R*,3'*RS*)-[3'-^{14}C,3'-^{3}H]Tryptophan	α-Cyclopiazonic acid (50% ret. of ^{3}H) β-Cyclopiazonic acid (94% ret. of ^{3}H)	
(2'*RS*)-[3'-^{14}C,2'-^{3}H]Tryptophan	α-Cyclopiazonic acid (49% ret. of ^{3}H. This result is explained by proposing that the *R*-isomer is incorporated by conversion to the *S*-isomer via the α-keto acid)	

(These results indicate that β-cyclopiazonic acid is converted to α-cyclopiazonic acid, possibly via a 1,4-didehydro derivative, but not via a 4,5-didehydro intermediate).

Penicillium oxalicum		
(2'*S*,3'*S*)-[2-^{14}C,3'-^{3}H]-Histidine 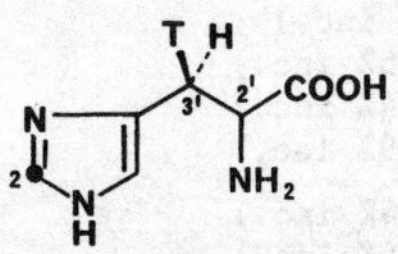	Oxaline (1-4% inc. 4.6% ret. of ^{3}H) 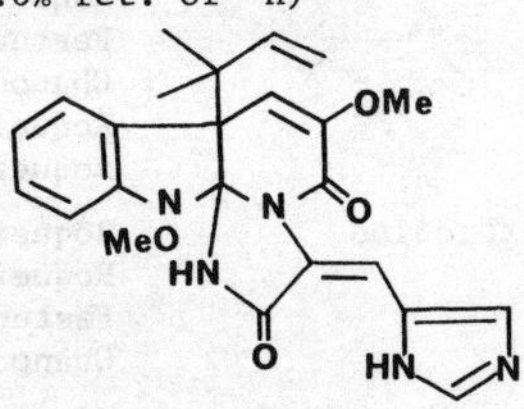	328
(2'*S*,3'*R*)-[2-^{14}C,3'-^{3}H]-Histidine	Oxaline (1-4% inc. 96% ret. of ^{3}H)	

(These results indicate that there is a *syn* elimination of the hydrogen at C-2' and the *pro-S* hydrogen at C-3' of the histidine).

Penicillium roqueforti		
DL-[3'-^{14}C]Tryptophan (All the reported inc. are for 2-day feedings, comparable inc. were obtained in 4-day incubations)	Roquefortine A (0.25% inc.) (see next page for structure)	329

Species *Compound Administered*	*Alkaloids labelled*	*Ref.*
Penicillium roqueforti (continued) -	Roquefortine A 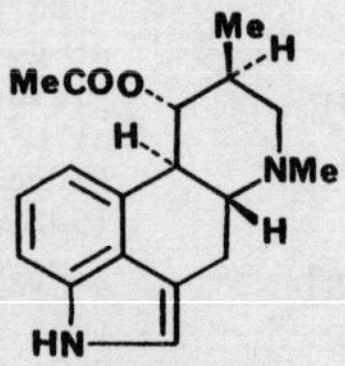 Roquefortine B (desacetyl-roquefortine A) (0.06% inc.) Festuclavine (0.21% inc.) 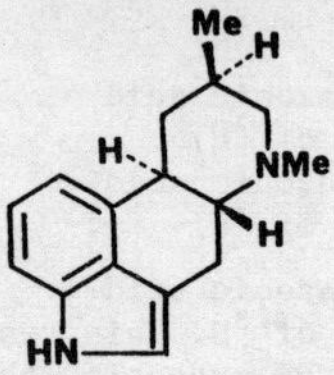 Chanoclavine-I (0.03% inc.) 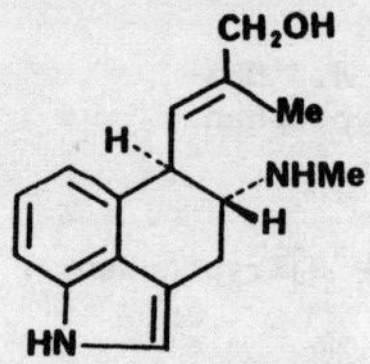 Roquefortine C (1.5% inc.) (see *Penicillium crustosum* for structure) Roquefortine D (0.17% inc.) (roquefortine C with $\Delta^{3,17}$ bond reduced)	329
DL-[1'-^{14}C]Tryptophan	Roquefortine C (0.91% inc) Roquefortine D (0.11% inc.)	
DL-[2-^{14}C]Mevalonic acid lactone	Roquefortine A (0.4% inc.) Roquefortine B (0.08 % inc.) Festuclavine (0.16% inc.) Chanoclavine-I (0.06% inc.) Roquefortine C (0.64% inc.) Roquefortine D (0.09% inc.)	330
L-[*methyl*-^{14}C]Methionine	Roquefortine A (0.14% inc.) Roquefortine B (0.05% inc.) Festuclavine (0.12% inc.) Chanoclavine-I (<0.01% inc.)	
L-[2-^{14}C]Histidine	Roquefortine C (3.9% inc.) Roquefortine D (0.38% inc.)	
[2-^{14}C]Mevalonic acid lactone	Roquefortine C (0.08% inc.)	
[3'-^{14}C]Tryptophan	Roquefortine C (0.15% inc.)	
[2-^{14}C]Histidine	Roquefortine C (1.12% inc.)	
[^{2}H]Tryptophan (a mixture of various deuterated species, but >95% at C-2 of indole nucleus)	Roquefortine C (No deuterium at C-5a. This result is consistent with initial substitution at C-2 of tryptophan with an isoprene unit, which then rearranges to the β-position).	

Species / *Compound Administered*	*Alkaloids labelled*	*Ref.*
Penicillium roqueforti (continued) -		
(2'*S*,3'*S*)-[2-^{14}C,3'-^{3}H]-Histidine	Roquefortine C (1-4% inc. 4.6% ret. of ^{3}H)	328
(2'*S*,3'*R*)-[2-^{14}C,3'-^{3}H]-Histidine	Roquefortine C (1-4% inc. 94.2% ret. of ^{3}H)	

(These results indicate that there is a *syn* elimination of the hydrogen at C-2' and the *pro-S* hydrogen at C-3' of histidine).

Phoma exigua		
Sodium [2-^{13}C,2-^{2}H$_3$]Acetate + inactive malonate	Cytochalasin B	331

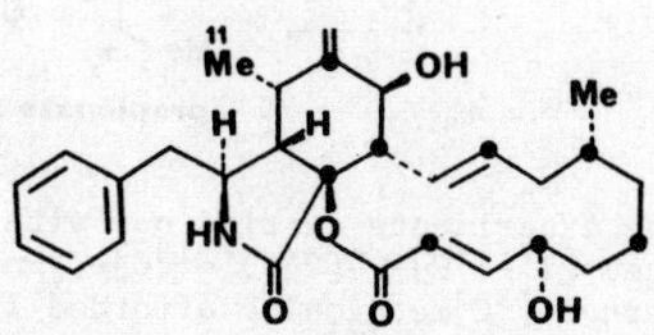

(^{13}C was incorporated into the positions indicated with heavy dots in the above structure. Enhancement of the signal (n.m.r.) arising from C-11 was not observed because of the presence of deuterium at this position . Thus this carbon is recognised as the starter unit for the polyketide chain).

Phoma lingam		
Sodium [1-^{13}C]Acetate	Sirodesmin PL (9.5% inc.	332

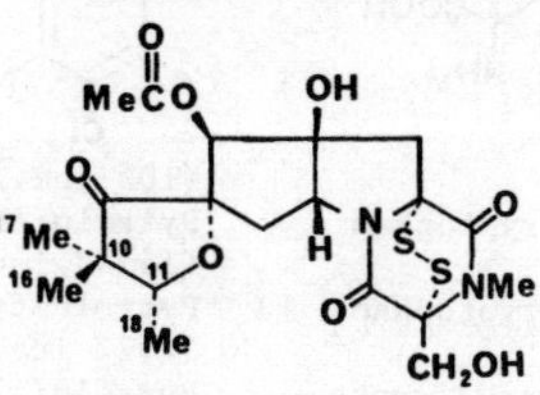

Enrichment at C-10, C-18, and the C=O of the acetyl group)

Sodium [1,2-^{13}C$_2$]Acetate	Sirodesmin PL (23% spec. inc. The ^{13}C n.m.r. indicated the following coupled carbons: C-10,16 and C-11,18, and the *O*-acetyl group)	
L-[U-^{14}C]Serine	Phomamide (0.29% inc.)	

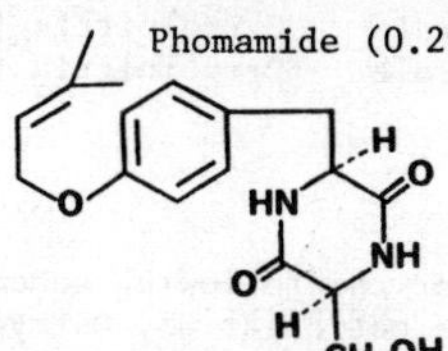

cyclo-L-[U-^{14}C]Tyrosine-L-serine	Sirodesmin PL (12.5% inc. 2.4% spec. inc.)	
Phomamide (*cyclo*-O-(γγ-Dimethyl-allyl)-L-[U-^{14}C]tyrosine-L-serine)	Sirodesmine PL (25.5% inc. 4.75% spec. inc.)	
L-[U-^{14}C]Tyrosine	Phomamide (0.49% inc.)	

Species — *Compound Administered* | *Alkaloids labelled* | *Ref.*

Pseudeurotium ovalis

Sodium [1-^{14}C]Acetate (0.17% inc.)
Sodium [2-^{14}C]Acetate (0.23% inc.)
Sodium [2-^{14}C]Malonate (1.4% inc.)
Sodium [1-^{14}C]Propionate (2.5% inc.)
L-[*methyl*-^{14}C]Methionine (7.2% inc.)
L-[U-^{14}C]Phenylalanine (10.5% inc.)
L-[4-^{3}H]Phenylalanine (8.7% inc.)

Pseurotin A — 333

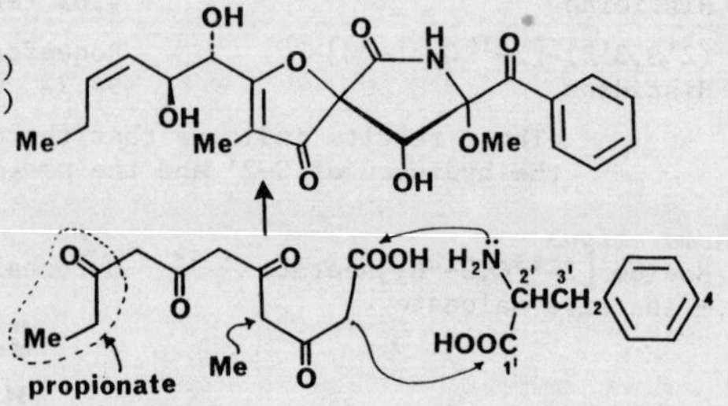

(Feeding experiments carried out with [1,2-^{13}C]propionate, [2-^{13}C]-malonate, [1'-^{13}C]-, [2',3'-^{13}C]-, and [2'-^{13}C,^{15}N]-phenylalanine, and [*methyl*-^{13}C]methionine afforded labelled pseurotin in which the location of the ^{13}C was determined by n.m.r. The results were consistent with its formation from one propionate starter unit (for the polyketide chain), four acetate units, a *C*-methyl and an *O*-methyl from methionine, and the rest from phenylalanine as depicted schematically above)

Pseudomonas aureofaciens

DL-[2-^{14}C,1-^{15}N]Tryptophan — Pyrrolnitrin — 334

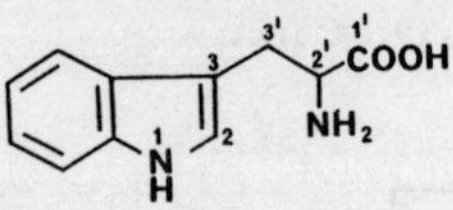

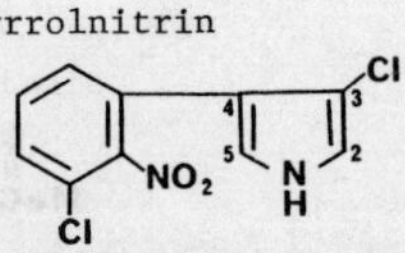

(10% inc., 107% ret. of ^{15}N)

DL-[3'-^{14}C,2'-^{15}N]Tryptophan — Pyrrolnitrin (15.6% inc., 19% ret. of ^{15}N)

DL-[3'-^{14}C,2'- ^{3}H]Tryptophan — Pyrrolnitrin (4.2% inc., 26% ret. of ^{3}H)

D-[3'-^{14}C,2'-^{3}H,2'-^{15}N]Tryptophan — Pyrrolnitrin (^{14}C: 18% inc.,2% ret. of ^{3}H, 10% ret. of ^{15}N)

L-[3'-^{14}C,2'-^{3}H]Tryptophan — Pyrrolnitrin (9.6% inc., 71% ret. of ^{3}H, all at C-2)

DL-[3'-^{13}C]Tryptophan — Pyrrolnitrin (^{13}C detected at C-3 by n.m.r.)

DL-[2-^{2}H]Tryptophan — Pyrrolnitrin (44% spec. inc.)

3-(*o*-Nitrophenyl)-[2-^{3}H]pyrrole — Pyrrolnitrin (0.36% inc.)

3-(*o*-Aminophenyl)-[2-^{3}H]pyrrole — Pyrrolnitrin (6.7% inc.)

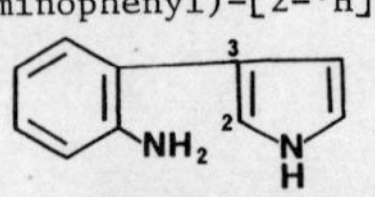

(These results favor the following scheme for pyrrolnitrin biosynthesis. It is not clear why D-tryptophan, which was a more efficient precursor than the L-isomer, loses tritium from C-2', and the side-chain amino nitrogen, prior to incorporation)

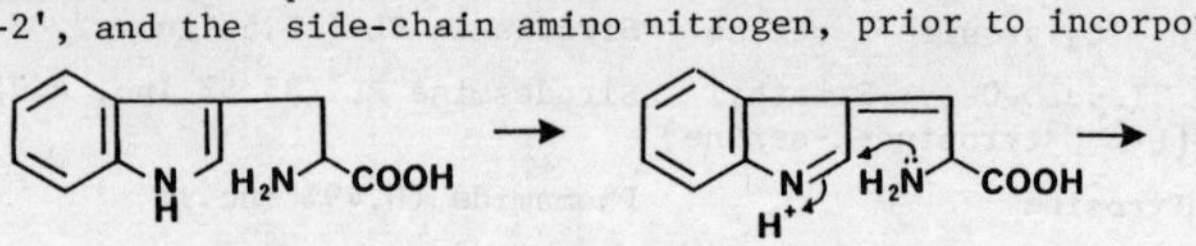

Species / *Compound Administered*	*Alkaloids labelled*	*Ref.*
Pseudomonas aureofaciens (continued)	-	334

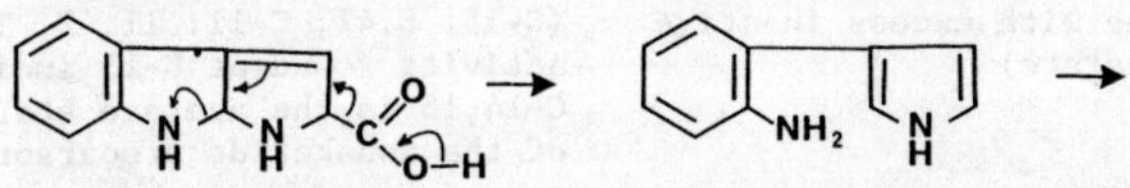

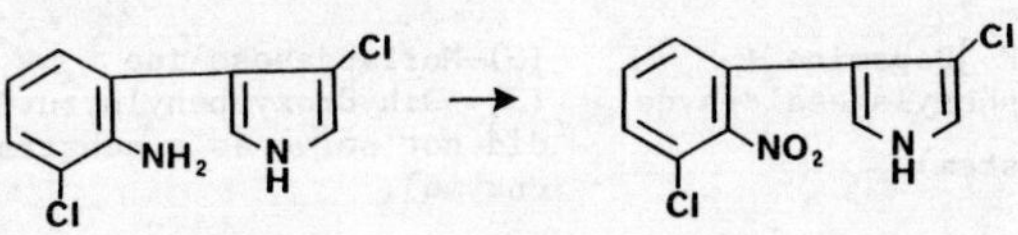

[*dicarboxy*-^{14}C]Phenazine-1,6-dicarboxylic acid	Phenazine-1-carboxylic acid	335

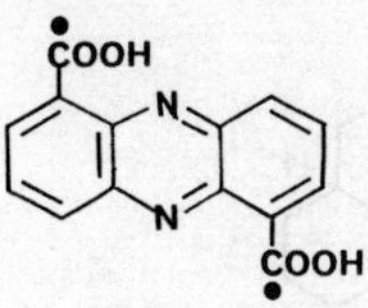

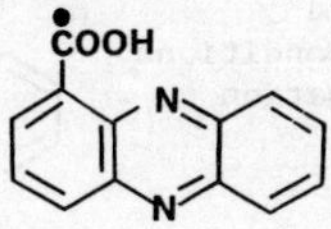

	(2.6% inc. in cells washed with Et_2O, 1.2% inc. in normal cells)	
	2-Hydroxyphenazine-1-carboxylic acid (0.2% inc. in Et_2O-treated cells)	
Dimethyl [*dicarboxy*-^{14}C]Phenazine-1,6-dicarboxylate	Phenazine-1-carboxylic acid (very low inc.)	
Methyl [*carboxy*-^{14}C]Phenazine-1-carboxylate	Phenazine-1-carboxylic acid (14, 16% inc. with Et_2O-treated cells, 4% inc. with normal cells).	
[1,6,7-^{14}C]Shikimic acid (equal labelling at each site)	Phenazine-1-carboxylic acid (The COOH had 21% of the total activity)	216
(for related work see under *Brevibacterium iodinum*)	2-Hydroxyphenazine-1-carboxylic acid (COOH: 21%)	
Pyrenochaeta terrestris		
Sodium [1-^{14}C]Acetate	Phomazarin (35.9% spec. inc.)	336
Sodium [2-^{13}C]Acetate Sodium [1,2-$^{13}C_2$]Acetate Diethyl [2-^{13}C]Malonate		

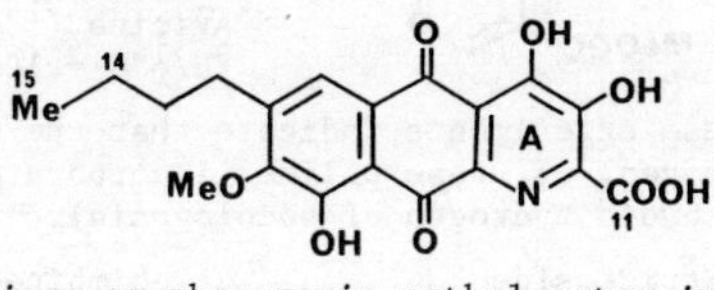

(^{13}C n.m.r, determinations on phomazarin methyl ester indicated enrichments consistent with the formation of this alkaloid from a nonaketide, with subsequent ring opening of ring A as illustrated below:

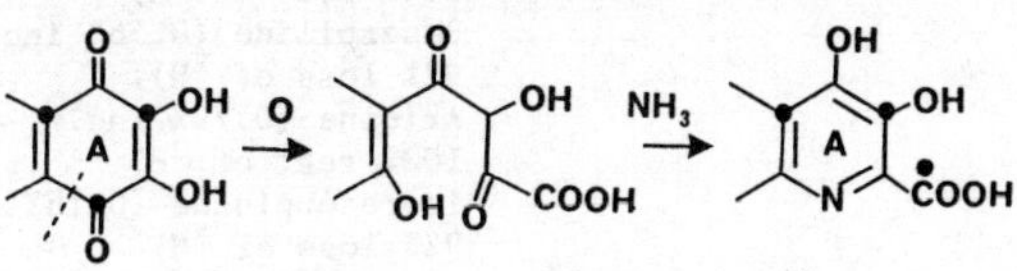

• labelled atoms derived from [2-^{13}C]acetate

Species *Compound Administered*	*Alkaloids labelled*	*Ref*

Pyrenochaeta terrestris (continued) -

Diethyl [2-^{14}C]Malonate (fed along with excess inactive sodium acetate)

Phomazarin (C-15: 8.4%, C-11: 11.4%. The lower activity found at C-15 indicates that C-14,15 is the acetate starter unit of the nonaketide precursor) — 336

Ranunculus flammula

[2'-^{14}C,2,6-$^{3}H_2$]Dopamine + 3,4-Dihydroxyphenylacetaldehyde (cell-free system)

(*S*)-Norlaudanosoline (3,4-Dihydroxyphenylpyruvic acid did not serve as a substrate for this enzyme). — 143

Rauwolfia canescens

[7-^{3}H]Secologanin + [2'-^{14}C]-Tryptamine. These were fed at the same time, under conditions where no chemical condensation occurs. $^{3}H/^{14}C$ = 4.21

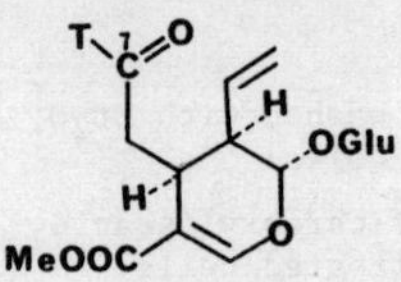

Secologanin

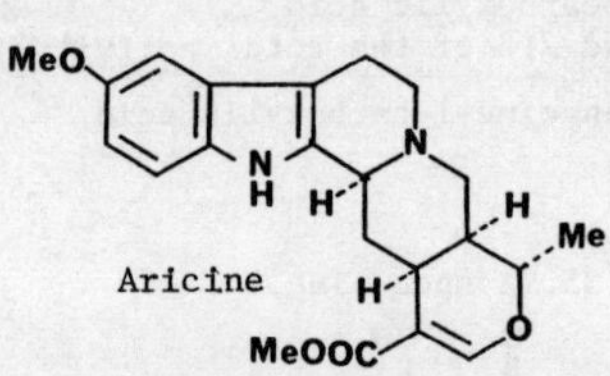

Aricine

α-Yohimbine (^{14}C: 0.14% inc. $^{3}H/^{14}C$ = 8.15) — 236

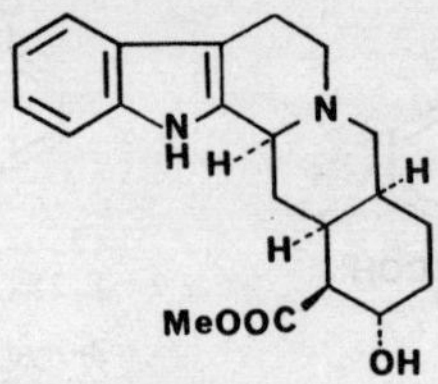

Isoreserpiline (^{14}C: 0.05% inc. $^{3}H/^{14}C$ = 14.82)

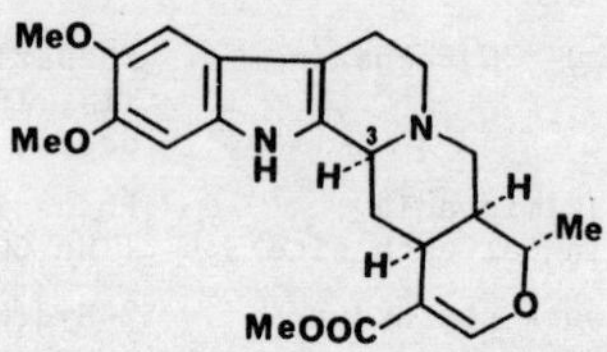

Reserpiline (3β-epimer of isoreserpiline) (^{14}C: 0.04% inc. $^{3}H/^{14}C$ = 0.98)

Aricine (^{14}C: 0.02% inc. $^{3}H/^{14}C$ = 10.49)

(These experiments indicate that the alkaloids having a 3β-hydrogen, eg. reserpiline, incorporate little of the aldehydic hydrogen of secologanin).

[6-^{14}C,3-^{3}H]Strictoside (see under *Catharanthus roseus* for structure)

α-Yohimbine (1.63% inc. 107% ret. of ^{3}H)
Isoreserpiline (1.46% inc. 104% ret. of ^{3}H)
Reserpiline (0.56% inc. 92% loss of ^{3}H).
Aricine (0.79% inc. 102% ret. of ^{3}H)
Isoreserpinine (0.16% inc. 92% loss of ^{3}H)

(see next page for structure)

Species *Compound Administered*	*Alkaloids labelled*	*Ref.*
Rauwolfia canescens (continued) -		
	Isoreserpinine 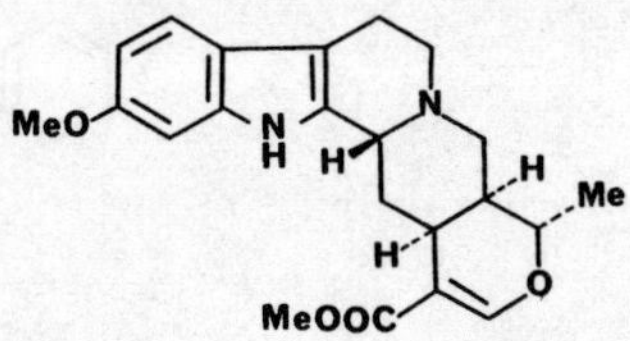	236
[6-^{14}C,3-^{3}H]Vincoside	Negligible (<0.001%) inc. into the alkaloids in this species.	
Rauwolfia verticillata [1'-^{14}C]Tryptamine + secologanin (a cell-free extract from a cell-suspension culture)	Cathenamine 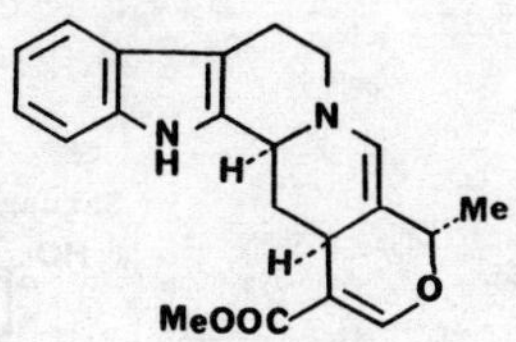	161
	Vallesiachotamine (*E*,Δ^{19}), Isovallesiachotamine (*Z*,Δ^{19}) 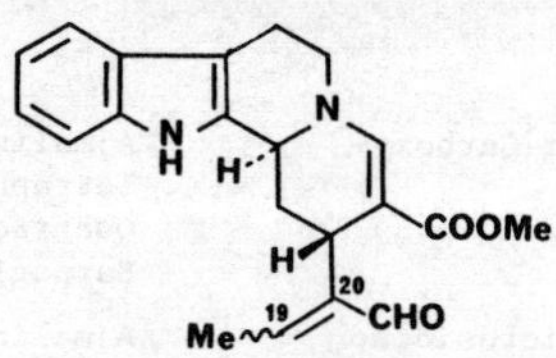	
	22-*N*-Tetrahydroalstonine 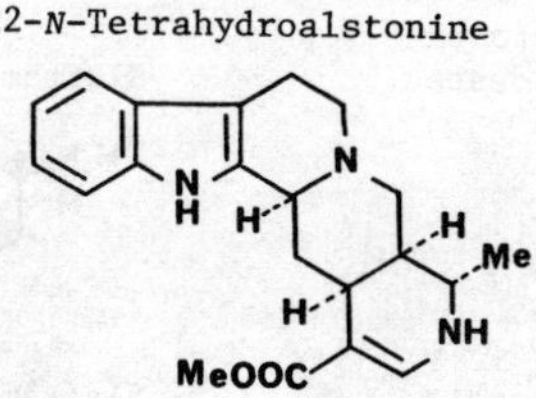 (This compound is probably an artifact, resulting from reaction of $^{+}NH_4$ with the corresponding oxygen heterocycle).	
Rauwolfia vomitoria [6-^{14}C,10-^{3}H]-5α-Carboxystrictosidine (see under *Gelsemium sempervirens*)	Ajmaline (no inc.) 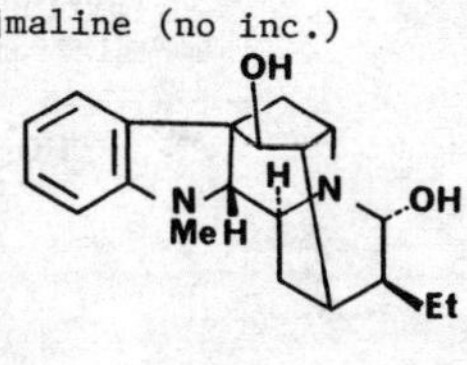	288

Species *Compound Administered*	*Alkaloids labelled*	*Ref.*
Rauwolfia vomitoria (continued) -		
	Tetraphyllicine (<0.01% inc.)	288

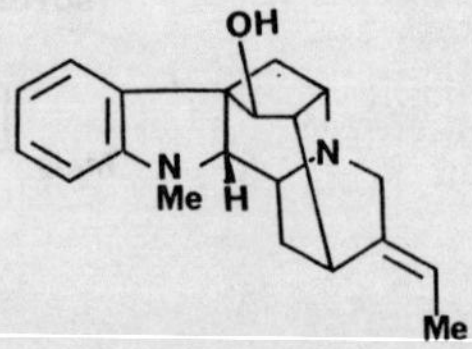

Quebrachidine (<0.01% inc.)

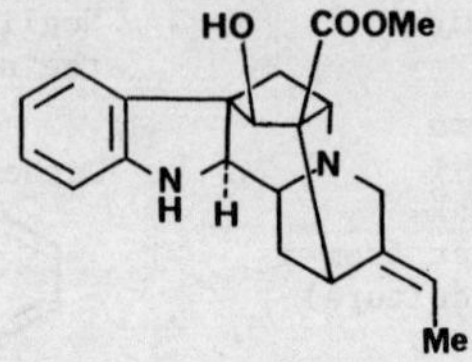

Sarpagine (<0.01% inc.)

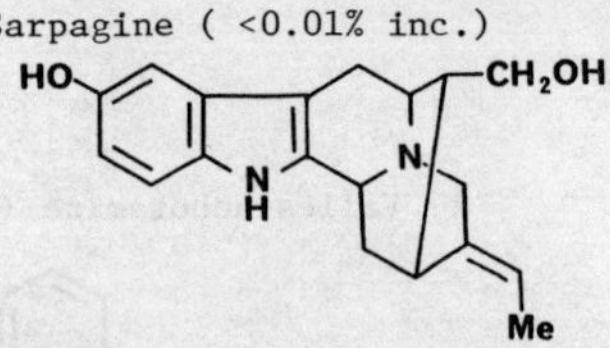

[6-^{14}C,10-^{3}H]-5α-Carboxy-vincoside	Ajmaline (inactive) Tetraphyllicine (<0.01% inc.) Quebrachidine (<0.01% inc.) Sarpagine (<0.01% inc.)	
[6-^{14}C,3-^{3}H]Strictosidine	Ajmaline (^{14}C: 0.72% inc. 92% ret. of ^{3}H)	236
Rhizoctonia leguminicola		
Sodium [1-^{14}C]Malonate	Slaframine (0.023-0.44% specific inc.)	337

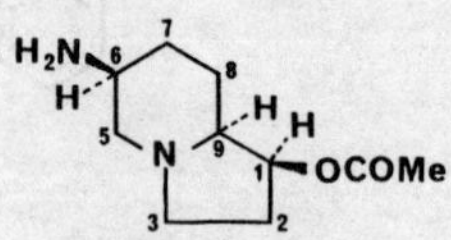

Sodium [2-^{14}C]Malonate — Slaframine (0.051-2.0% spec. inc.)

(Activity was equally divided between the acetyl group and the heterocyclic ring system. The incorporation was dependent on the concentration of the precursor in the medium).

3,4,5-Trihydroxy-octahydro-1-pyrindine(inc. about the same as into slaframine).

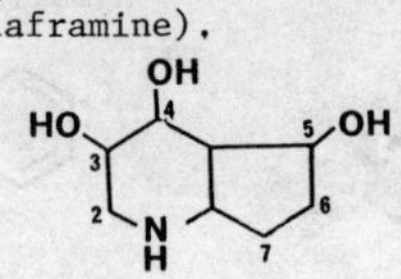

Species — *Alkaloids labelled* — *Ref.*
Compound Administered

Rhizoctonia leguminicola (continued) -

Sodium [1-^{14}C]Acetate — Slaframine (0.02-0.62% spec. inc.) (This precursor was somewhat less efficient as a precursor than [2-^{14}C]-acetate, when fed under the same conditions). — 337

Sodium [2-^{2}H$_3$]Acetate — Slaframine (Mass spectrometry on the desacetylslaframine indicated the presence of two deuterium atoms at C-2).
3,4,5-Trihydroxy-octahydro-1-pyrindine (Mass spectrometry indicated the presence of only one deuterium atom, presumably at C-6, not at C-7 as postulated in the original publication).

(*RS*)-[^{3}H]Pipecolic acid — Slaframine (labelled)
3,4,5-Trihydroxy-octahydro-1-pyrindine
(It is not clear whether the (*R*) or (*RS*)-pipecolic acid was fed. At certain times the (*R*)- isomer is mentioned, but this seems an unlikely precursor since slaframine has the (*S*) configuration at C-9).

[3-^{14}C]-1,3-Dioxo-octahydro-indolizine — Slaframine (0.29% spec. inc. The inc. was not significantly different from that of acetate or malonate. It is presumed that no direct incorporation takes place, but it is degraded to pipecolic acid and [1-^{14}C]acetate prior to incorporation). — 338

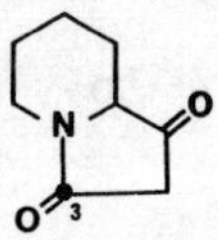

[1,3,3-^{2}H$_3$]-1-Hydroxy-octahydroindolizine — Slaframine
(As expected, the (1*S*) isomer was a more efficient precursor than the (1*R*) epimer , which was presumably incorporated via the 1-keto derivative).

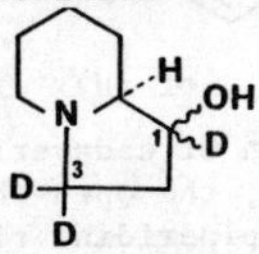

[U-^{2}H$_9$]Pipecolic acid — Swainsonine — 339

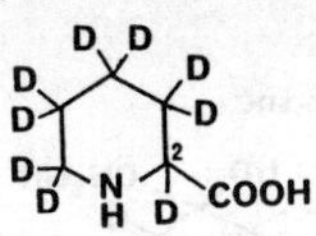

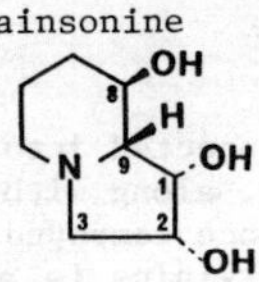

(The major isotopically labelled species was d$_7$, deuterium not being present at C-8 or C-9)

Slaframine
(The major isotopically labelled species was d$_7$).

[2-^{2}H,G-^{3}H]Pipecolic acid — Slaframine (^{2}H and ^{3}H incorporated)
Swainsonine (no ^{2}H incorporated)

Species / *Compound Administered*	*Alkaloids labelled*	*Ref.*
Rhizoctonia leguminicola (continued)-		
[1-$^2H_2$]Ethanol	Swainsonine (Deuterium incorporated into the C-9 position)	339
Rhodopseudomonas viridis		
[1,4-^{14}C]Putrescine (fed to cell-free system)	Homospermidine (It is proposed that the putrescine is oxidized to 4-aminobutanal, which forms a Schiff base with putrescine. Reduction with NADH then yields the homospermidine)	84

Homospermidine

Ruta graveolens		
Sodium [1-^{13}C]Acetate Sodium [2-^{13}C]Acetate Sodium [1,2-$^{13}C_2$]Acetate (fed to cell-suspension culture)	Rutacridone	340

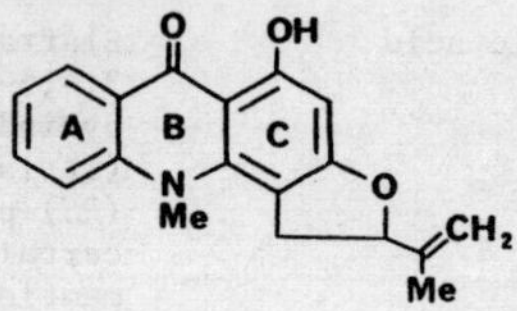

(^{13}C n.m.r. of the alkaloid derived from [1,2-^{13}C]acetate showed six bonded ^{13}C-^{13}C pairs arising from the carbons of ring C. This result indicates that a free aminobenzophenone is probably an intermediate in the biosynthesis).

Sedum acre		
(1*S*)-[1,5-^{14}C,1-^3H]Cadaverine	Sedamine (48% loss of ^3H)	341

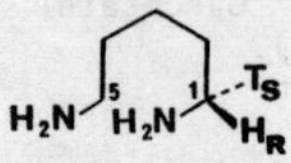

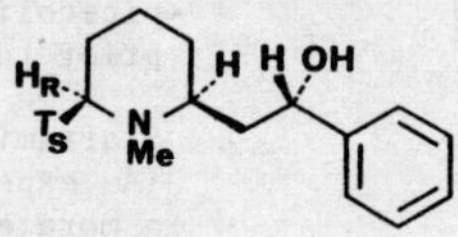

(1*R*)-[1,5-^{14}C,1-^3H]Cadaverine	Sedamine (100% ret. of ^3H)	

(These results indicate that the 1*S*-hydrogen of cadaverine is lost in the formation of 5-aminopentanal, the open-chain form of Δ^1-piperideine, a precursor of the piperidine ring of this alkaloid).

Senecio isatideus		
(Potential precursors were injected into the xylem through stem punctures, along with L-[5-^3H]arginine as a reference compound. The ratio of $^3H/^{14}C$ = 5.0. Arginine is an established precursor of retronecine).	Retrorsine	86,87

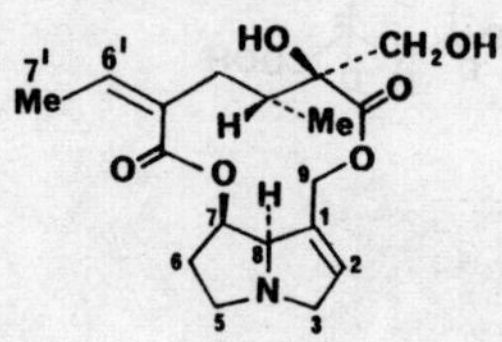

Retronecine

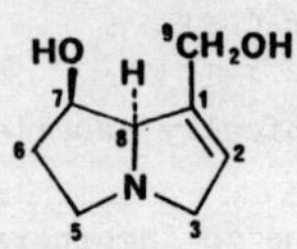

Species / *Compound Administered*	*Alkaloids labelled*	*Ref.*
Senecio isatideus (continued) -		86,87
L-[U-^{14}C]Proline	Retrorsine (^{14}C: 0.04% inc. ^{3}H/^{14}C = 77, 68% of ^{14}C in retronecine, 30% in isatinecic acid).	
[U-^{14}C]-4-Aminobutanoic acid	Retrorsine (^{14}C: 0.04% inc. ^{3}H/^{14}C = 37, 52% of ^{14}C in retronecine, 48% in isatinecic acid).	
L-[U-^{14}C]Glutamic acid	Retrorsine (^{14}C: 0.12% inc. ^{3}H/^{14}C = 10, 48% of ^{14}C in retronecine, 58% in isatinecic acid).	
L-[U-^{14}C]Arginine	Retrorsine (^{14}C: 0.46% inc. ^{3}H/^{14}C = 5.5, 99% of ^{14}C in retronecine).	
DL-[5-^{14}C]Ornithine	Retrorsine (^{14}C: 0.25% inc. ^{3}H/^{14}C = 6.2, 97% of ^{14}C in retronecine, C-9: 25%, C-5,6,7: 22%).	
[1,4-^{14}C]Putrescine	Retrorsine (^{14}C: 1.6% inc. ^{3}H/^{14}C = 0.60, 94% of ^{14}C in retronecine, C-9: 24%, C-5,6,7: 22%).	
[6,9-^{14}C]Spermine	Retrorsine (^{14}C: 5.2% inc. ^{3}H/^{14}C = 0.73, C-9: 23%, C-5,6,7: 23%).	

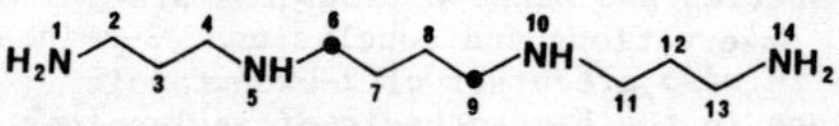

[6,9-^{14}C]Spermidine	Retrorsine (^{14}C: 2.0% inc. ^{3}H/^{14}C = 0.80, C-9: 22%, C-5,6,7: 24%).	

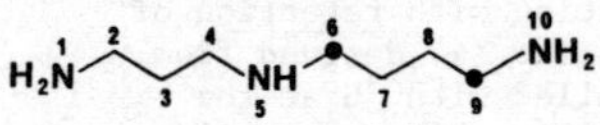

The following compounds were fed alone, ie. not with [^{3}H]arginine.

[1,4-$^{13}C_2$,^{14}C]Putrescine	Retronecine (equal enrichments, establish- by ^{13}C n.m.r. at C-3,5,8, and 9).	90
[2,3-$^{13}C_2$,1,4-^{14}C]Putrescine	Retronecine (^{14}C: 0.5% spec. inc. Contiguous ^{13}C atoms were detected in the n.m.r. (satellites) at C-1,2 and C-6,7, with enrichments of 0.16% and 0.18% respectively).	
[1-^{13}C,1,4-^{14}C,1-*amino*-^{15}N]-Putrescine	Retronecine (2.2% spec. inc. In its ^{13}C n.m.r. spectrum, C-3 and C-5 exhibited satellites due to the contiguous ^{13}C and ^{15}N atoms. $^{1}J_{3,N}$ = 5 Hz, $^{1}J_{5,N}$ = 4.5 Hz).	92
[2,10-^{14}C]Homospermidine	Retrorsine (0.5% inc., 0.3% spec. inc. Degradation of the retronecine produced on hydrolysis had 44±4% at C-9, consistent with the direct incorporation of this compound).	

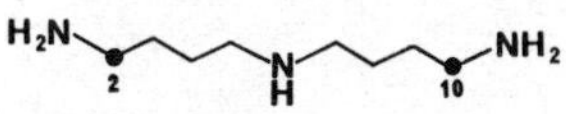

DL-[5-^{14}C]Ornithine	Homospermidine (isolated by dilution, 0.5% inc.)	
		342,343
(2*RS*)-[3-^{14}C,3-^{3}H]-2-Amino-butanoic acid (1)	Retrorsine (^{14}C: 3.4% inc., 98% in the isatinecic acid, 56% ret. of ^{3}H in the C-6',7' portion, which contained 43% of the ^{14}C. Similar results were obtained with the (*R*) and (*S*)-1).	

Species *Compound Administered*	*Alkaloids labelled*	*Ref.*
Senecio isatideus (continued) -		
(2*S*)-[U-^{14}C]Isoleucine + (2*S*,4*R*)-[4-^{3}H]Isoleucine	Retrorsine (^{14}C: 2.2% inc., 95% ret. of ^{3}H in the isatinecic acid)	342, 343
Senecio magnificus		
(2*RS*,3*S*)-[3-^{14}C,3-^{3}H]-2-Amino-butanoic acid	Senecionine (^{14}C: 0.015% inc., 95.3% loss of ^{3}H)	342, 343
(2*S*)-[U-^{14}C]Isoleucine + (2*S*,4*S*)-[3,4-^{3}H]Isoleucine	Senecionine (^{14}C: 0.004% inc., 93.5% loss of ^{3}H)	
(2*S*)-[U-^{14}C]Isoleucine + (2*RS*,4*RS*)-[3,4-^{3}H]Isoleucine	Senecionine (^{14}C: 0.05% inc., 94.9% loss of ^{3}H).	

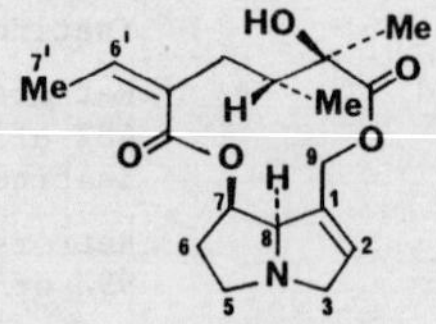

(The results obtained in this species and *Senecio isatideus* are consistent with the following observations and conclusions. 2-Amino-butanoic acid (1) serves as an *in vivo* precursor of 2-oxobutanoic acid (2), the normal intermediate in the biosynthesis of isoleucine. Reaction of this keto-acid with thiamine-bound acetaldehyde yields (3), which then undergoes an ethyl migration with retention of configuration. Thus the ultimate isoleucine (4) derived from (3*S*)-[3-^{3}H]-2-aminobutanoic acid is labelled with ^{3}H at the 4*S*- position. In the formation of the necic acids, two molecules of isoleucine condense, with loss of the carboxyl group, oxidation of C-2 to COOH, and loss of the C-4 *pro-S* hydrogen. These reactions are illustrated below:

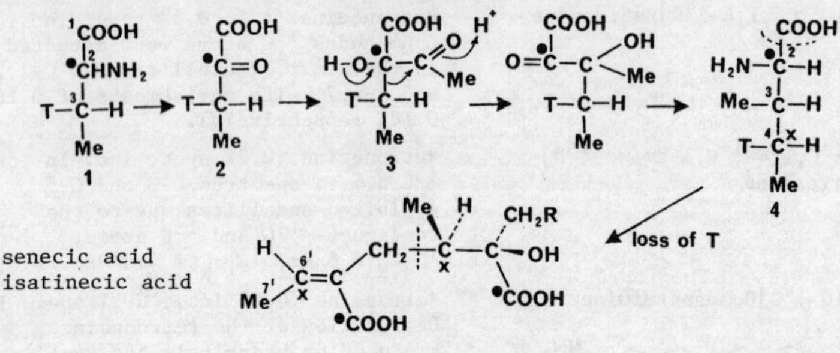

R = H senecic acid
R = OH isatinecic acid

Senecio vulgaris		
[1-^{13}C,1,4-^{14}C,1-*amino*-^{15}N]-Putrescine	Retronecine (6.6% spec. inc. This necine base was obtained by hydrolysis of senecionine, seneciphylline, and	88,91

retrorsine, the major alkaloids of this species. Enrichment was observed at C-3,5,8, and 9. Both C-3 and C-5 exhibited satellites in the ^{13}C n.m.r. due to coupling with adjacent ^{15}N; $^{1}J_{3,N}$ = 5.1 Hz, $^{1}J_{5,N}$ = 5.8 Hz. See Highlights,2.2. for discussion of these results)

Species / *Compound Administered*	*Alkaloids labelled*	*Ref.*
Senecio vulgaris (continued) -		
L-[4-^{3}H]Ornithine	Retronecine (Activities of degradation products are consistent with the following distribution of activity: C-2: 25%, C-6: 50%, C-7: 25%)	88,91
DL-[5-^{3}H]Ornithine (This material contained 14% of [4-^{3}H]ornithine)	Retronecine (Activity in degradation products consistent with one ^{3}H at C-8 and C-9, and three at C-3 and C-5)	
DL-[5-^{14}C]Ornithine	Retronecine (C-8: 23%, C-9: 25%)	
[2-^{14}C]-Δ^1-Pyrroline	Retronecine (no inc.)	
[5-^{14}C,5-^{3}H]-Δ^1-Pyrroline 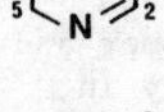	Retronecine (no inc.)	
[6,9-^{14}C,2,9-^{3}H]Spermidine 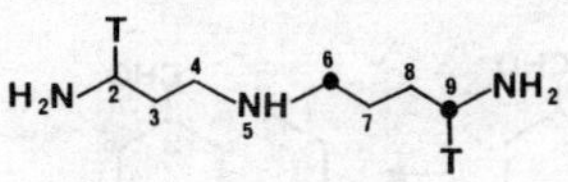	Retronecine (^{14}C: C-8: 24%, C-9: 24%, ^{3}H: C-8: 21%, C-9: 34%. This distribution of tritium is different from that observed with [5-^{3}H]ornithine, and may be explained by the production of [1-^{3}H]-4-aminobutanal from the spermidine)	
Solanum tuberosum		
[^{3}H]Solanidine (fed to tuber disks and a cell-suspension culture) 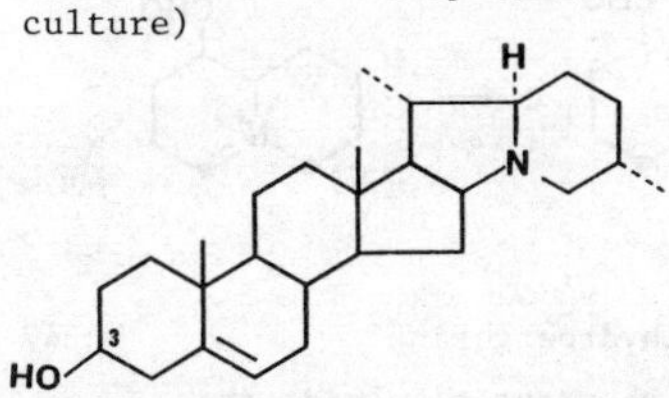	α-Chaconine (3-β-*O*-glucosylsolanidine) (The subsequent formation of 3-β-*O*-glucosyl(glucosyl)solanidine was also observed)	344
Sophora griffithii		
[^{14}C]Cytisine (obtained biosynthetically from [^{14}C]lysine) 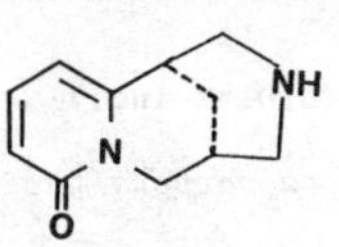	Cytisine (recovered, 3.5% spec. inc.) *N*-Methylcytisine (0.8% spec. inc.) Argentine (1.0% spec. inc.) 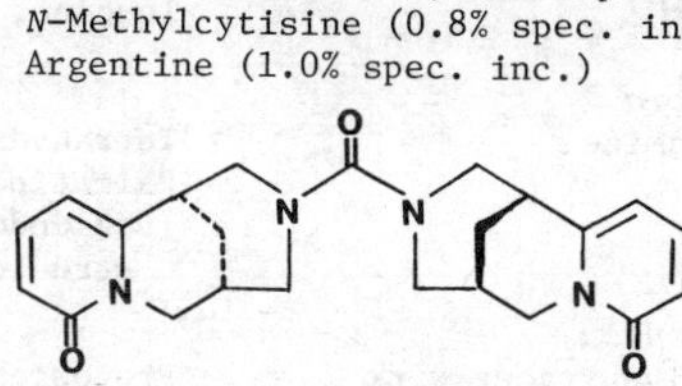	345
Sophora microphylla		
[6-^{14}C]-Δ^1-Piperideine	Matrine (see next page for structure) (~0.01% inc. C-2: 25%, C-10: 25%, C-15: 17%. See next page for discussion of these results).	346

Species | *Alkaloids labelled* | *Ref.*
Compound Administered

Sophora tetraptera
[6-^{14}C]-Δ^1-Piperideine | Matrine | 346

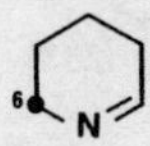

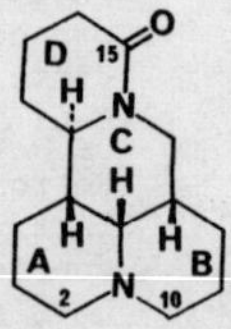

($\sim$0.01% inc. C-2: 26%, C-10: 25%, C-15: 16% . In this experiment and the one carried out in *Sophora microphylla*, only 2/3 of the activity was recovered in C-2,10, and 15. The higher level of activity found at C-2 and C-10 is rationalized on the basis of the following scheme, in which rings A and B of matrine are derived from a dimer of Δ^1-piperideine (1) . This then later condenses with the monomer to yield the carbon skeleton of matrine . Different pool sizes of the monomer and dimer could then result in different levels of activity in ring D and rings A/B.

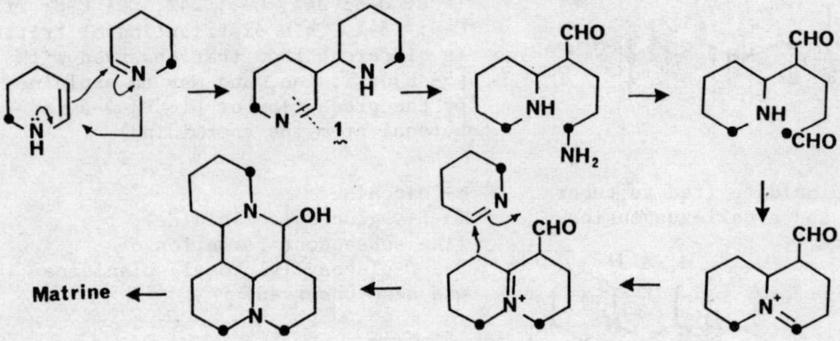

Sphacelia sorghi
DL-[2-^{14}C]-4-Hydroxyproline | 9'-Hydroxydihydroergosine | 347

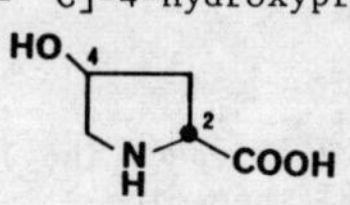

(This is an aberrant alkaloid; the normal dihydroergosine contains the tripeptide moiety derived from proline, leucine, and phenylalanine).

Stephania glabra
[2'-^{14}C]Dopamine | Tetrahydropalmatine (0.018% inc.) | 248
Palmatine (0.02% inc.)
(see under *Cissampelos pariera* for structures)

Stephania japonica
(The extensive tracer work reported in these definitive publications has been previously described in preliminary communications, and reviewed : *Biosynthesis*, 1976, 4, 168 ; 1980, 6, 194) | Protostephanine | 348-351
Hasubanonine

Species / *Compound Administered*	*Alkaloids labelled*	*Ref.*
Stizolobium hassjoo		
(*RS*)-[2'-^{14}C]-3,4-Dihydroxyphenylalanine (fed to callus tissue)	3-Carboxy-6,7-dihydroxy-1,2,3,4-tetrahydroisoquinoline 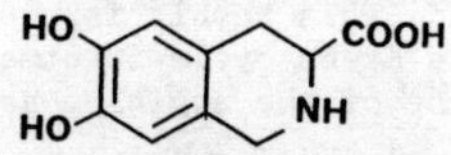 1-Methyl-3-carboxy-6,7-dihydroxy-1,2,3,4-tetrahydroisoquinoline 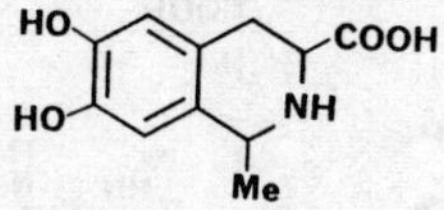 Stizolobic acid 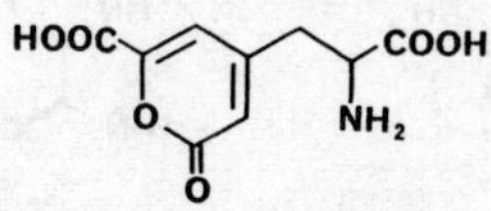 (A significant incorporation of activity was obtained into the above amino-acids)	352, 353
[2-^{14}C]Acetate [2-^{14}C]Pyruvate	Negligible incorporation into the above amino-acids; a surprising result.	
Streptomyces species (strain E/784)		
[*carboxy*-^{14}C]-3-Amino-5-hydroxybenzoic acid 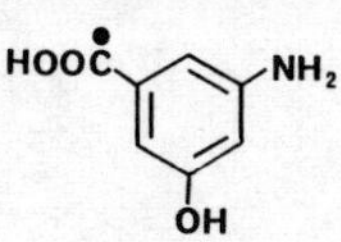	Actamycin (7.9% inc., 8.6% spec. inc.) 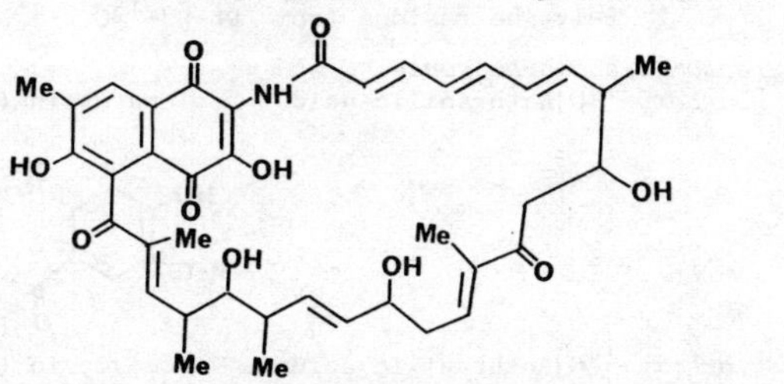	118

(Activity only present in the naphthaquinone moiety)

[2,3,4,5-^{14}C]Shikimic acid (see under *Brevibacterium iodinum*)	Actamycin (no incorporation)	
Streptomyces species (strain L-1689-23)		
[1,2-$^{13}C_2$]Acetate	Streptothricin F 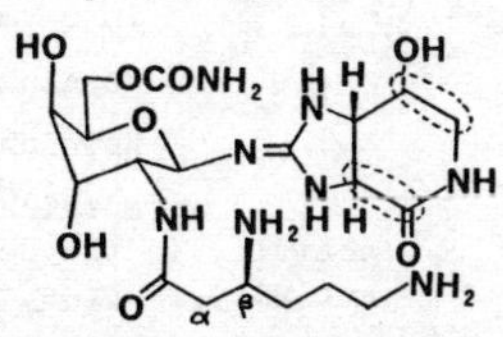	354

Species / *Compound Administered*	*Alkaloids labelled*	*Ref.*

Streptomyces species (strain L-1689-23) *(continued)*-

(^{13}C n.m.r. established the presence of coupled carbons, indicated with dotted lines, in the streptolidine moiety of the compound. This result is consistent with the formation of arginine via Krebs cycle intermediates, and subsequent rearrangement of the arginine as illustrated below: 354

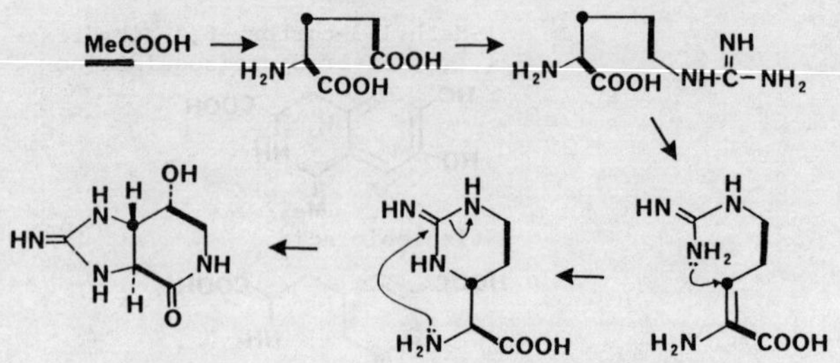

L-[*guanido*-^{13}C,^{15}N] Arginine + DL-[1-^{14}C]Arginine	Streptothricin F (11.8% spec. inc. of ^{14}C, 25% spec. inc. of ^{13}C into ring B, indicating the incorporation of only L-arginine. The	355

^{13}C n.m.r. exhibited coupling between the ^{15}N and the guanido carbon, $^1J_{CN}$ = 20 Hz. A small enrichment (~0.4%) was also observed in the carbamate group on the sugar residue, $^1J_{CN}$ = 26 Hz. Work has also been carried out on the formation of the β-lysine residue from DL-[3-^{13}C,2-^{15}N]lysine[356]).

Streptomyces achromogenes

[*carboxy*-^{14}C]Anthranilic acid	Tomaymycin (10.4% inc.)	357

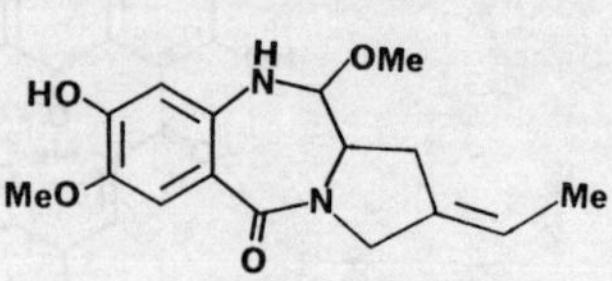

[*carboxy*-^{14}C]Anthranilic acid + inactive DL-Tryptophan	Tomaymycin (1.9% inc.)	
DL-[7a-^{14}C]Tryptophan	Tomaymycin (4.0% inc.)	

Streptomyces bernensis

DL-[1-^{14}C]Serine	Berninamycin (2.8% inc)	358
(48 hr incubations reported; somewhat lower incorporations after 72 hr feedings)	(for structure, see next page) (Degradations indicated incorporation into the dehydroalanine residues)	
L-[U-^{14}C]Serine	Berninamycin (4.2% inc.)	
DL-[1-^{14}C]Alanine	Berninamycin (0.042% inc.)	
L-[U-^{14}C]Cysteine	Berninamycin (1.0% inc. Negligible activity in the dehydroalanine residues, so the activity is presumably in the thiazole moiety)	

Species / *Compound Administered*	*Alkaloids labelled*	*Ref.*
Streptomyces bernensis (continued) -	Berninamycin	358

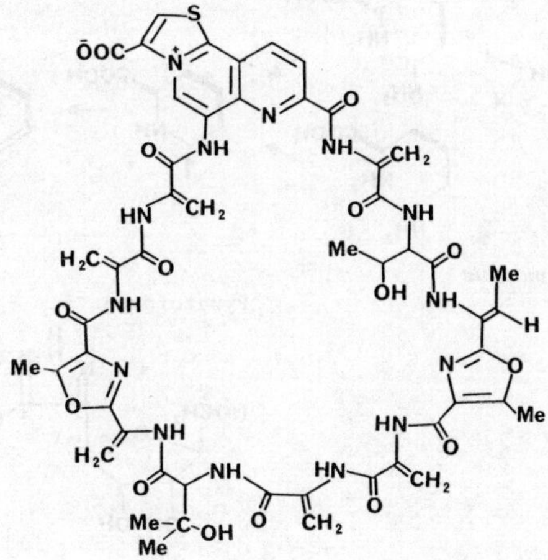

Streptomyces caeruleus	Caerulomycin A	359

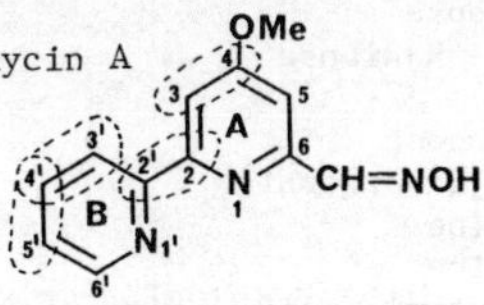

	Specific inc. (%)
Sodium [1-^{14}C]Acetate	3.8
D-[U-^{14}C]Glucose	3.2
DL-[4-^{14}C]Aspartic acid	2.3
L-[U-^{14}C]Alanine	1.8
DL-[1-^{14}C]Aspartic acid	0.9
L-[3'-^{14}C]Tryptophan	0.6
DL-[*benzene ring*-^{14}C]Tryptophan	0.2
Sodium [^{14}C]Formate	0.13
L-[U-^{14}C]Lysine	40.6
DL-[3-^{14}C]Serine	12.0
[1-^{14}C]Glycine	0.6
Sodium [1-^{13}C]Acetate	(enrichments at C-2, 4, 4')
Sodium [1,2-$^{13}C_2$]Acetate	(The coupled carbons observed in the ^{13}C n.m.r. are indicated with dotted lines in the above structure)
DL-[3-^{13}C]Serine	(enrichment of the OMe group)

(These results are consistent with the formation of ring B from lysine (labelled as indicated on the next page from [1,2-$^{13}C_2$]-acetate). The lysine then affords pipecolic acid (1) . Chain extension then takes place with an acetate unit and a three-carbon unit (possibly from the glycolytic pathway).)

Species *Compound Administered*	*Alkaloids labelled*	*Ref.*

Streptomyces caeruleus (continued) - 359

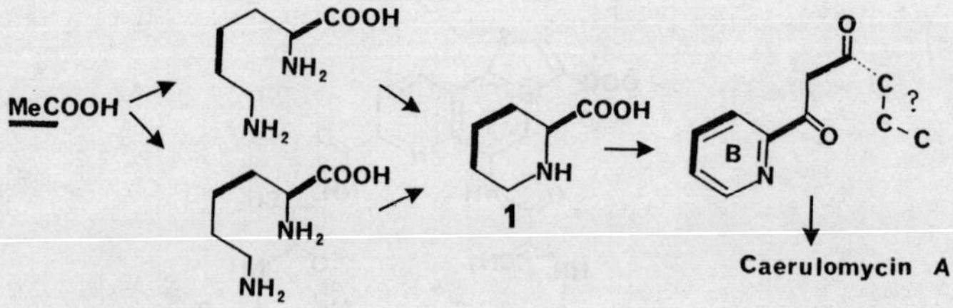

Streptomyces candidus

Pyrazofurin 360

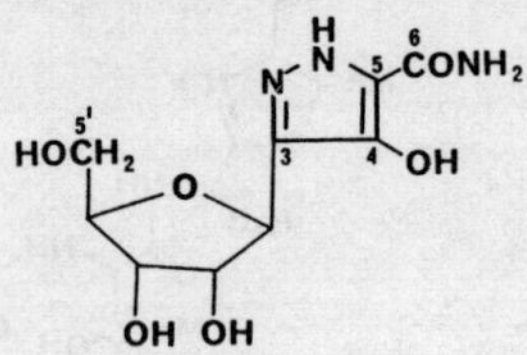

	Incorporation (%)
D-[1-^{14}C]Ribose	6.0
D-[1-^{14}C, 5-^{2}H]Ribose	7.9 (spec. inc.) with all the deuterium at C-5'
D-[1-^{14}C]Glucose	1.3
Sodium [^{14}C]Bicarbonate	0.014
[1-^{14}C]Glycine	0.08
[2-^{14}C]Glycine	0.7
L-[U-^{14}C]Glutamic acid	1.1
L-[1-^{14}C]Glutamic acid	0.37
L-[U-^{14}C]Lysine	0.07

(It is proposed that carbons C-3 to C-6 are derived from glutamic acid or α-ketoglutaric acid, with C-1 of these acids giving rise to C-6 of pyrazofurin)

Streptomyces chartreusis

Antibiotic A23187 361

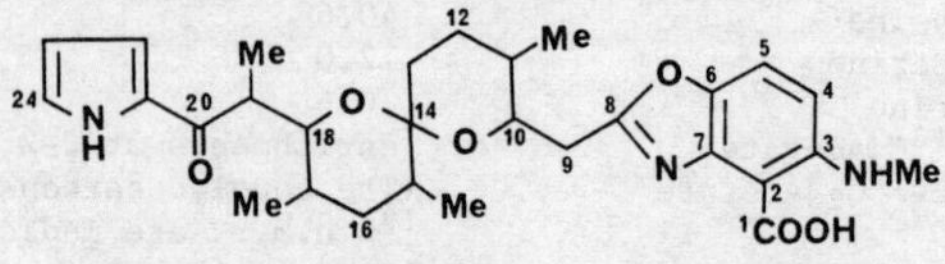

	Incorporation (%)
[3-^{14}C]Propionate	28.4
[2-^{14}C]Propionate	18.0
[1-^{14}C]Propionate	8.1

(Degradations indicated that all the *C*-methyl groups were derived from C-2 and C-3 of propionate)

[1-^{14}C]Acetate	1.8
[2-^{14}C]Acetate	1.6

(Degradations indicated that the acetate was incorporated via propionate)

Species / *Compound Administered*	*Alkaloids labelled*	*Ref.*
Streptomyces chartreusis (continued) -	Incorporation (%) into Antibiotic A23187	
L-[U-^{14}C]Proline	24.2	361
DL-[*carboxy*-^{13}C]Proline	7.0 (excess ^{13}C at C-20)	
DL-[1-^{14}C]Ornithine	0.55	
D-[U-^{14}C]Glucose	1.3	
[U-^{14}C]Glycerol	0.5	
DL-[7a-^{14}C]Tryptophan	no inc.	
L-[*methyl*-^{14}C]Methionine	22.0 (no activity in *C*-methyl groups)	
DL-[*methyl*-^{13}C]Methionine	16.4 (all excess ^{13}C in the NMe group)	
D-[U-$^{13}C_6$]Glucose	(The ^{13}C n.m.r. indicated that C-1,2,7 were derived from an intact three-carbon unit (phosphoenolpyruvate), and C-3,4,5,6 were derived from an intact four-carbon unit (erythrose-4-phosphate). The observation of these discrete units eliminates from the biosynthetic sequence symmetrical compounds such as 2,6-diaminobenzoic acid. Coupling was also observed between C-8 and C-9, these two carbons being the acetate starter unit for the polyketide-derived chain).	362
[U-^{14}C]Shikimic acid	Antibiotic A23187 (Incorporated into the C_7N_2 unit. 2,6-Diaminobenzoic acid did not reduce the degree of incorporation of the shikimic acid into the alkaloid)	
Sodium [1-^{13}C]Propionate	Antibiotic A23187 (Enrichments at C-10,14,16, and 18)	363
Sodium [1-^{13}C]Acetate	Antibiotic A23187 (Enrichments at C-8 and 12. Enrichment was also observed at C-24 due to the production of [5-^{13}C]glutamate via the Krebs cycle)	

(Addition of tryptophan or anthranilic acid to the culture medium caused the production of a new compound: Cezomycin (3-desmethyl-antibiotic A23187))

Streptomyces clavuligerus		
Sodium [2-^{14}C]Acetate	Clavulanic acid	364

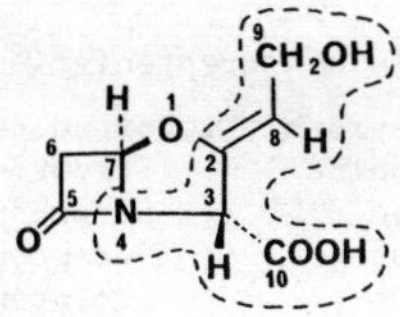

(10.4% of the activity was lost when C-8 and C-9 are removed in a degradation. C-5,6,7 contained 38.6% of activity)

[U-^{14}C]Glycerol	Clavulanic acid (95% of the activity at C-5,6,7. These results are in agreement with previously reported studies using ^{13}C-labelled precursors [365].)	

Species *Compound Administered*	*Alkaloids labelled*	*Ref.*
Streptomyces clavuligerus (continued) -		
DL-[3,4-$^{13}C_2$]Glutamic acid	Clavulanic acid	366

(Spin-spin couplings indicative of contiguous ^{13}C atoms were observed for the following atoms: 2-8, 2-3, 3-10, and 5-6. The direct incorporation of glutamic acid, as indicated by the dotted line on the structure, would account for the 2-8 coupling. Metabolism of the glutamate via the Krebs cycle could yield [2,3-$^{13}C_2$]phosphoenol pyruvate, which is considered to be a precursor of C-5,6,7. Reformed glutamic acid after operation of the Krebs cycle could result in the observed couplings at 2-3 and 3-10.)

Streptomyces flocculus

[U-$^{13}C_6$]Glucose	Streptonigrin	367

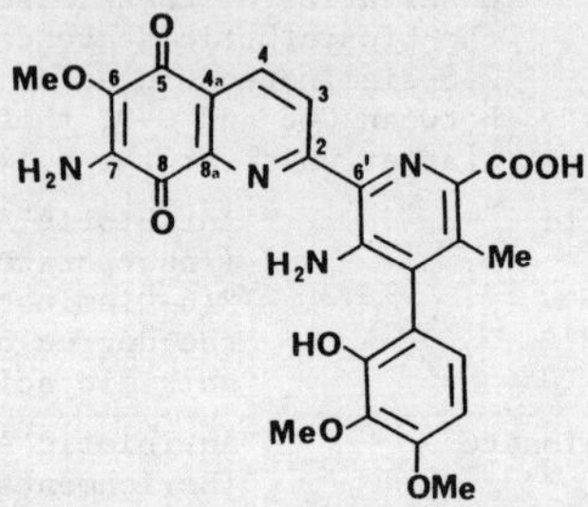

(The observed couplings in the ^{13}C n.m.r. indicated that the quinoline moiety is plausibly formed from 2,4-diaminobenzoic acid (arising from phosphoenol pyruvate and erythrose-4-phosphate) by condensation with a second molecule of erythrose-4-phosphate according to the following scheme:

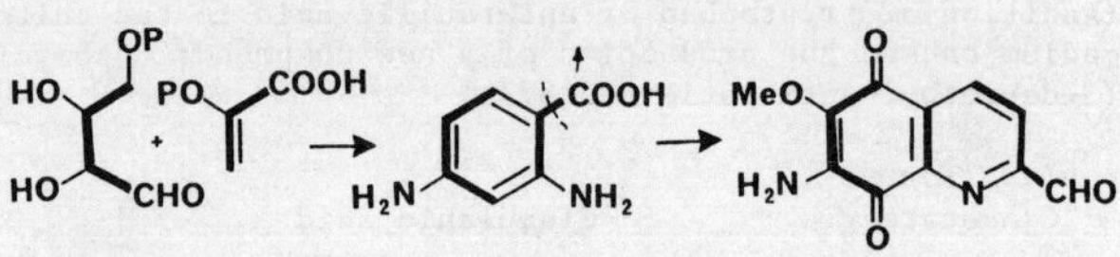

The biosynthesis of streptonigrin has been reviewed[368].)

(For details of work previously reported, see *Biosynthesis*, 1980, 6, 279.)

L-[*methyl*-^{13}C,^{14}C]Methionine	Streptonigrin (2.74% inc)	369
DL-[3'-^{13}C,^{14}C]Tryptophan	Streptonigrin (2.08% inc.)	
DL-[3'-^{14}C,5-^{3}H]Tryptophan	Streptonigrin (2.19% inc. 94% ret. of ^{3}H)	
L-[3'-^{14}C]Tryptophan + DL-[5-^{3}H]Tryptophan	Streptonigrin (2.78% inc. 50% ret. of ^{3}H)	
DL-[7a,3'-^{14}C]Tryptophan	Streptonigrin (1.34% inc.)	

(Little activity in the quinoline moiety, indicating that this nucleus is not derived from tryptophan or its degradation products such as anthranilic acid)

[U-^{14}C]Shikimic acid	Streptonigrin (0.005% inc.)	
[*carboxy*-^{14}C]Anthranilic acid	Streptonigrin (0.005% inc.)	

Species	*Alkaloids labelled*	*Ref.*
Streptomyces griseus		
L-[*methyl*-^{3}H]Methionine	Cycloheximide (25% inc)	370

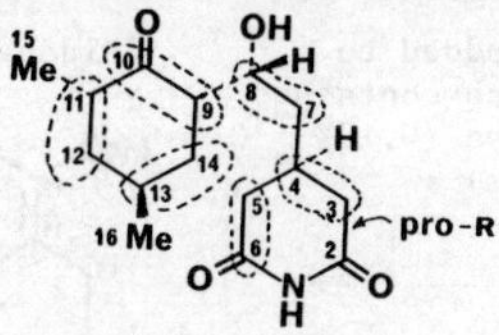

L-[*methyl*-$^2H_3$]Methionine	Cycloheximide (>95% contained either three or six deuterium atoms, with 55% of the d_6 species)	
L- [*methyl*-^{13}C]Methionine	Cycloheximide (About 45% ^{13}C was found on the methyl groups. It has been previously established that this compound is derived from acetate and malonate units[371-374]).	
[1-^{13}C]Acetate	Cycloheximide (Enrichments at C-2,4,6,8,10,12,14)	375
[2-^{13}C]Acetate	Cycloheximide (Enrichments at C-3,5,7,9,11,13)	
Sodium [^{13}C]Carbonate	Cycloheximide (Slight enrichment at the amide C=O)	
[1,2-$^{13}C_2$]Acetate	Cycloheximide	

(The coupled carbons are indicated with dotted lines on the structure above. Differentiation between the *pro-R* and *pro-S* sides of the glutarimide ring was achieved by a chemical degradation. Malonate is considered to be a stereospecific precursor of C-2,3,4.)

(This publication reports details of experiments previously described in a communication[377], and reviewed in *Biosynthesis*, 1980, 6, 280.)

(2*S*,*methyl*-*S*)-[*methyl*-^{14}C,^{2}H,^{3}H]-Methionine	Indolmycin (19.8% inc.)	376

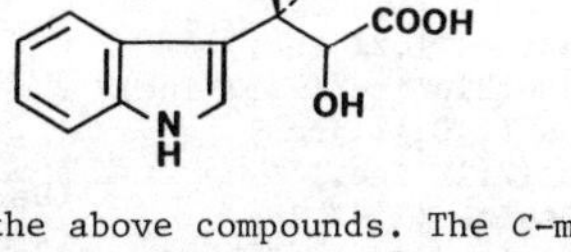

Indolmycenic acid (34.4% inc.)

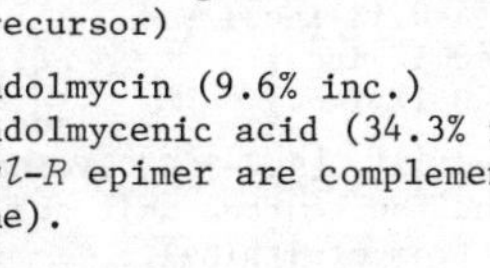

(~100% ret. of ^{3}H in both the above compounds. The *C*-methyl and *N*-methyl groups had an opposite configuration from the methyl group in the administered precursor)

(2*S*,*methyl*-*R*)-[*methyl*-^{14}C,^{2}H,^{3}H]-Methionine	Indolmycin (9.6% inc.) Indolmycenic acid (34.3% inc.)	

(The results obtained with this *methyl*-*R* epimer are complementary to those obtained with the *methyl*-*S*-methionine).

Species / *Compound Administered*	*Alkaloids labelled*	*Ref.*
Streptomyces hygroscopicus var. *geldanus* var. *nova*		
D-[U-$^{13}C_6$]Glucose added to a culture medium which contained ^{13}C-depleted glucose (0.12% ^{13}C at each carbon atom) + D-[U-^{14}C]Glucose	Geldanamycin (^{14}C: 1.8% inc.)	123

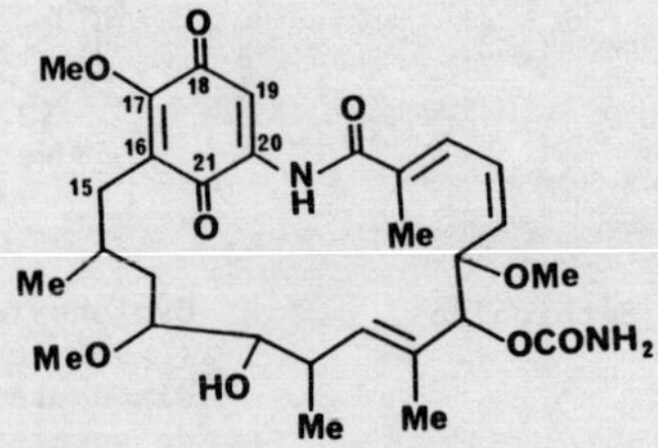

(The pattern of labelling in the C_7N unit was consistent with C-15,16, and 21 being derived from phosphoenol pyruvate, and C-17,18,19, and 20 from erythrose-4-phosphate)

Streptomyces lavendulae		
Sodium [^{14}C]Cyanide	Saframycin	378

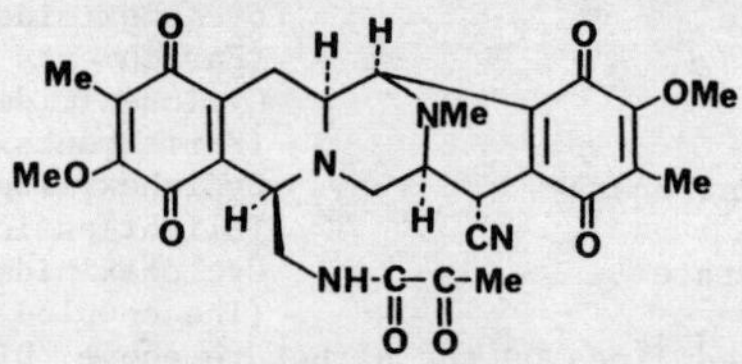

(The direct incorporation of the cyanide group was confirmed by feeding [^{13}C]cyanide, which resulted in enhancement of the CN group in the ^{13}C n.m.r.)

L-[U-^{14}C]Tyrosine	Saframycin (Presumably two molecules of tyrosine are incorporated into this alkaloid, but no degradations were carried out to locate the ^{14}C)	
Streptomyces lusitanus		
L-[U-^{14}C]Tyrosine (1.8% inc.)	Naphthyridinomycin	379

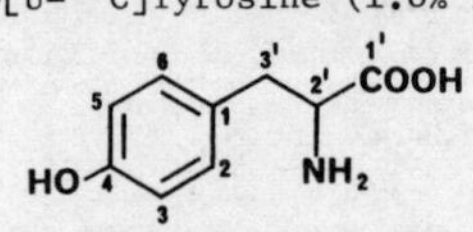

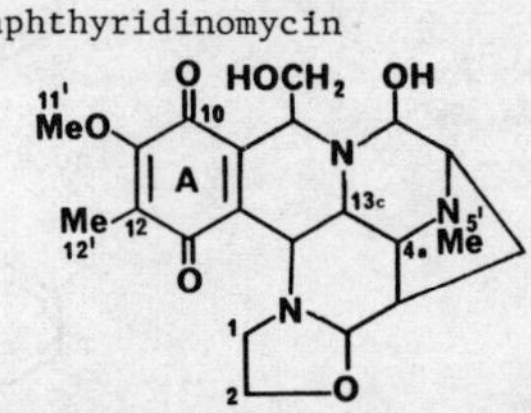

L-[1'-^{14}C]Tyrosine (1.2% inc.)
L-[*methyl*-^{14}C]Methionine (2.5% inc)
DL-[3'-^{14}C]Dopa (<0.1% inc.)
[2-^{14}C]Glycine (2.1% inc.)
DL-[1-^{14}C]Ornithine (1.5% inc.)
DL-[1-^{14}C]Glutamic acid (<0.1% inc.)
D-[U-^{14}C]Glucose (<0.1% inc.)
[1-^{14}C]Acetate (<0.1% inc.)
[2-^{14}C]Acetate (<0.1% inc.)

(Degradations indicated that tyrosine is a precursor of ring A, confirmed by feeding [1'-^{13}C]tyrosine which labelled 4a, [2'-^{13}C]tyrosine which labelled 13c, [3,5-$^{13}C_2$]tyrosine which labelled 10 and 12. [*methyl*-^{13}C]Methionine labelled 5',11',and 12'. [1,2-^{13}C]Glycine labelled 1 and 2 and the methyl groups.The four-carbon unit not derived from tyrosine may come directly from ornithine).

Species — *Compound Administered* | *Alkaloids labelled* | *Ref.*

Streptomyces naraensis

[1,2,3-$^{13}C_3$]Malonate — Cyclohexieximide — 380

(The results obtained are in conflict with those obtained by Jeffs in *Streptomyces griseus*[375]. In this work the results seem to indicate unequivocally that the malonate serves as a precursor of *pro-S* side of the glutarimide ring).

Streptomyces pactum

D-[U-$^{13}C_6$]Glucose — Pactamycin — 381

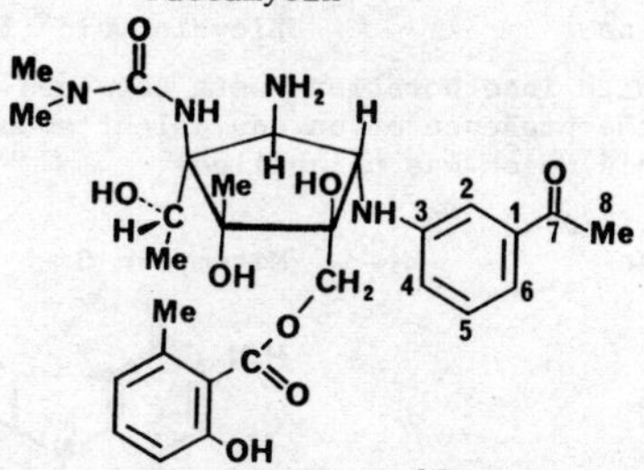

(The coupling patterns observed in the ^{13}C n.m.r. spectrum of the *m*-aminoacetophenone moiety indicated that C-1,6, and 7 were derived from phosphoenol pyruvate, and C-2,3,4, and 5 from erythrose-4-phosphate. Earlier work was carried out with [1-^{14}C]glucose[382]).

[*ring*-U-^{14}C]-*m*-Aminobenzoic acid — Pactamycin (11.5% inc.)

Streptomyces refuineus

L-[5-^{3}H, *indole*-^{15}N]Tryptophan + DL-[1'-^{13}C]Tyrosine (fed together) — Anthramycin (^{15}N: 55% spec. inc. ^{13}C: 44% spec. inc. ^{3}H: 64% spec. inc.) — 383

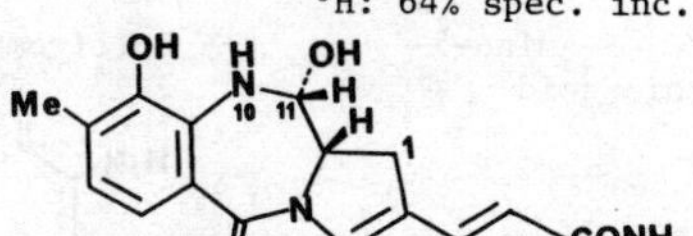

(C-11 is derived from C-1' of tyrosine, and N-10 is derived from the indole nitrogen of tryptophan. Intramolecular labelling with ^{13}C and ^{15}N was observable in the ^{13}C n.m.r. spectrum of a derivative of anthramycin : didehydroanhydroanthramycin, $^{1}J_{CN}$ = 7 .7)

[*carboxy*-^{14}C]Anthranilic acid — Anthramycin (<0.1% inc.) — 357

3-Hydroxy-4-methyl-[2-^{14}C]-anthranilic acid — Anthramycin (<0.1% inc.)

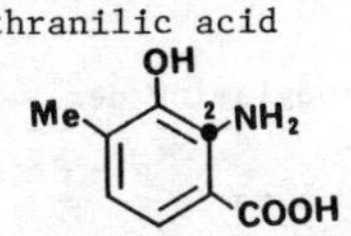

(The poor incorporation obtained in the last two feedings is attributed to poor permeability of the labelled substrates)

DL-[7a-^{14}C]Tryptophan — Anthramycin (18.0% inc.)

DL-[(3'*RS*)-3'-^{3}H]Tyrosine — Anthramycin — 384

DL-[(3'*S*)- 3'-^{3}H]Tyrosine — Anthramycin

DL-[(3'*R*)- 1'-^{14}C,3'-^{3}H]Tyrosine — Anthramycin (Degradations carried out on the anthramycin from the last three experiments indicated that all the tritium was at C-1).

Species / *Compound Administered*	*Alkaloids labelled*	*Ref.*
Streptomyces sapporonensis		
L-[U-^{14}C]Leucine	Bicyclomycin (8.8 - 11.1% inc.)	385

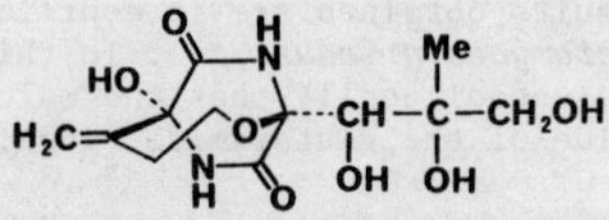

L-[U-^{14}C]Isoleucine	Bicyclomycin (10.9 - 13.2% inc.)	

(These high incorporations were obtained in feedings carried out in the presence of an equivalent amount of the other amino-acid which was unlabelled).

Streptomyces verticillatus		
D-[4-^{14}C]Erythrose	Mitomycin C	122

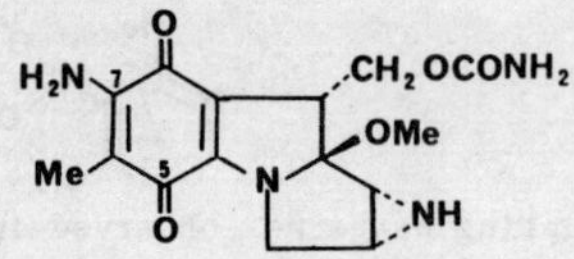

(0.8% inc. into the mixture of mitomycins . C-7 contained 37% of the total activity)

[3-^{14}C]Pyruvate	Mitomycin C (0.2% inc. Degradations consistent with the major amount of activity at C-5)	
[7-^{14}C]-3-Dehydroquinic acid	Mitomycins + Porfiromycin (<0.01% inc.)	
D-[1,6-^{14}C]Shikimic acid methyl ester	Mitomycins + Porfiromycin (<0.01% inc.)	
[*carboxy*-^{13}C]-3-Amino-5-hydroxybenzoic acid	Porfiromycin (8.5% spec. inc.	121

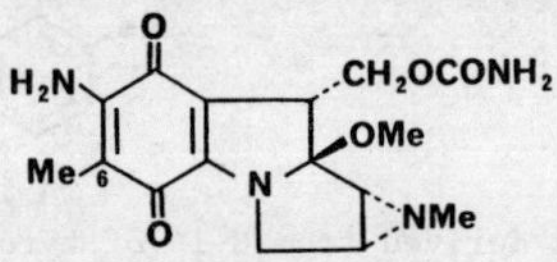

^{13}C was detected at the C-6 methyl group by proton and ^{13}C n.m.r)

Streptomyces verticillus (Labelled substrates were added to the culture medium which contained 3-morpholinopropylamine)		
L-[*methyl*-^{13}C]Methionine	Bleomycin-morpholinopropylamine deriv. (Bleomycin-MPO)	386
	See next page for structure	

(~3.6% spec. inc. In the ^{13}C n.m.r. spectrum, enhanced signals were observed for the methyl group in the pyrimidine ring and the methyl at C-2 of the 4-amino-3-hydroxy-2-methylpentanoic acid moiety. These positions are indicated with heavy dots on the structure).

Species / *Compound Administered*	*Alkaloids labelled*	*Ref.*
Streptomyces verticillus (continued) -		
DL-[3-^{13}C]Alanine	Bleomycin-MPO	386

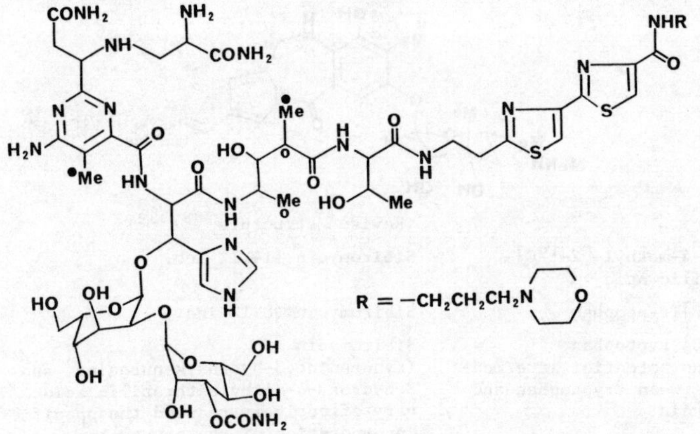

(~5% spec. inc. The labelled atoms are indicated with open circles in the above structure. This result suggests that the pentanoic acid moiety is derived from two molecules of pyruvate).

Species / *Compound Administered*	*Alkaloids labelled*	*Ref.*
Streptomyces virginiae		
Sodium [2-^{14}C]Acetate (5.0% inc.)	Virginiamycin M_1	387
L-[*methyl*-^{3}H]Methionine (4.0%)		
DL-[3-^{14}C]Serine (1.4%)		
D-[U-^{14}C]Glucose (0.6%)		
L-[3,4-^{3}H]Proline (5.0%)		
[2-^{14}C]Glycine (0.4%)		
L-[U-^{14}C]Alanine (no inc.)		

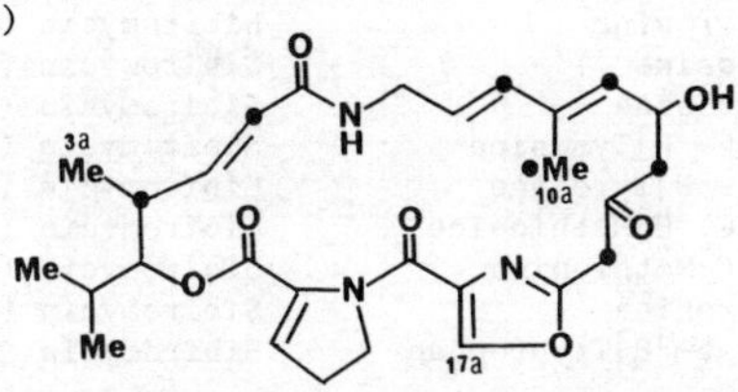

DL-[3-^{13}C]Serine — Virginiamycin M_1

(^{13}C n.m.r. indicated that the most enriched carbon was C-17a, indicating that the oxazole ring was partially formed from serine. A lower degree of enrichment was observed at C-3a, which is derived from the one-carbon pool)

L-[*methyl*-^{13}C]Methionine — Virginiamycin M_1 (enrichment at C-3a)

Sodium [2-^{13}C]Acetate — Virginiamycin M_1

(Enrichment at positions indicated with heavy dots)

Sodium [1,2-$^{13}C_2$]Acetate — Virginiamycin M_1

(The lack of coupling of C-10a to its adjacent carbon indicates that this methyl is formed from acetate which undergoes decarboxylation)

(These results indicate that this compound is formed from the following units : isobutyric acid (from valine), (acetate)$_2$,glycine, (acetate)$_4$,serine,proline, with peripheral methylation from methionine and acetate).

Species *Compound Administered*	*Alkaloids labelled*	*Ref.*
Streptosporangium sibiricum		
[*carboxy*-^{14}C]Anthranilic acid	Sibiromycin (<0.1% inc.)	357

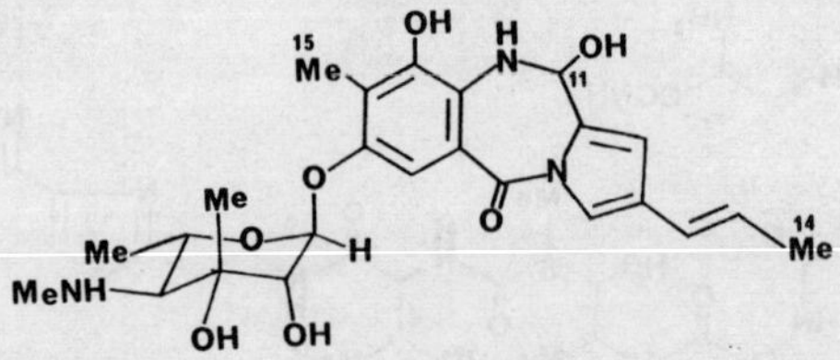

(Revised structure [388,389])

3-Hydroxy-4-methyl-[2-^{14}C]-anthranilic acid	Sibiromycin (14.1% inc.)	
DL-[7a-^{14}C]Tryptophan	Sibiromycin (8.1% inc.)	
DL-[7a-^{14}C]Tryptophan + various potential intermediates between tryptophan and sibiromycin	Sibiromycin (Kynurenine,3-hydroxykynurenine, and 3-hydroxy-4-methylanthranilic acid very effectively reduced the specific incorporation of the tryptophan. Anthranilic acid had a lesser effect and 3-hydroxyanthranilic acid had no effect).	
D-[6-^{14}C]Glucose	Sibiromycin (2.1% inc.)	390
L-[1'-^{14}C]Dopa	Sibiromycin (18.4% inc.)	
L-[*methyl*-^{14}C]Methionine	Sibiromycin (24.5% inc.)	
DL-[7a-^{14}C]Tryptophan	Sibiromycin (7.7% inc.)	
DL-[1'-^{14}C]Tyrosine	Sibiromycin (15.3% inc.)	
L-[U-^{14}C]Tyrosine	Sibiromycin (6.5% inc.)	
L-[1'-^{14}C]Tyrosine	Sibiromycin (27.9% inc.)	
L-[1'-^{14}C,2,6-^{3}H]Tyrosine	Sibiromycin (34.6% ret. of ^{3}H)	
L-[U-^{14}C,2,6-^{3}H]Tyrosine	Sibiromycin (45.1% ret. of ^{3}H)	
L-[*methyl*-^{14}C,^{3}H]Methionine	Sibiromycin (100% ret. of ^{3}H)	
L-[*methyl*-^{13}C]Methionine	Sibiromycin (enrichment at C-14 and C-15)	
L-[1'-^{13}C]Tyrosine	Sibiromycin (enrichment at C-11)	
DL- [7a-^{14}C, 5-^{3}H]Tryptophan	Sibiromycin (100% ret. of ^{3}H)	
Strychnos nux-vomica		
[6-^{14}C]Strictosidine	Strychnine (0.12% inc.)	236

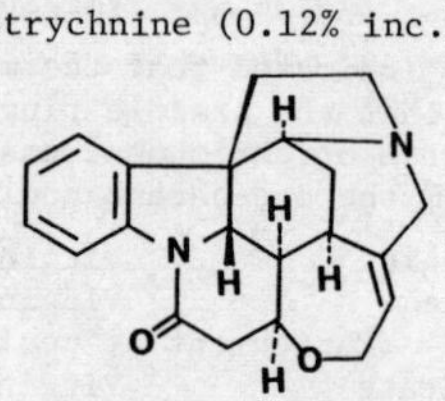

[6-^{14}C,3-^{3}H]Vincoside	Strychnine (<0.001% inc.)	
Thalicitrum minus var.*elatum*		391
(*RS*)-[1-^{14}C]Reticuline	Thalicarpine (see under *Cocculus laurifolius* for structure)	

(0.76% inc. Activity found in both halves of the alkaloid;however, a radiochemical balance was not obtained on degradation to yield the two halves)

Species *Compound Administered*	*Alkaloids labelled*	*Ref.*
Thalicitrum minus var. *elatum (continued)* -		
(*RS*)-[1-^{14}C]Reticuline	Thalicarpine (1.76% inc.)	392
(*RS*)-[1-^{14}C]Norreticuline	Thalicarpine (1.12% inc.)	
(*RS*)-[1-^{14}C]Protosinomenine	Thalicarpine (0.16% inc.)	
(*RS*)-[1-^{14}C]Norprotosinomenine	Thalicarpine (0.01% inc.)	
(*RS*)-[1-^{14}C]Orientaline	Thalicarpine (0.02% inc.)	
(*RS*)-[1-^{14}C]Nororientaline	Thalicarpine (0.01% inc.)	
(*S*)-[8-^{3}H]Isoboldine	Thalicarpine (0.50% inc.)	
[^{14}C]Carbon dioxide	Activity was detected in reticuline and isoboldine by dilution.	
Thalictrum tuberosum		
[2'-^{14}C,2,6-^{3}H]Dopamine + 3,4-Dihydroxyphenylacetaldehyde (cell-free system)	(*S*)-Norlaudanosoline (3,4-Dihydroxyphenylpyruvic acid did not serve as a substrate for the enzyme)	143
Tiliacora racemosa		
(*RS*)-[1-^{3}H]Norcoclaurine (R = H)	Tiliacorine (0.10% inc.)	393

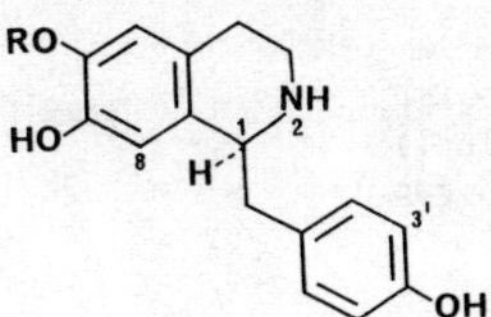

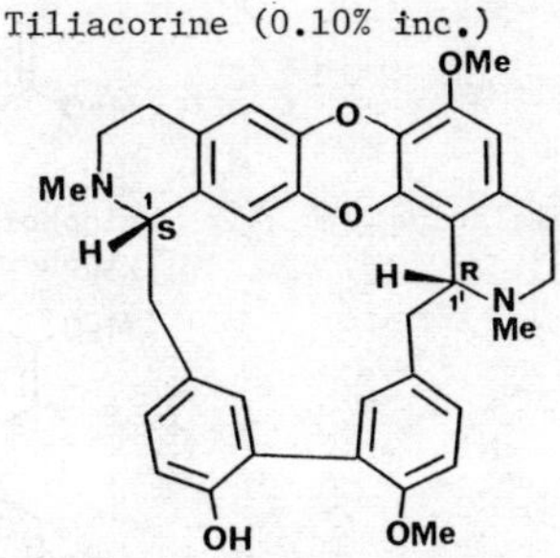

Tiliacorinine (0.09% inc.)

(The structure is the same as tiliacorine except that 1' is (*S*))

L-[U-^{14}C]Tyrosine	Tiliacorine (0.09% inc.) Tiliacorinine (0.06% inc.)	
(*RS*)-[3',5',8-^{3}H]Coclaurine (R = Me)	Tiliacorine (0.14% inc.) Tiliacorinine (0.12% inc.)	
(*RS*)-[3',5',8-^{3}H]-*N*-Methylcoclaurine	Tiliacorine (0.18% inc.) Tiliacorinine (0.15% inc.) (Activity found in both halves of both alkaloids)	
(*RS*)-[3',5',8-^{3}H]-*N,O,O*-Trimethylcoclaurine	Tiliacorine (no inc.) Tiliacorinine (no inc.)	
[*N-methyl*-^{14}C]-*N*-Methyl-1,2-dehydrococlaurine	Tiliacorine (0.12% inc.) Tiliacorinine (0.10% inc.)	
(*RS*)-[*N-methyl*-^{14}C,1-^{3}H]-*N*-Methylcoclaurine	Tiliacorine (0.15% inc.) Tiliacorinine (0.12% inc.) (~100% ret. of ^{3}H in both alkaloids)	
(*RS*)-[6-*O-methyl*-^{14}C,1-^{3}H]-*N*-Methylcoclaurine	Tiliacorine (0.19% inc.) Tiliacorinine (0.16% inc.)	

(There was a 50% loss of ^{14}C relative to ^{3}H. This result is consistent with the loss of the methyl group from the 6-methoxy group during the formation of the dioxin ring).

Species / *Compound Administered*	*Alkaloids labelled*	*Ref.*
Tiliacora racemosa (continued) –		
(*S*)-[3',5',8-^{3}H]-*N*-Methylcoclaurine	Tiliacorine (0.15% inc.) (Activity found on the left side of the alkaloid as drawn on the previous page)	393
	Tiliacorine (0.28% inc. Activity found in both halves of the alkaloid).	
(*R*)-[3',5',8-^{3}H]-*N*-Methylcoclaurine	Tiliacorine (0,16% inc. Activity on the right side of the molecule)	
	Tiliacorinine (no inc.)	
Tylophora asthmatica		
DL-[U-^{14}C]Tyrosine	Tylophorine (0.33% inc.) 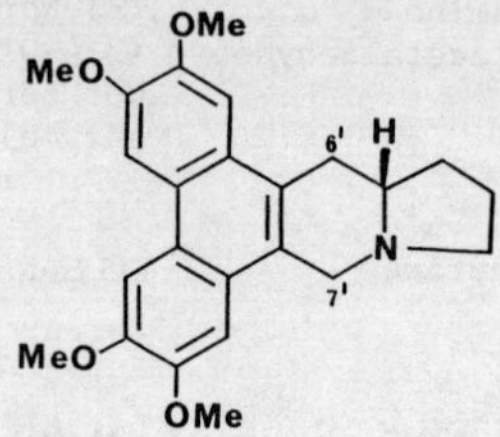	394
DL-[2'-^{14}C]Phenylalanine	Tylophorine (2,6% inc.)	
	Tylophorinine (2.9% inc.) 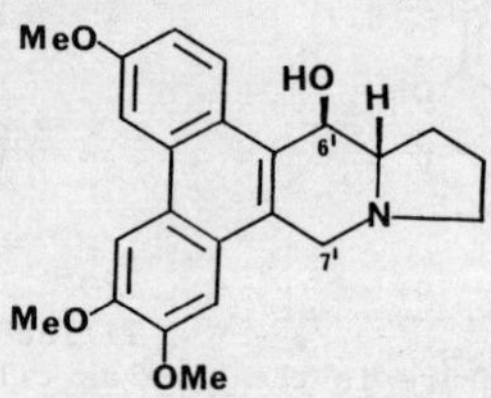	
DL-[2'-^{14}C]Tyrosine	Tylophorine (3.4% inc. All activity at C-7', a result in agreement with previous work[395].)	
	Tylophorinine (2.1% inc.)	
DL-[2'-^{14}C]Dopa	Tylophorine (4.2% inc., essentially all the activity being at C-7')	
	Tylophorinine (2.7% inc., mostly at C-7')	
[3,5-^{3}H]-4-Hydroxyphenylpyruvic acid	Tylophorine (0.6% inc.)	
	Tylophorinine (0.65% inc.)	
(*RS*)-6-(3-Hydroxy-4-methoxyphenyl)-7-(4-hydroxy-3-methoxyphenyl)-hexahydro-[5-^{14}C]indolizine 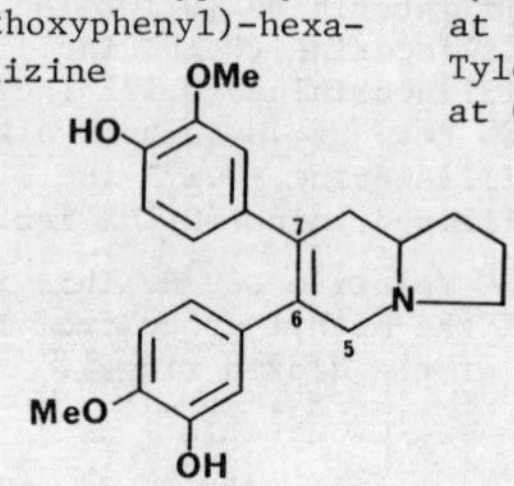	Tylophorine (13.1% inc. all activity at C-7')	
	Tylophorinine (4.2% inc. all activity at C-7')	

Species / *Compound Administered*	*Alkaloids labelled*	*Ref.*
Tylophora asthmatica (continued) -		
(*RS*)-6,7-Bis(3,4-dihydroxyphenyl)-hexahydro-[5-^{14}C]indolizine	Tylophorine (2.6% inc.) Tylophorinine (1.4% inc.)	394

(Indolizines with a different methylation pattern in the dihydroxyphenyl rings failed to serve as efficient precursors of tylophorine or tylophorinine).

Species / *Compound Administered*	*Alkaloids labelled*	*Ref.*
Vallesia glabra		
[6-^{14}C,10-^{3}H]-5α-Carboxystrictosidine (see under *Gelsemium sempervirens*)	Akuammidine (<0.01% inc.)	288

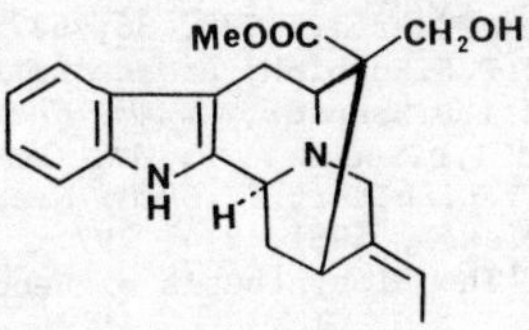

Species / *Compound Administered*	*Alkaloids labelled*	*Ref.*
[6-^{14}C,10-^{3}H]-5α-Carboxy-vincoside	Akuammidine (<0.002% inc.)	
Vinca major		
[6-^{14}C]Strictosidine	Vincadifformine (0.14% inc.)	236

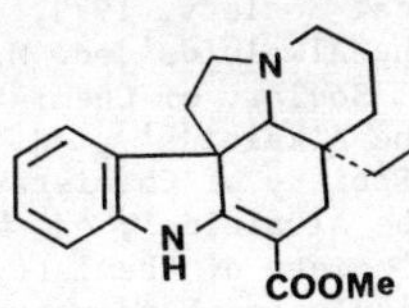

Species / *Compound Administered*	*Alkaloids labelled*	*Ref.*
[6-^{14}C]Vincoside	Vincadifformine (no inc.)	
Voacanga africana		
[6-^{14}C,10-^{3}H]-5α-Carboxy-strictosidine (see under *Gelsemium sempervirens*)	Akuammidine (<0.01% inc.) (A similar negligible incorporation was obtained with the corresponding vincoside derivative).	288
Zygosporium masonii		
Sodium [2-^{13}C,2-^{2}H$_3$]Acetate + inactive malonate	Cytochalasin D	331

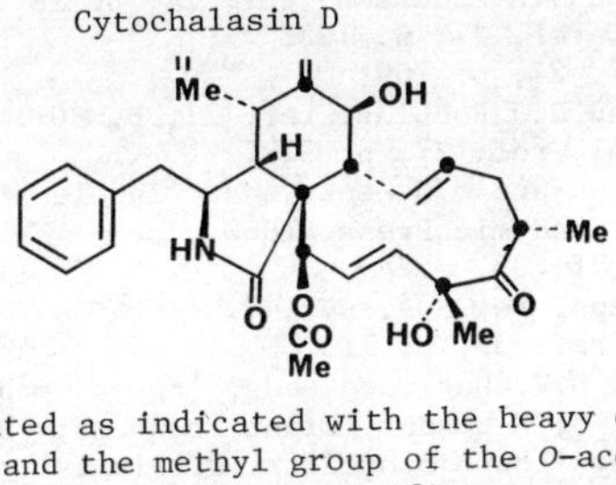

(The ^{13}C was incorporated as indicated with the heavy dots in the above structure. C-11 and the methyl group of the *O*-acetyl group were the main sites of deuterium . No significant amount of deuterium was detected (by ^{2}H n.m.r.)elsewhere, indicating that the methylene groups of the polyketide exchange with the media)

Species / *Compound Administered*	*Alkaloids labelled*	*Ref.*
[U-^{14}C]Phenylalanine [^{3}H]Phenylalanine	Cytochalasin D (Yields as high as 514 mg/liter were obtained on a special medium)	396

References

1 S.W.Pelletier, in 'Alkaloids: Chemical and Biological Perspectives', Vol.1, ed. S.W.Pelletier, John Wiley and Sons, 1983, p.2.
2 R.J.Parry and M.V.Naidu, *Tetrahedron Lett.*, 1980, 21, 4783.
3 D.A.Trainer, R.J.Parry, and A.Gitterman, *J.Am.Chem.Soc.*, 1980, 102, 1467.
4 J.Arditti and J.B.Tarr, *Am. J. Bot.*, 1979, 66, 1105.
5 J.L.Cooper, B.L.Hilton, J.Arditti, and J.B.Tarr, *New Phytol.*, 1982, 91, 621.
6 G.J.Vella, R.E.Hill, B.S.Mootoo, and I.D.Spenser, *J. Biol. Chem.*, 1980, 255, 3042.
7 G.J.Vella, R.E.Hill, and I.D.Spenser, *J. Biol. Chem.*, 1981, 256, 10469.
8 S.David, B.Estramareix, J.-C.Fischer, and M.Therisod, *J. Am. Chem. Soc.*, 1981, 103, 7431; idem, *J. Chem. Soc., Perkin Trans.1.*, 1982, 2131.
9 R.L.White and I.D.Spenser, *Biochem. J.*, 1979, 179, 315.
10 R.H.White, *Experientia*, 1980, 36, 637.
11 R.H.White and F.B.Rudolph, *Biochemistry*, 1979, 18, 2632.
12 R.L.White and I.D.Spenser, *J. Am. Chem. Soc.*, 1979, 101, 5102.
13 R.L.White and I.D.Spenser, *J. Am. Chem. Soc.*, 1982, 104, 4934.
14 G.Lowe and B.V.L.Potter, *J. Chem. Soc., Perkin Trans 1.*, 1980, 2026.
15 R.H.White, *Science*, 1981, 214, 797.
16 R.B.Herbert, 'The Biosynthesis of Secondary Metabolites', Chapman and Hall, 1981.
17 R.B.Herbert, in 'Rodd's Chemistry of Carbon Compounds', 2 edn. ed. S.Coffey, Elsevier, Vol. IV, Part L, 1980, p. 291.
18 R.B.Herbert, in 'Comprehensive Organic Chemistry', Vol. 5, ed. E. Haslam, Pergamon, Oxford, 1979, p. 1045.
19 R.B.Herbert, in 'The Alkaloids', ed. M.F.Grundon (Specialist Periodical Report), The Chemical Society, 1979, 9, 1.
20 R.B.Herbert, in 'The Alkaloids',ed. M.F.Grundon (Specialist Periodical Report), The Royal Society of Chemistry, 1980, 10, 1.
21 R.B.Herbert, in 'The Alkaloids' , ed. M.F.Grundon (Specialist Periodical Report) The Royal Society of Chemistry, 1981, 11, 1.
22 R.B.Herbert, in 'The Alkaloids', ed. M.F.Grundon (Specialist Periodical Report) The Royal Society of Chemistry, 1982, 12, 1.
23 D.R.Dalton, 'The Alkaloids: Fundamental Chemistry-A Biogenetic Approach', Marcel Dekker, New York and Basel, 1979.
24 M.L.Vickery and B.Vickery, 'Secondary Plant Metabolism',McMillan, London, 1981.
25 G.A.Cordell, 'Introduction to Alkaloids, A Biogenetic Approach', John Wiley, 1981.
26 T.Robinson, 'The Biochemistry of Alkaloids', 2nd edn. Springer Verlag, 1981.
27 P.S.Steyn, ed. 'The Biosynthesis of Mycotoxins', Academic Press, 1980.
28 H.G.Floss and J.A.Anderson, ref. 27, p. 18.
29 M. Yamazaki, ref. 27, p. 193.
30 Ch.Tamm, ref. 27, p. 269.
31 G.W.Kirby and D.J.Robbins, ref. 27, p. 301.
32 C.W.Holzapfel, ref. 27, p. 327.
33 J.D.Phillipson and M.H.Zenk, eds. 'Indole and Biogenetically Related Alkaloids', Academic Press, 1980.
34 H.G.Floss, ref. 33, p.249.
35 C.R.Hutchinson, ref. 33, p. 143.
36 J.Stöckigt, ref. 33, p. 113.
37 E.A.Bell and B.V.Charlwood, eds. 'Encyclopedia of Plant Physiology', New Series, Vol. 8, Secondary Plant Products, Springer Verlag, 1980.
38 E.E.Conn, ed. 'The Biochemistry of Plants', Vol. 7, Secondary Plant Products, Academic Press, 1981.
39 G.R.Waller and O.C.Dermer, ref. 38, p. 317.
40 J.W.Corcoran, ed. 'Antibiotics', Vol. IV, Biosynthesis, Springer Verlag, 1981.

41 G.Lancini and M.Grandi, ref. 40, p. 12.
42 J.O'Sullivan and E.P.Abraham, ref. 40, p. 101.
43 L.H.Hurley and M.K.Speedie, ref. 40, p. 262.
44 U.Hornemann, ref. 40, p. 295.
45 R.J.Suhadolnik, ref. 40, p. 353.
46 H.R.Schütte, *Prog. Bot.*, 1980, 42, 96.
47 R.J.Parry, in 'Recent Advances in Phytochemistry, Topics in the Biochemistry of Natural Products', Plenum Press, N.Y., 1979, 13, 55.
48 K.M.Madyastha and C.J.Coscia, ref. 47, p. 85.
49 K.L.Rinehart, D.D.Weller, and C.J.Pearce, *J. Nat. Prod.*, 1980, 43, 1.
50 M.Luckner, *J. Nat. Prod.*, 1980, 43, 21.
51 J.Staunton, *Planta Med.*, 1979, 36, 97.
52 E.McDonald, in 'The Chemistry of Heterocyclic Compounds, Isoquinolines', Part I, ed. G.Grethe, Vol. 38, 1981, p. 275.
53 P.Adriaens, B.Meeschaert, L.Dumon, and H.Eyssen, *J. Pharm. Belg.*, 1979, 34, 222.
54 J.E.Baldwin, M.Jung, P.Singh, T.Wan, S.Haber, S.Herchen, J.Kitchin, A.L.Demain, N.A.Hunt, M.Kohsaka, T.Konomi and M.Yoshida, *Philos. Trans. R. Soc. London Ser. B*, 1980, 289, 169.
55 E.P.Abraham, J.A.Huddleston, G.S.Jayatilake, J.O'Sullivan, and R.L.White, in 'Recent Advances in the Chemistry of β-Lactam Antibiotics', ed. G.I.Gregory, Special Publication, The Royal Society of Chemistry, 1981.
56 M.H.Zenk, *Planta Med.*, 1980, 43, 438.
57 E.Leete, *Planta Med.*, 1979, 36, 97.
58 E.Leete, ref. 1, p. 85.
59 L.H.Hurley, *Acc. Chem. Res.*, 1980, 13, 263.
60 P.A.Stadler, *Planta Med.*, 1982, 46, 131.
61 E.Leete, *Rev. Latinoam. Quim.*, 1980, 11, 8.
62 E.Leete, *J. Chem. Soc., Chem. Commun.*, 1980, 1170.
63 E.Leete, *J. Am. Chem. Soc.*, 1981, 104, 1403.
64 E.Leete, *Phytochemistry*, 1983, 22, 699.
65 A.Ahmad and E.Leete, *Phytochemistry*, 1970, 9, 2345.
66 F.E.Baralle and E.G.Gros, *Chem. Commun.*, 1969, 721.
67 S.H.Hedges and R.B.Herbert, *Phytochemistry*, 1981, 20, 2064.
68 H.W.Liebisch, W.Maier, and H.R.Schütte, *Tetrahedron Lett.*, 1966, 4079.
69 H.W.Liebisch, A.S.Radwan, and H.R.Schütte, *Liebigs Ann. Chem.*, 1969, 721, 163.
70 E.Leete and J.A.McDonell, *J. Am. Chem. Soc.*, 1981, 103, 658.
71 S.Mizusaki, T.Kisaki, and E.Tamaki, *Plant Physiol.*, 1967, 43, 93.
72 H.W.Liebisch, K.Pelsker, A.S.Radwan, and H.R.Schütte, *Z. Pflanzenphysiol.*, 1972, 67, 1.
73 B.A.McGaw and J.G.Woolley, *Phytochemistry*, 1978, 17, 257.
74 D.G.O'Donovan and M.F.Keogh, *J. Chem. Soc. C*, 1969, 223.
75 P.Tamprateep, E.H.Taylor, and E.Ramstad, *Lloydia*, 1963, 26, 203.
76 D.Gross and H.R.Schütte, *Arch. Pharm. (Weinheim, Ger.)*, 1963, 296, 1.
77 D.Gross, Dissertation, Halle, DDR, 1963; cited in ref. 96, p. 198.
78 E.Leete, *J. Org. Chem.*, 1976, 41, 3438.
79 E.Nowacki and R.U.Byerrum, *Life Sci.*, 1962, 5, 157.
80 W.Bottomley and T.A.Geissman, *Phytochemistry*, 1964, 3, 357.
81 R.Robinson, 'Structural Relations of Natural Products', Clarendon Press, Oxford, 1955.
82 K.Babor, I.Jezo, V.Kalac, and M.Karvas, *Chem. Zvesti*, 1959, 13, 163.
83 N.J.Leonard and S.W.Blum, *J. Am. Chem. Soc.*, 1960, 82, 503.
84 G.H.Tait, *Biochem. Soc. Trans.*, 1979, 7, 199.
85 C.A.Hughes, R.Letcher, and F.L.Warren, *J. Chem. Soc.*, 1964, 4974.
86 D.J.Robins and J.R.Sweeney, *J. Chem. Soc., Chem. Commun.*, 1979, 120.
87 D.J.Robins and J.R.Sweeney, *J. Chem. Soc. Perkin Trans 1*, 1981, 3083.
88 G.Grue-Sørensen and I.D.Spenser, *Can. J. Chem.*, 1982, 60, 643.
89 T.A.Smith, *Phytochemistry*, 1972, 11, 899.
90 H.A.Khan and D.J.Robins, *J. Chem. Soc., Chem. Commun.*, 1981, 146.

91 G.Grue-Sørensen and I.D.Spenser, *J. Am. Chem. Soc.*, 1981, 103, 3208.
92 H.A.Khan and D.J.Robins, *J. Chem. Soc., Chem. Commun.*, 1981, 554.
93 N.M.Bale and D.H.G.Crout, *Phytochemistry*, 1975, 14, 2617.
94 H.Birecka and J.L.Catalfamo, *Phytochemistry*, 1982, 21, 2645.
95 M.Wink, T.Hartmann, and H.-M.Schiebel, *Z. Naturforsch.*, 1979, 34c, 704.
96 H.R.Schütte, in 'Biosynthese der Alkaloide', eds. K.Mothes and H.R.Schütte, VEB: Berlin, 1969, p. 324.
97 T.Hartmann, G.Schoofs, and M.Wink, *FEBS Lett.*, 1980, 115, 35.
98 M.Wink, T.Hartmann, and L.Witte, *Z. Naturforsch.*, 1980, 35c, 93.
99 M.Wink and T.Hartmann, *Plant Physiol.*, 1982, 70, 74.
100 M.Wink and T.Hartmann, *Plant Cell Rep.*, 1981, 1, 6.
101 M.Wink and T.Hartmann, *Z. Naturforsch.*, 1982, 37c, 369.
102 G.Schoofs, S.Teichmann, T.Hartmann, and M.Wink, *Phytochemistry*, 1983, 22, 65.
103 M.Wink and T.Hartmann, *Z. Pflanzenphysiol.*, 1981, 102, 337.
104 M.Wink and T.Hartmann, *FEBS Lett.*, 1979, 101, 343.
105 M.Wink, T.Hartmann, and L.Witte, *Planta Med.*, 1980, 40, 31.
106 M.Wink and T.Hartmann, *Planta Med.*, 1980, 40, 149.
107 V.D.Cho, R.O.Martin, and J.N.Anderson, *J. Am. Chem. Soc.*, 1971, 93, 2087.
108 V.D.Cho and R.O.Martin, *Korean Biochem. J.*, 1977, 10, 147.
109 W.M.Golebiewski and I.D.Spenser, *J. Am. Chem. Soc.*, 1976, 98, 6726.
110 For review: E. Leete, *Biosynthesis*, 1977, 5, 145.
111 R.J.White, E.Martinelli, G.G.Gallo, G.Lancini, and P.Beynon, *Nature (London)*, 1973, 243, 273.
112 R.D.Johnson, A.Haber, and K.L.Rinehart, *J. Am. Chem. Soc.*, 1974, 96, 3316.
113 A.Haber, R.D.Johnson, and K.L.Rinehart, *J. Am. Chem. Soc.*, 1977, 99, 3541.
114 (a) M.S.Allen, I.A.McDonald, and R.W.Rickards, *Tetrahedron Lett.*, 1981, 22, 1145; (b) I.A.McDonald and R.W.Rickards, *Tetrahedron Lett.*, 1981, 22, 1149.
115 U.Hornemann, J.T.Kehrer, C.S.Nunez, and K.L.Ranieri, *J. Am. Chem. Soc.*, 1974, 96, 320.
116 R.W.Frank, *Prog. Chem. Nat. Prod.*, 1979, 38, 1.
117 O.Ghisalba and J.Nuesch, *J. Antibiot.*, 1981, 34, 64.
118 J.J.Kibby, I.A.McDonald, and R.W.Rickards, *J. Chem. Soc., Chem. Commun.*, 1980, 768.
119 O.Ghisalba and J.Nuesch, *J. Antibiot.*, 1978, 31, 202, 215.
120 O.Ghisalba, H.Fuhrer, W.J.Richter, and S.Moss, *J. Antibiot.*, 1981, 34, 58.
121 M.G.Anderson, J.J.Kibby, R.W.Rickards, and J.M.Rothschild, *J. Chem. Soc., Chem. Commun.*, 1980, 1277.
122 U.Hornemann, J.H.Eggert, and D.P.Honor, *J. Chem. Soc., Chem. Commun.*, 1980, 11.
123 K.L.Rinehart, M.Potgieter, and D.A.Wright, *J. Am. Chem. Soc.*, 1982, 104, 2649.
124 S.M.Kupchan, Y.Komoda, W.A.Court, G.J.Thomas, R.M.Smith, A.Karim, C.J.Gilmore, R.C.Haltiwanger, and R.F.Bryan, *J. Am. Chem. Soc.*, 1972, 94, 1354.
125 M.Asai, E.Mizuta, M.Izawa, K.Hailbara, and T.Kishi, *Tetrahedron*, 1979, 35, 1079.
126 K.Hatano, S.Akiyama, M.Asai, and R.W.Rickards, *J. Antibiot.*, 1982, 35, 1415.
127 J.P.Kutney, M.H.Beale, P.J.Salisbury, K.L.Stuart, B.R.Worth, P.M.Townsley, W.T.Chambers, K.Nilsson, and G.G.Jacoli, *Phytochemistry*, 1981, 20, 653.
128 W.Barz, E.Reinhard, and M.H.Zenk, eds., 'Plant Tissue Culture and its Biotechnological Application', Springer-Verlag, Berlin, 1977.
129 A.W.Alfermann and E.Reinhard, eds. 'Production of Natural Compounds by Cell Culture Methods', Proceedings of an International Symposium, Munich, 1978.
130 E.J.Staba, ed. 'Plant Tissue Culture as a Source of Biochemicals', C.R.C. Press, Boca Raton, Florida, 1980.
131 L.R.Wetter and F.Constable, eds. 'Plant Tissue Culture Methods', N.R.C. Canada, Saskatoon, 1982.
132 J.Stöckigt and H.J.Soll, *Planta Med.*, 1980, 40, 22.

133 W.G.W.Kurz, K.B.Chatson, F.Constabel, J.B.Kutney, L.S.L.Choi, P.Kolodziejczyk, S.K.Sleigh, K.L.Stuart, and B.R.Worth, *Phytochemistry*, 1980, 19, 2583.
134 W.G.W.Kurz, K.B.Chatson, F.Constabel, J.P.Kutney, L.S.L.Choi, P.Kolodziejczyk, S.K.Sleigh, K.L.Stuart, B.R.Worth, *Planta Med.*, 1980, 39, 284.
135 J.B.Kutney, L.S.Choi, P.Kolodiejczk, S.K.Sleigh, K.L.Stuart, B.R.Worth, W.G.W.Kurz, K.B.Chatson, and F.Constabel, *Phytochemistry*, 1980, 19, 2589.
136 F.Constabel, S.Rambold, J.P.Shyluk, D.LeTourneau, W.G.W.Kurz, and J.B.Kutney, *Z. Pfanzenphysiol.*, 1981, 105, 53.
137 F.Constabel, S.Rambold, K.B.Chatson, W.G.M.Kurz, and J.P.Kutney, *Plant Cell Rep.*, 1981, 1, 3.
138 W.G.W.Kurz, K.B.Chatson, F.Constabel, J.P.Kutney, L.S.L.Choi, P.Kolodziejczyk, S.K.Sleigh, K.L.Stuart, and B.R.Worth, *Planta Med.*, 1981, 42, 22.
139 S.-L.Lee, K.-D.Chang, and A.I.Scott, *Phytochemistry*, 1981, 20, 1841.
140 H.Hinz and M.H.Zenk, *Naturwissenschaften*, 1981, 68, 620.
141 W.Barz, H.Herzbeck, W.Husemann, G.Schneiders, and H.K.Mangola, *Planta Med.*, 1980, 40, 137.
142 W.J.Kelleher, A.Rother, E.Wellmann, and H.Grisebach, *Planta Med.*, 1980, 40, 127.
143 M.Rueffer (Rüffer), H.El-Shagi, N.Nagakura, and M.H.Zenk, *FEBS Lett.*, 1981, 129, 5.
144 G.J.Kapadia, G.S.Rao, E.Leete, M.B.E.Fayez, Y.S.Vaishar, and H.M.Fales, *J. Am. Chem. Soc.*, 1970, 92, 6943.
145 M.L.Wilson and C.J.Coscia, *J. Am. Chem. Soc.*, 1975, 97, 431.
146 A.R.Battersby, R.C.F.Jones and R.Kazlauskas, *Tetrahedron Lett.*, 1975, 1873.
147 A.I.Scott, S.-L.Lee, and T.Hirata, *Heterocycles*, 1978, 11, 159.
148 L.K.Klyshev and L.A.Zhesakova, *Izv. Akad. Nauk Kaz. SSR, Ser. Biol.*, 1980, No.5, p.12.
149 H.Fujiwake, T.Suzuki, S.Oka, and K.Iwai, *Agric. Biol. Chem.*, 1980, 44, 2907.
150 K.Iwai, T.Suzuki, and H.Fujiwake, *Agric. Biol. Chem.*, 1979, 43, 2493.
151 J.Stöckigt and M.H.Zenk, *J. Chem. Soc., Chem. Commun.*, 1977, 646.
152 J.F.Treimer and M.H.Zenk, *FEBS Lett.*, 1979, 97, 159.
153 J.F.Treimer and M.H.Zenk, *Eur. J. Biochem.*, 1979, 101, 225.
154 J.Stöckigt, *Phytochemistry*, 1979, 18, 965.
155 H.Mizukami, H.Nordlov, S.-L.Lee, A.I.Scott, *Biochemistry*, 1979, 18, 3760.
156 U.Pfitzner and M.H.Zenk, *Planta Med.*, 1982, 46, 10.
157 A.I.Scott, S.-L.Lee, and W.Wan, *Biochem. Biophys. Res. Commun.*, 1977, 75, 1004.
158 T.Hemscheidt and M.H.Zenk, *FEBS Lett.*, 1980, 110, 187.
159 M.Rueffer, C.Kan-Fan, H.P.Husson, J.Stöckigt, and M.H.Zenk, *J. Chem. Soc., Chem. Commun.*, 1979, 1016.
160 J.Stöckigt, G.Höfle, and A.Pfitzner, *Tetrahedron Lett.*, 1980, 21, 1925.
161 P.Heinstein, G.Höfle, and J. Stöckigt, *Planta Med.*, 1979, 37, 349.
162 P.Heinstein, J.Stöckigt, and M.H.Zenk, *Tetrahedron Lett.*, 1980, 21, 141.
163 A.I.Scott, S.-L.Lee, T.Hirata, and M.G.Calver, *Rev. Latinoam. Quim.*, 1978, 9, 131.
164 S.-L.Lee, T.Hirata, and A.I.Scott, *Tetrahedron Lett.*, 1979, 691.
165 A.Pfitzner and J.Stöckigt, *Phytochemistry*, 1982, 21, 1585.
166 T.Hirata, S.-L.Lee, and A.I.Scott, *J. Chem. Soc., Chem. Commun.*, 1979, 1081.
167 R.T.Brown and J.Leonard, *J. Chem. Soc., Chem. Commun.*, 1979, 877.
168 J.Stöckigt, M.Rueffer, C.Kan-Fan, and H.P.Husson, *Planta Med.*, 1980, 39, 73.
169 J.P.Kutney, L.S.L.Choi, T.Honda, N.G.Lewis, T.Sato, K.L.Stuart, and B.R.Worth, *Helv. Chim. Acta*, 1982, 65, 2088.
170 R.L.Baxter, C.A.Dorschel, S.-L.Lee, and A.I.Scott, *J. Chem. Soc., Chem. Commun.*, 1979, 257.
171 R.L.Baxter, M.Hasam, N.E.Mackenzie, and A.I.Scott, *J. Chem. Soc., Chem. Commun.*, 1982, 791.
172 J.P.Kutney, B.Aweryn, L.S.L.Choi, P.Kolodziejczyk, W.G.W.Kurz, K.B.Chatson, and F.Constabel, *Helv. Chim. Acta*, 1982, 65, 1271.

173 A.I.Scott, S.-L.Lee, M.G.Culver, W.Wan, T.Hirata, F.Guéritte, R.L.Baxter, H.Nordlöv, C.A.Dorschel, H.Mizukami, and N.E.MacKenzie, *Heterocycles*, 1981, 15, 1257.
174 J.O'Sullivan, R.C.Bleaney, J.A.Huddleston, and E.P.Abraham, *Biochem. J.*, 1979, 184, 421.
175 T.Konomi, S.Herchen, J.E.Baldwin, M.Yoshida, N.A.Hunt, and A.L.Demain, *Biochem. J.*, 1979, 184, 427.
176 J.E.Baldwin, B.L.Johnson, J.J.Usher, E.P.Abraham, J.A.Huddleston, and R.L.White, *J. Chem. Soc., Chem. Commun.*, 1980, 1271.
177 J.E.Baldwin, M.Jung, J.J.Usher, E.P.Abraham, J.A.Huddleston, and R.L.White, *J. Chem. Soc., Chem. Commun.*, 1981, 246.
178 R.M.Adlington, R.T.Aplin, J.E.Baldwin, L.D.Field, E.-M.M.John, E.P.Abraham, and R.L.White, *J. Chem. Soc., Chem. Commun.*, 1982, 137.
179 B.Meesschaert, P.Adriaens, and H.Eyssen, *J. Antibiot.*, 1980, 33, 722.
180 G.A.Bahadur, J.E.Baldwin, J.J.Usher, E.P.Abraham, G.S.Jayatilake, and R.L.White, *J. Am. Chem. Soc.*, 1981, 103, 7650.
181 G.Bahadur, J.E.Baldwin, T.Wan, M.Jung, E.P.Abraham, J.A.Huddleston, and R.L.White, *J. Chem. Soc., Chem. Commun.*, 1981, 1146.
182 E.P.Abraham, R.M.Adlington, J.E.Baldwin, M.J.Crimmin, L.D.Field, G.S.Jayatilake, and R.L.White, *J. Chem. Soc., Chem. Commun.*, 1982, 1130.
183 J.E.Baldwin and T.S.Wan, *J. Chem. Soc., Chem. Commun.*, 1979, 249.
184 H.R.Felix, J.Nuesch, and W.Wehrli, *FEMS Microbiol. Lett.*, 1980, 8, 55.
185 G.S.Jayatilake, J.A.Huddleston, and E.P.Abraham, *Biochem. J.*, 1981, 194, 645.
186 J.E.Baldwin, P.D.Singh, M.Yoshida, Y.Sawada, and A.L.Demain, *Biochem. J.*, 1980, 186, 889.
187 S.E.Jensen, D.W.S.Westlake, R.J.Bowers, and S.Wolfe, *J. Antibiot.*, 1982, 35, 1351.
188 H.Otsuka, F.R.Quigley, D.Gröger, J.A.Anderson, and H.G.Floss, *Planta Med.*, 1980, 40, 109.
189 S.-L.Lee, H.G.Floss, and P.F.Heinstein, *Arch. Biochem. Biophys.*, 1976, 177, 84.
190 D.Erge, W.Maier, and D.Gröger, *Biochem. Physiol. Pflanzen*, 1973, 164, 234.
191 J.C.Hsu and J.A.Anderson, *Biochim. Biophys. Acta*, 1971, 230, 518.
192 M.F.Roberts and G.R.Waller, *Phytochemistry*, 1979, 18, 451.
193 M.F.Roberts, *Plant Cell Rep.*, 1981, 1, 10.
194 K.J.Koelen and G.G.Gross, *Planta Med.*, 1982, 44, 227.
195 V.R.Villanueva, R.C.Adlakha, and R.Calvayrac, *Phytochemistry*, 1980, 19, 787.
196 E.Meyer, *Plant Cell Rep.*, 1982, 236.
197 C.R.Bird and T.A.Smith, *Phytochemistry*, 1981, 20, 2345.
198 D.C.Billington, B.T.Golding, and I.K.Nassereddin, *J. Chem. Soc., Chem. Commun.*, 1980, 90.
199 K.S.Srivenugopal and P.R.Adiga, *FEBS Lett.*, 1980, 112, 260.
200 G.H.Tait, *Biochem. Soc. Trans.*, 1976, 4, 610.
201 J.W.Saunders and L.P.Bush, *Plant Physiol.*, 1979, 64, 236.
202 C.C.Hodges and H.Rapoport, *Phytochemistry*, 1980, 19, 1681.
203 C.C.Hodges and H.Rapoport, *Biochemistry*, 1982, 16, 3729.
204 D.Vágújfalvi and M.Petz-Stifter, *Phytochemistry*, 1982, 21, 1533.
205 J.P.G.Mack and M.Slaytor, *Phytochemistry*, 1979, 18, 1921.
206 S.Ōmura, A.Nakagawa, K.Shibata, and H.Sano, *Tetrahedron Lett.*, 1982, 23, 4713.
207 J.O'Sullivan, A.M. Gillum, C.A.Aklonis, M.L.Souser, and R.B.Sykes, *Antimicrob. Agents Chemother.*, 1982, 21, 558.
208 B.A.Abdusalamov, *Khim. Prir. Soedin.*, 1980, 16, 56.
209 O.Prakash, D.S.Bhakuni, and R.S.Kapil, *J. Chem. Soc., Perkin Trans. 1*, 1979, 1515.
210 V.Sharma, S.Jain, D.S.Bhakuni, and R.S.Kapil, *J. Chem. Soc., Perkin Trans. 1*, 1982, 1153.
211 J.K.Allen, K.D.Barrow, and A.J.Jones, *J. Chem. Soc., Chem. Commun.*, 1979, 280.

212 M.F.Grundon, M.R.Hamblin, D.M.Harrison, J.N.D.Logue, M.Maguire, and J.A.McGrath, *J. Chem. Soc., Perkin Trans. 1*, 1980, 1294.

213 C.Fuganti, P.Grasselli, and G.Pedrocchi-Fantoni, *Tetrahedron Lett.*, 1979, 2453.

214 M.Yamazaki, K.Suzuki, H.Fujimoto, T.Akiyama, U.Sankawa, and Y.Iitaka, *Chem. Pharm. Bull.*, 1980, 28, 861.

215 M.I.P.Boente, G.W.Kirby, and D.J.Robins, *J. Chem. Soc., Chem. Commun.*, 1981, 619.

216 R.B.Herbert, F.G.Holliman, P.N.Ibberson, and J.B.Sheridan, *J. Chem. Soc., Perkin Trans. 1*, 1979, 2411.

217 T.Etherington, R.B.Herbert, F.G.Holliman, and J.B.Sheridan, *J. Chem. Soc., Perkin Trans. 1*, 1979, 2416.

218 U.Hollstein, D.L.Mock, and R.R.Sibbitt, *Tetrahedron Lett.*, 1978, 2987.

219 R.B.Herbert, J.Mann, and A.Römer, *Z. Naturforsch.*, 1982, 37c, 159.

220 A.Römer, and R.B.Herbert, *Z. Naturforsch.*, 1982, 37c, 1070.

221 C.R.Hutchinson, A.H.Heckendorf, J.L.Straughn, P.E.Daddona, and D.E.Cane, *J. Am. Chem. Soc.*, 1979, 101, 3358.

222 C.R.Hutchinson, *Tetrahedron*, 1981, 37, 1047.

223 T.Suzuki, T.Kawada, and K.Iwai, *Plant Cell Physiol.*, 1981, 22, 23.

224 T.Suzuki, T.Kawada, and K.Iwai, *Agric. Biol. Chem.*, 1981, 45, 535.

225 D.J.Bennette and G.W.Kirby, *J. Chem. Soc. C.*, 1968, 442.

226 E.Leete and M.C.Louden, *J. Am. Chem. Soc.*, 1968, 90, 6837.

227 B.Kopp and J.Jurenitsch, *Planta Med.*, 1981, 43, 272.

228 I.N.Sharipov, I.V.Terentjeva, and G.V.Lazurjevskii, *Izv. Akad. Nauk Mold. SSR, Ser. Biol. Khim. Nauk*, 1979, No.1, p.86.

229 M.Ya.Lovkova, N.I.Klimenteva, and G.V.Lazurjevskii, *Prikl. Biokhim. Mikrobiol.*, 1979, 15, 775.

230 E.Leete, *J. Nat. Prod.*, 1980, 43, 130.

231 A.I.Scott, H.Mizukami, T.Hiratu, and S.-L.Lee, *Phytochemistry*, 1980, 19, 488.

232 J.Balsevich, F.Constabel, and W.G.W.Kurz, *Planta Med.*, 1982, 44, 231.

233 A.R.Battersby, M.Thompson, K.-H.Glüsenkamp, and L.-F.Tietze, *Chem. Ber.*, 1981, 114, 3430.

234 A.R.Battersby, N.D.Westcott, K.-H.Glüsenkamp, and L.-F.Tietze, *Chem. Ber.*, 1981, 114, 3439.

235 H.Inouye, *Planta Med.*, 1978, 33, 193.

236 N.Nagakura, M.Rüffer, and M.H.Zenk, *J. Chem. Soc., Perkin Trans. 1*, 1979, 2308.

237 F.Guéritte, N.V.Bac, Y.Langlois, and P.Potier, *J. Chem. Soc., Chem. Commun.*, 1980, 452.

238 J.E.Baldwin and T.S.Wan, *Tetrahedron*, 1981, 37, 1589.

239 R.L.Baxter, A.I.Scott, and M.Fukumura, *J. Chem. Soc., Chem. Commun.*, 1982, 66.

240 J.E.Baldwin, B.Chakravarti, M.Jung, N.J.Patel, P.D.Singh, J.J.Usher, and C.Vallejo, *J. Chem. Soc., Chem. Commun.*, 1981, 934.

241 J.O'Sullivan, R.T.Aplin, C.M.Stevens, and E.P.Abraham, *Biochem. J.*, 1979, 179, 47.

242 R.J.Parry, D.D.Sternbach, and M.D.Cabelli, *J. Am. Chem. Soc.*, 1976, 98, 6380.

243 A.Gitterman, R.J.Parry, R.F.Dufresne, D.D.Sternbach, and M.D.Cabelli, *J. Am. Chem. Soc.*, 1980, 102, 2074.

244 R.J.Parry, M.N.T.Chang, J.M.Schwab, and B.M.Foxman, *J. Am. Chem. Soc.*, 1980, 102, 1099.

245 A.Probst and Ch.Tamm, *Helv. Chim. Acta*, 1981, 64, 2065.

246 A.R.Battersby, J.Staunton, M.C.Summers, and R.Southgate, *J. Chem. Soc., Perkin Trans. 1*, 1979, 45.

247 N.Takao, N.Bessho, M.Kamigauchi, K.Iwasa, K.Tomita, T.Fujwara, and S.Fujii, *Tetrahedron Lett.*, 1979, 495.

248 D.S.Bhakuni, S.Jain, and S.Gupta, *Tetrahedron*, 1980, 36, 2491.

249 H.Otsuka, J.A.Anderson, and H.G.Floss, *J. Chem. Soc., Chem. Commun.*, 1979, 660.

250 H.Otsuka, F.R.Quigley, D.Gröger, J.A.Anderson, and H.G.Floss, *Planta Med.*, 1980, 40, 109.
251 J.E.Robbers, H.Otsuka, H.G.Floss, E.V.Arnold, and J.Clardy, *J. Org. Chem.*, 1980, 45, 1117.
252 S.B.Hassam and H.G.Floss, *J. Nat. Prod.*, 1981, 44, 756.
253 W.Aklin, F.Weibel, and D.Arigoni, *Chimia*, 1977, 31, 63.
254 M.Tanabe, Abstracts 26th IUPAC Symposium, Tokyo, Sept. 1977, No. 8C206.
255 U.Keller, R.Zocher, and H.Kleinkauf, *J. Gen. Microbiol.*, 1980, 118, 485.
256 N.Crespi-Perellino, A.Guicciardi, A.Minghetti, and C.Spalla, *Experientia*, 1981, 37, 217.
257 C.M.Belzecki, F.R.Quigley, H.G.Floss, N.Crespi-Perellino, and A.Guicciardi, *J. Org. Chem.*, 1980, 45, 2215.
258 A.Baument, D.Erge, and D.Gröger, *Planta Med.*, 1982, 44, 122.
259 D.S.Bhakuni, S.Jain, and R.S.Singh, *Tetrahedron*, 1980, 36, 2525.
260 D.S.Bhakuni and S.Jain, *Tetrahedron*, 1980, 36, 2153.
261 D.S.Bhakuni and S.Jain, *Tetrahedron*, 1981, 37, 3171.
262 D.S.Bhakuni, S.Jain, and A.N.Singh, *Phytochemistry*, 1980, 19, 2347.
263 D.S.Bhakuni, A.N.Singh, and S.Jain, *Tetrahedron*, 1980, 36, 2149.
264 D.S.Bhakuni and S.Jain, *Tetrahedron*, 1982, 38, 729.
265 P.M.Frischknecht and T.W.Baumann, *Planta Med.*, 1980, 40, 245.
266 P.M.Frischknecht, B.M.Eller, and T.W.Baumann, *Planta*, 1982, 156, 295.
267 H.L.Holland, P.W.Jeffs, T.M.Capps, and D.B.MacLean, *Can. J. Chem.*, 1979, 57, 1588.
268 W.J.Keller, *J. Pharm. Sci.*, 1979, 68, 85.
269 W.J.Keller, *Phytochemistry*, 1980, 19, 413.
270 W.J.Keller, *Phytochemistry*, 1981, 20, 2165.
271 P.G.Rao, U.Zatshi, A.Soni, and C.K.Atal, *Planta Med.*, 1979, 35, 279.
272 D.S.Bhakuni, S.Jain, and R.Chaturvedi, *Tetrahedron*, 1979, 35, 2323.
273 D.S.Bhakuni and S.Jain, *Tetrahedron*, 1981, 37, 3175.
274 B.A.McGaw and J.G.Woolley, *Phytochemistry*, 1979, 18, 1647.
275 K.Basey and J.G.Woolley, *Phytochemistry*, 1973, 12, 2883.
276 R.K.Hill, S.-W.Rhee, E.Leete, and B.A.McGaw, *J. Am. Chem. Soc.*, 1980, 102, 7344.
277 E.Leete and J.A.McDonell, *J. Am. Chem. Soc.*, 1981, 103, 658.
278 B.A.McGaw and J.G.Woolley, *Phytochemistry*, 1982, 21, 2653.
279 V.Koppel, *Tartu Riikliku Ulik. Toim.*, 1980, 523, 53; *Chem. Abs.*, 1981, 94, 80309.
280 R.N.Gupta, P.Horsewood, S.H.Koo, and I.D.Spenser, *Can. J. Chem.*, 1979, 57, 1606.
281 P.Horsewood, W.M.Golebiewski, J.T.Wrobel, I.D.Spenser, J.F.Cohen, and F.Comer, *Can. J. Chem.*, 1979, 57, 1615.
282 W.J.Keller, *Phytochemistry*, 1982, 21, 2851.
283 O.Luanratana and W.J.Griffin, *J. Nat. Prod.*, 1982, 45, 551.
284 R.B.Herbert and J.Mann, *J. Chem. Soc., Perkin Trans. 1*, 1982, 1523, prelim. account: *J. Chem. Soc., Chem. Commun.*, 1980, 841.
285 B.T.Golding and I.K.Nassereddin, *J. Am. Chem. Soc.*, 1982, 104, 5815.
286 M.A.Akhtar, W.G.Brouwer, J.A.D.Jeffreys, C.W.Gemenden, W.I.Taylor, R.N.Seelye, and D.W.Stanton, *J. Chem. Soc. C*, 1967, 859.
287 J.C.Powers and I.Ponticello, *J. Am. Chem. Soc.*, 1968, 90, 7102.
288 J.Stöckigt, *Tetrahedron Lett.*, 1979, 2615.
289 G.W.Kirby, D.J.Robins, M.A.Sefton, and R.R.Talekar, *J. Chem. Soc., Perkin Trans. 1*, 1980, 119.
290 C.A.Russo and E.G.Gros, *Phytochemistry*, 1981, 20, 1763.
291 C.A.Russo and E.G.Gros, *Phytochemistry*, 1982, 21, 609.
292 A.A.Ghini, G.Burton, and E.G.Gros, *Phytochemistry*, 1982, 21, 605.
293 J.Schallenberg and E.Meyer, *Planta Med.*, 1981, 42, 133.
294 E.Leistner and I.D.Spenser, *Pol. J. Chem.*, 1979, 53, 49.
295 K.K.Chexal, C.Snipes, and Ch.Tamm, *Helv. Chim. Acta*, 1980, 63, 761.

296 R.R.Arndt, C.W.Holzapfel, N.P.Ferreira, and J.J.Marsh, *Phytochemistry*, 1974, 13, 1865.
297 B.A.McGaw and J.G.Woolley, *Tetrahedron Lett.*, 1979, 3135.
298 E.Leete, *J. Org. Chem.*, 1979, 44, 165.
299 E.Leete, *J. Nat. Prod.*, 1982, 45, 197.
300 N.Fangmeier and E.Leistner, *J. Chem. Soc., Perkin Trans. 1*, 1981, 1769.
301 E.Leete and M.-L.Yu, *Phytochemistry*, 1980, 19, 1093.
302 E.Leete, *Phytochemistry*, 1981, 20, 1037.
303 I.D.Wigle, L.J.J.Mestichelli, and I.D.Spenser, *J. Chem. Soc., Chem. Commun.*, 1982, 662.
304 J.Berlin, *Phytochemistry*, 1981, 20, 53, see also: J.Berlin and L.Witte, *J. Nat. Prod.*, 1982, 45, 88.
305 J.Berlin, K.G.Kukoschke, and K.-H.Knobloch, *Planta Med.*, 1981, 42, 173.
306 O.Ghisalba, R.Roos, T.Schupp, and J.Nüesch, *J. Antibiot.*, 1982, 35, 74.
307 P.Traxler and O.Ghisalba, *J. Antibiot.*, 1982, 35, 1361.
308 M.A.O'Leary and J.R.Hanson, *Tetrahedron Lett.*, 1982, 23, 1855.
309 H.Wyler and U.Meuer, *Helv. Chim. Acta.*, 1979, 62, 1330.
310 E.Brochmann-Hanssen and C.Y.Chang, *J. Nat. Prod.*, 1982, 45, 434.
311 A.R.Battersby, E.J.T.Chrystal, and J.Staunton, *J. Chem. Soc., Perkin Trans. 1*, 1980, 31.
312 A.R.Battersby, P.W.Sheldrake, J.Staunton, and M.C.Summers, *Bioorg. Chem.*, 1977, 6, 43.
313 A.-F.Hsu, *J. Nat. Prod.*, 1981, 44, 408.
314 J.W.Fairbairn and M.J.Steele, *Phytochemistry*, 1981, 20, 1031.
315 W.H.J.Tam, W.G.W.Kurz, F.Constabel, and K.B.Chatson, *Phytochemistry*, 1982, 21, 253.
316 W.H.J.Tam, F.Constabel, and W.G.W.Kurz, *Phytochemistry*, 1980, 19, 486.
317 E.Brochmann-Hanssen, C.Y.Cheng, and H.C.Chiang, *J. Nat. Prod.*, 1982, 45, 629.
318 E.Brochmann-Hanssen and C.Y.Cheng, *J. Nat. Prod.*, 1982, 45, 437.
319 J.S.Horn, A.G.Paul, and H.Rapoport, *J. Am. Chem. Soc.*, 1978, 100, 1895.
320 E.Brochmann-Hanssen and Y.Okamoto, *J. Nat. Prod.*, 1980, 43, 731.
321 E.Brochmann-Hanssen, C.-Y.Chen and E.E.Linn, *J. Nat. Prod.*, 1980, 43, 736.
322 S.Wolfe, R.J.Bowers, D.A.Lowe, and R.B.Morin, *Can. J. Chem.*, 1982, 60, 355.
323 A.E.deJesus, P.S.Steyn, F.R.vanHeerden, R.Vleggaar, P.L.Wessels, and W.E.Hall, *J. Chem. Soc., Chem. Commun.*, 1981, 289.
324 A.E.deJesus, W.E.Hull, P.S.Steyn, F.R.vanHeerden, R.Vleggaar, and P.L.Wessels, *J. Chem. Soc., Chem. Commun.*, 1982, 837.
325 C.P.Gorst-Allman, P.S.Steyn, and R.Vleggaar, *J. Chem. Soc., Chem. Commun.*, 1982, 652.
326 A.A.Chalmers, C.P.Gorst-Allman, and P.S.Steyn, *J. Chem. Soc., Chem. Commun.*, 1982, 1367.
327 A.E.deJesus, P.S.Steyn, R.Vleggaar, G.W.Kirby, M.J.Varley, and N.P.Ferreira, *J. Chem. Soc., Perkin Trans. 1*, 1981, 3292.
328 R.Vleggaar and P.L.Wessels, *J. Chem. Soc., Chem. Commun.*, 1980, 160.
329 S.Ohmomo, T.Ohashi, and M.Abe, *Agric. Biol. Chem.*, 1979, 43, 2035.
330 K.D.Barrow, P.W.Colley, and D.E.Tribe, *J. Chem. Soc., Chem. Commun.*, 1979, 225.
331 R.Wyss, Ch.Tamm, and J.C.Vederas, *Helv. Chim. Acta*, 1980, 63, 1538.
332 J.-P.Ferezou, A.Quesneau-Thierry, C.Servy, E.Zissmann, and M.Barbier, *J. Chem. Soc., Perkin Trans. 1*, 1980, 1739.
333 P.Mohr and Ch.Tamm, *Tetrahedron*, 1981, 37, Suppl. 1, 201.
334 C.J.Chang, H.G.Floss, D.J.Hook, J.A.Mabe, P.E.Manni, L.L.Martin, K.Schröder, and T.L.Shieh, *J. Antibiot.*, 1981, 34, 555.
335 P.R.Buckland, R.B.Herbert, and F.G.Holliman, *Tetrahedron Lett.*, 1981, 22, 595.
336 A.J.Birch and T.J.Simpson, *J. Chem. Soc., Perkin Trans. 1*, 1979, 816.
337 E.C.Clevenstine, H.P.Broquist, and T.M.Harris, *Biochemistry*, 1979, 18, 3658.
338 E.C.Clevenstine, P.Walter, T.M.Harris, and H.P.Broquist, *Biochemistry*, 1979, 18, 3663.

339 M.J.Schneider, F.S.Ungemach, H.P.Broquist, and T.M.Harris, *J. Am. Chem. Soc.*, 1982, 104, 6863.
340 A.Zschunke, A.Baumert, and D. Gröger, *J. Chem. Soc., Chem. Commun.*, 1982, 1263.
341 H.J.Gerdes and E.Leistner, *Phytochemistry*, 1979, 18, 771.
342 R.Cahill, D.H.G.Crout, M.B.Mitchell, and U.S.Müller, *J. Chem. Soc., Chem. Commun.*, 1980, 419.
343 R.Cahill, D.H.G.Crout, M.V.M.Gregorio, M.B.Mitchell, and U.S.Müller, *J. Chem. Soc., Perkin Trans. 1*, 1983, 173.
344 S.F.Osman, R.M.Zacharius, and D.Naglak, *Phytochemistry*, 1980, 19, 2599.
345 B.A.Abdusalamov, *Khim. Prirod. Soedin.*, 1979, 15, 737.
346 F.J.Leeper, G.Grue-Sørensen, and I.D.Spenser, *Can. J. Chem.*, 1981, 59, 106.
347 S.M.Atwell and P.G.Mantle, *Experientia*, 1981, 37, 1257.
348 A.R.Battersby, A.K.Bhatnagar, P.Hackett, C.W.Thornber, and J.Staunton, *J. Chem. Soc., Perkin Trans. 1*, 1981, 2002.
349 A.R.Battersby, R.C.F.Jones, R.Kazlauskas, A.P.Ottridge, C.Poupat, and J.Staunton, *J. Chem. Soc., Perkin Trans. 1*, 1981, 2010.
350 A.R.Battersby, R.C.F.Jones, R.Kazlauskas, C.W.Thornber, S.Ruchirawat, and J.Staunton, *J. Chem. Soc., Perkin Trans. 1*, 1981, 2016.
351 A.R.Battersby, R.C.F.Jones, A.Minta, A.P.Ottridge, and J.Staunton, *J. Chem. Soc., Perkin Trans. 1*, 1981, 2030.
352 H.Obata-Sasamoto, A.Komamine, and K.Saito, *Z.Naturforsch.*, 1981, 36c, 921.
353 K.Saito, H.Obata-Sasamoto, S.-I.Hatanaka, H.Noguchi, U.Sankawa, and A.Komamine, *Phytochemistry*, 1982, 21, 474.
354 S.J.Gould, K.J.Martinkus, and C.-H.Tann, *J. Am. Chem. Soc.*, 1981, 103, 2871.
355 S.J.Gould, K.J.Martinkus, and C.-H.Tann, *J. Am. Chem. Soc.*, 1981, 103, 4639.
356 S.J.Gould and T.K.Thiruvengadem, *J. Am. Chem. Soc.*, 1981, 103, 6752.
357 L.H.Hurley and C.Gairola, *Antimicrob. Agents Chemother.*, 1979, 15, 42.
358 C.J.Pearce and K.L.Rinehart, *J. Am. Chem. Soc.*, 1979, 101, 5069.
359 A.G.McInnes, D.G.Smith, J.A.Walter, L.C.Vining, and J.L.C.Wright, *Can. J. Chem.*, 1979, 57, 3200.
360 J.G.Buchanan, M.R.Hamblin, G.R.Sood, and R.H.Wightman, *J. Chem. Soc., Chem. Commun.*, 1980, 917.
361 M.J.Zmijewski, *J. Antibiot.*, 1980, 33, 447.
362 M.J.Zmijewski, *Tetrahedron Lett.*, 1982, 23, 1775.
363 L.David and S.Emadzadeh, *J. Antibiot.*, 1982, 35, 1616.
364 I.Stirling and S.W.Elson, *J. Antibiot.*, 1979, 32, 1125.
365 S.W.Elson and R.S.Oliver, *J. Antibiot.*, 1978, 31, 586, cf. *Biosynthesis*, 1980, 6, 278.
366 S.W.Elson, R.S.Oliver, B.W.Bycroft, and E.A.Faruk, *J. Antibiot.*, 1982, 35, 81.
367 S.J.Gould and D.E.Cane, *J. Am. Chem. Soc.*, 1982, 104, 343.
368 S.J.Gould and S.M. Weinreb, *Prog. Chem. Nat. Prod.*, 1982, 41, 77.
369 S.J.Gould, C.C.Chang, D.S.Darling, J.D.Roberts, and M.Squillacote, *J. Am. Chem. Soc.*, 1980, 102, 1707.
370 R.S.P.Hsi, D.F.Witz, W.T.Stolle, L.Baczynskyj, and T.A.Scahill, *J. Labelled Compd. Radiopharm.*, 1980, 17, 613.
371 S.Kharatyzn, M.Puza, J.Spizek, L.Dolezilova, Z.Vanek, M.Vondracek, and R.W.Rickards, *Chem. Ind. (London)*, 1963, 1038.
372 Z.Vanek, M.Puza, J.Cudlin, and L.Dolezilova, *Biochem. Biophys. Res. Commun.*, 1964, 17, 532.
373 Z.Vanek, M.Puza, J.Cudlin, M.Vandracek, and R.W.Rickards, *Folia Microbiol. (Prague)*, 1969, 14, 388.
374 J.Cudlin, M.Puza, M.Vandracek, Z.Vanek, and R.W.Rickards, *Folia Microbiol. (Prague)*, 1967, 12, 376.
375 P.W.Jeffs and D.McWilliams, *J. Am. Chem. Soc.*, 1981, 103, 6185.
376 R.W.Woodward, L.Mascaro, R.Hörhammer, S.Eisenstein, and H.G.Floss, *J. Am. Chem. Soc.*, 1980, 102, 6314.

377 L.Mascaro, R.Hörhammer, S.Eisenstein, L.K.Sellers, K.Mascaro, and H.G.Floss, *J. Am. Chem. Soc.*, 1977, 99, 273.
378 K.Ishiguro, K.Takahashi, K.Yazawa, S.Sakiyama, and T.Arai, *J. Biol. Chem.*, 1981, 256, 2162.
379 M.J.Zmijewski, M.Mikolajczak, V.Viswanatha, and V.J.Hruby, *J. Am. Chem. Soc.*, 1982, 104, 4969.
380 H.Shimada, H.Noguchi, Y.Iitaka and U.Sankawa, *Heterocycles*, 1981, 15, 1141.
381 K.L.Rinehart, H.Potgieter, D.L.Delaware, and H.Seto, *J. Am. Chem. Soc.*, 1981, 103, 2099.
382 D.D.Weller and K.L.Rinehart, *J. Am. Chem. Soc.*, 1978, 100, 6757.
383 J.M.Ostrander, L.H.Hurley, A.G.McInnes, D.G.Smith, J.A.Walter, and J.L.C.Wright, *J. Antibiot.*, 1980, 33, 1167.
384 R.K.Malhotra, J.M.Ostrander, L.H.Hurley, A.G.McInnes, D.G.Smith, J.A.Walter, and J.L.C.Wright, *J. Nat. Prod.*, 1981, 44, 38.
385 M.Iseki, T.Miyoshi, T.Konomi, and H.Imanaka, *J. Antibiot.*, 1980, 33, 488.
386 T.Nakatani, A.Fujii, H.Naganawa, T.Takita, and H.Umezawa, *J. Antibiot.*, 1980, 33, 717.
387 D.G.I.Kingston and M.X.Kolpak, *J. Am. Chem. Soc.*, 1980, 102, 5964.
388 K.A.Parker and R.E.Babine, *Tetrahedron Lett.*, 1982, 23, 1763.
389 K.A.Parker and R.E.Babine, *J. Am. Chem. Soc.*, 1982, 104, 7330.
390 L.H.Hurley, W.L.Lasswell, R.K.Malhotra, and N.V.Das, *Biochemistry*, 1979, 18, 4225.
391 N.L.Marekov and A.K.Sidjimov, *Tetrahedron Lett.*, 1981, 22, 2311
392 A.K.Sidjimov and N.L.Marekov, *Phytochemistry*, 1982, 21, 871.
393 D.S.Bhakuni and S.Jain, *J. Chem. Soc., Perkin Trans. 1*, 1981, 2598.
394 D.S.Bhakuni and V.K.Mangla, *Tetrahedron*, 1981, 37, 401.
395 B.Mulchandani, S.S.Iyer, and L.P.Badheka, *Phytochemistry*, 1969, 8, 1931.
396 R.A.Zabel, C.A.Miller, and S.W.Tanenbaum, *Appl. Env. Microbiol.*, 1979, 37, 208.
397 I.D.Spenser and H.P.Tiwari, *Chem. Commun.*, 1966, 55.

Acknowledgments

My work on the biosynthesis and metabolism of natural products has been supported by a research grant GM-13246 from the National Institutes of Health, U. S. Public Health Service. A preliminary survey of the literature with the key words: 'alkaloid' and 'biosynthesis' using the data base of Chemical Abstracts, generated only 57 citations for the 1979-1982 period. I thank Professor Norman R. Farnsworth, Department of Pharmacognosy, University of Illinois, Chicago, for providing me with considerably more references relating to alkaloid biosynthesis from his personal data base (NAPRALERT) on natural products of medicinal interest. I thank my wife, Sheila Anne Leete, for preparing the index to the Table, and for proof reading the whole manuscript, which was typed by Mary Sende.

NATURAL PRODUCT REPORTS

The contents of issues in 1984 are expected to be:

February (Volume 1, No.1) - 'Rotenoids' (L. Crombie), 'Indole Alkaloids and Mould Metabolites' (J.E. Saxton), 'Triterpenoids' (R.B. Boar), 'Carotenoids and Polyterpenoids' (G. Britton), 'Microbial Metabolites' (R.C.F. Jones).

April (Volume 1, No.2) - 'Alkaloid Biosynthesis' (R.B. Herbert), 'Aporphinoid Alkaloids' (M. Shamma), 'Quinoline, Quinazoline and Acridone Alkaloids' (M.F. Grundon), 'Sesquiterpenoids' (J.S. Roberts and I. Bryson), 'Diterpenoids' (J.R. Hanson).

June (Volume 1, No.3) - 'Biosynthesis of Polyketides' (T.J. Simpson), 'Steroids-Physical Methods' (D.N. Kirk), 'Steroids - Reactions and Partial Synthesis' (B.A. Marples), 'Insect Pheromones' (R. Baker and R. Herbert), plus a collection of chapters on alkaloids from amongst those regularly contributed by A.R. Pinder, G. Fodor and R. Dharanipragada, D.J. Robins, J.A. Lamberton, M.F. Grundon, K.W. Bentley, A.S. Chawla and A.H. Jackson, S.W. Pelletier and S.W. Page, D.M. Harrison, and J.R. Lewis
(the rest of this group to be in Volume 1, No. 4)

August (Volume 1, No.4) - 'Biosynthesis of Porphyrins and Vitamin B_{12}' (D.G. Buckley), 'Arachidonic Acid Metabolites' (S.M.F. Lai), 'Enzyme Inhibitors in Medicine' (J. Saunders), 'Monoterpenoids' (D.H. Grayson), 'Tissue Culture'.

October (Volume 1, No.5) - 'Biosynthesis of C_5 - C_{20} Terpenoids' (J.R. Hanson), 'Fatty Acids' (F.D. Gunstone), 'Lipids - Physical Properties' (D.P. Quinn), 'Lipids-Chemical Properties' (W.W. Christie), 'Diterpenoids' (J.R. Hanson), 'Unusual Long-chain compounds from Mycobacteria and Related Organisms', (D.E. Minnikin).

December (Volume 1, No.6) - 'Biosynthesis of Shikimate Metabolites' (P.M. Dewick), 'N.M.R. as Applied to Structural Elucidation of Compounds' (I.H. Sadler), 'Marine Natural Products' (D.J. Faulkner), 'Biosynthesis of Non-protein Amino-acids' (P. Olesen Larsen), 'Triterpenoid Biosynthesis'.

There will be approximately 120 pages per issue. The ISSN is 0265-0568